Springer Praxis Books

Space Exploration

This book series presents the whole spectrum of Earth Sciences, Astronautics and Space Exploration. Practitioners will find exact science and complex engineering solutions explained scientifically correct but easy to understand.Various subseries help to differentiate between the scientific areas of Springer Praxis books and to make selected professional information accessible for you.

Thomas Lund

Spacecraft Exploring Outer Planets

From Jupiter to Neptune

Springer

Thomas Lund
San Diego, CA, USA

ISSN 2945-7475 ISSN 2945-7483 (electronic)
Springer Praxis Books
ISSN 2731-5401 2731-541X (electronic)
Space Exploration
ISBN 978-3-031-96543-2 ISBN 978-3-031-96544-9 (eBook)
https://doi.org/10.1007/978-3-031-96544-9

This Springer imprint is published by the registered company Springer Nature Switzerland AG
The registered company address is: Gewerbestrasse 11, 6330 Cham, Switzerland

This book is dedicated to my wife, Barbara, and to my children, Ann, Tom, and Colin.

Credits for Figures in Book

All of the pictures in the book, except one, are from NASA and are in public domain. The exception is Fig. 6.20 in Chap. 6 of the book. The caption on the figure reads: "Model of descent module for Huygens probe residing in Salon Européen de la Recherche, Paris (Wikimedia posting by David Monniaux)." Monniaux gives permission to copy and distribute the work.

Most of the line drawing figures are from NASA, and they are so noted. A few line drawings were prepared by the author, and those do not have any credits listed.

Competing Interests The author has no competing interests to declare that are relevant to the content of this manuscript.

Introduction

The four outer planets, Jupiter, Saturn, Uranus, and Neptune, are each unique and fascinating to observe. The advent of sophisticated space vehicles has allowed astronomers to get up close and personal to each of the planets by exploring them with inquisitive spacecraft. Jupiter and Saturn have been explored in most detail by orbiting spacecraft. The planets Uranus and Neptune and the dwarf planet Pluto have been explored by flyby spacecraft. This book describes the eight capable spacecraft that have explored the outer planets to date.

The best known of the outer planets are the gas giant planets Jupiter and Saturn. Both are visible to the naked eye as bright points of reflected sunlight. Their incredible features come alive when viewed with a medium-power telescope. Jupiter is by far the largest planet in the solar system. It weighs more than all of the other planets put together. Spectacular color bands circle the planet, and its Great Red Spot has intrigued astronomers for centuries. The planet Saturn is famous for its spectacular rings. A medium-power telescope displays the colorful banded gaseous body and its distinct, broad rings.

The planets Uranus and Neptune are ice giant planets, about four times larger than Earth. A large portion of these planets is made up of a dense icy mixture of water and other ices. They do not appear to have a firm surface. Their atmospheres, as seen from afar, have a blue color due to the presence of methane. The dwarf planet Pluto is a small icy world that orbits well beyond Neptune in the Kuiper Belt of icy bodies.

The outer planets indeed orbit far out from the sun. Using for reference the average radius of the orbit of Earth from the sun (one AU), the average orbital distances of the outer planets from the sun are Jupiter 5.2 AU, Saturn 9.6 AU, Uranus 19.2 AU, Neptune 30.1 AU, and Pluto 39.4 AU. Neptune orbits about 4.5 billion km from the sun. For perspective, it took the New Horizons spacecraft 9.5 years to reach Pluto after launching from Earth.

Eight spacecraft have explored the outer planets in detail up to the year 2024. All were developed and launched by the United States. Several spacecraft have benefited from international assistance, often with science instruments. The names of the spacecraft and the planets explored are listed below:

Spacecraft	Launch date	Main planets explored
Pioneer 10	24 March 1972	Jupiter
Pioneer 11	6 April 1973	Jupiter, Saturn
Voyager 2	20 August 1977	Jupiter, Saturn, Uranus, Neptune
Voyager 1	5 September 1977	Jupiter, Saturn
Cassini/Huygens	15 October 1997	Saturn
Galileo	18 October 1989	Jupiter
New Horizons	19 January 2006	Pluto, Arrokoth (KBO)
Juno	5 August 2011	Jupiter

The two Pioneer spacecraft, the two Voyager spacecraft, and the New Horizons spacecraft were flyby-type spacecraft. The Cassini spacecraft orbited Saturn and deployed the European Space Agency Huygens probe that landed on Saturn's moon, Titan. Galileo conducted equatorial orbits of Jupiter, and it dispatched an instrumented probe into the dense atmosphere of Jupiter. Juno conducted polar orbits of Jupiter. All of the spacecraft carried out extensive exploration with several scientific instruments. The investigation included color imaging from wide- and narrow-angle cameras. Many of the images are spectacular.

The human element, embodied by scientists and engineers, made these impressive missions possible. A case in point, the Galileo mission was a success in spite of failure of the critical high-gain antenna. Work-arounds to succeed after failure of the antenna were testament to human ingenuity and perseverance.

Four new missions to the moons of Jupiter are planned. Arrival times are several years from now. Two of the missions have already been launched. They are the European Space Agency *Jupiter Icy Moons Explorer (JUICE)* and the United States *Europa Clipper*. Two other missions, the United States *Dragonfly* and the Chinese *Tianwen-4*, are in work with planned launches a few years from now.

Contents

Chapter 1
The Planets of the Sun

Our star, the sun, holds eight orbiting planets and several dwarf planets in a firm gravitational grasp. It provides life-sustaining light and warmth to a special planet, third from the sun, named Earth. The sun is immense with an equatorial diameter of 1,392,000 km and mass of 1.9885×10^{30} kg. The temperature of the outer surface (photosphere) is 5500 °C. Powered by nuclear fusion, the total power radiated by the sun in all directions (luminosity) is 3.83×10^{26} W.

Four of the planets, referred to as the inner planets, are grouped closer to the sun. Those planets, in order from the sun, are Mercury, Venus, Earth, and Mars. The average distances of the inner planets from the sun range from 57.9 million km for Mercury to 227.9 million km for Mars. There are four outer planets: Jupiter, Saturn, Uranus, and Neptune. The average distances of the outer planets from the sun range from 779 million km for Jupiter to 4495 million km for Neptune. Pluto was considered a planet until 2006 when it was reclassified as a dwarf planet. It orbits an average distance of 5900 million km from the sun. Pluto has also been explored by a flyby spacecraft, and it is included in this book.

A summary of parameters associated with the inner and outer planets is shown in Tables 1.1 and 1.2.

The relative orbits of the outer planets are shown in Fig. 1.1. The distance marker in the figure represents 10^9 km. The orbits shown are proportional to their average distances from the sun. The orbits are elliptical, but the eccentricities of the orbits are small. At the scale of the drawing, the difference between the average distances plotted and the actual orbits would be small.

Pluto is not included in the list of outer planets since it was reclassified as a dwarf planet in 2006. Pluto was explored by a flyby of the New Horizons spacecraft in 2015, and that mission is included in this book. A summary of parameters associated with Pluto is given in Table 1.3. The orbit of Pluto extends far beyond the outer planets. The relative orbit of Pluto along with that of the outer planets is given in

T. Lund, *Spacecraft Exploring Outer Planets*, Springer Praxis Books,
https://doi.org/10.1007/978-3-031-96544-9_1

Table 1.1 Parameters of the inner planets

Parameter	Mercury	Venus	Earth	Mars
Average distance from sun, 10^6 km	57.9	108.2	149.6	227.9
Eccentricity of orbit	0.205	0.007	0.017	0.094
Orbital period, days	88	224.7	365.2	687.0
Orbital inclination from ecliptic, °	7.0	3.4	0	5.1
Mass, 10^{24} kg	0.330	4.87	5.97	0.642
Diameter, km	4879	12,104	12,756	6792
Gravity, m/s^2	3.7	8.9	9.8	3.7
Rotation period/length of day, h	1407.6/4222.6	5832.5/2802.0	23.9/24.0	24.6/24.7
Mean temperature, °C	167	464	15	−65

Table 1.2 Parameters of the outer planets

Parameter	Jupiter	Saturn	Uranus	Neptune
Average distance from sun, 10^6 km	778.6	1433.5	2872.5	4495.1
Eccentricity of orbit	0.049	0.057	0.046	0.011
Orbital period, days	4331	10,747	30,589	59,800
Orbital inclination from ecliptic, °	1.3	2.5	0.8	1.8
Mass, 10^{24} kg	1898	568	86.8	102
Diameter, km	142,984	120,536	51,118	49,529
Gravity, m/s^2	23.1	9.0	8.7	11.0
Rotation period and length of day, h	9.9	10.7	17.2	16.1
Mean temperature, °C	−110	−140	−195	−200

Fig. 1.2. The orbit of Pluto has a large eccentricity of 0.25, and the orbit is inclined 17.1° to the ecliptic (plane of the Earth's orbit).

The planets orbit the sun in a relatively flat plane. Writing in Astronomy magazine, Roen Kelly stated: "The planets orbit in a plane because the solar system formed out of a relatively flat disk of gas and dust swirling around our infant sun." The plane of the orbit of Earth is referred to as the ecliptic. The orbital planes of the other planets are within a few degrees of the ecliptic as can be seen in Tables 1.1 and 1.2.

Characteristics of Jupiter

Jupiter is the largest and most massive planet in the solar system. Its mass is about 2.5 times the mass of the other planets combined. The diameter of Jupiter is about 11 times that of Earth. A classic photograph of Jupiter taken by the Hubble Space Telescope is shown in Fig. 1.3.

Jupiter is a gas giant planet composed primarily of hydrogen and helium. The composition of the atmosphere is 89.8% molecular hydrogen and 10.2% helium. There are trace amounts of methane and ammonia present. The pressure is thought

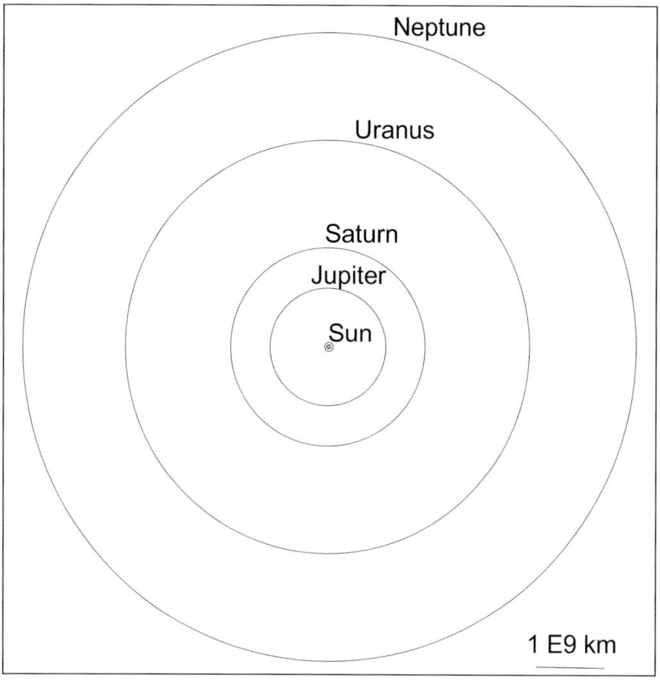

Fig. 1.1 Relative average orbits of outer planets

Table 1.3 Parameters of the dwarf planet Pluto

Parameter	Value
Average distance from sun	5.9×10^9 km
Eccentricity of orbit	0.25
Orbital period, earth years	248 Earth years
Orbital inclination from ecliptic	17.1°
Mass	1.2×10^{22} kg
Diameter, km	2370 km
Rotation period (length of day)	153 h
Mean temperature	−233 °C

to be sufficiently high in the lower regions of the gaseous structure to compress hydrogen into liquid form. The result is an ocean of liquid hydrogen deep within the planet. Even deeper, the immense pressure is thought to be sufficient to strip electrons from atoms forming an electrically conducting liquid. The rapid rotation of the conducting liquid is likely the source of the very high magnetic field around Jupiter. It is not known whether Jupiter has a solid core. The magnetic field of Jupiter is about 420,000 nanoTesla (nT) at the equator compared to about 30,000 nT at the equator for Earth. Jupiter's dipole magnetic field is tilted about 10° from the spin axis.

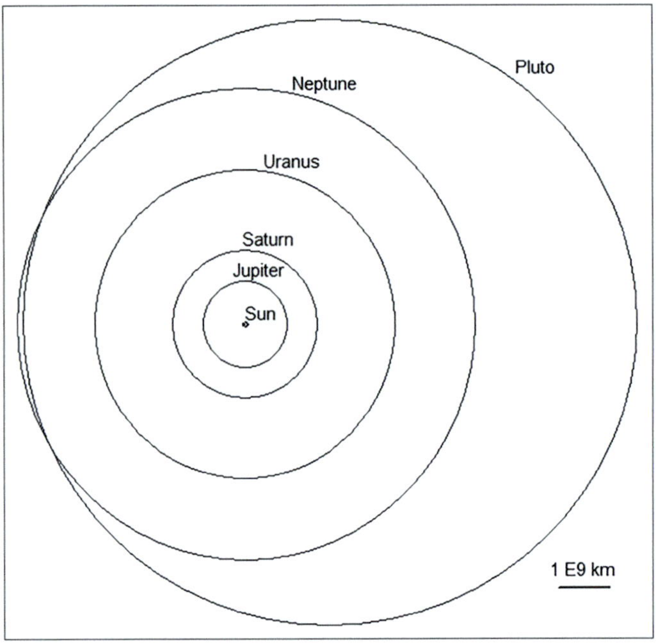

Fig. 1.2 Orbits of Pluto and outer planets

There are three cloud layers above the helium-hydrogen structure of the planet. The upper cloud layer is likely ammonia ice, the center layer is likely ammonium hydrosulfide crystals, and the lower layer may be water vapor and ice. The total thickness of the three layers together is about 44 km. The clouds are driven by strong winds blowing in the east-west direction. The distinctive alternating bands of white and reddish clouds of the planet are separated by winds blowing in opposite directions. The light-colored bands are likely ammonia ice. The darker-colored bands may be upwelling of sulfur and phosphorus gases rising from Jupiter's warm interior.

The famous Great Red Spot, visible in the photograph above, is a giant cyclonic storm, larger in diameter than Earth. This striking feature has been observed since the first use of telescopes over 300 years ago. Several smaller storms persist on the planet.

Jupiter has a faint ring system: an inner ring called the halo, the somewhat brighter main ring, and two outer faint rings called Amalthea Gossamer and Thebe Gossamer. The halo ring extends from about 92,000 to 122,500 km from the center of Jupiter. The main ring extends 122,000 to 129,130 km from the center.

Jupiter has over 80 moons and thousands of very small orbiting objects. The largest four moons are Io, Europa, Ganymede, and Callisto. They are called the Galilean satellites because they were discovered by Galileo Galilei in 1610. They are by far the largest moons orbiting Jupiter. The size and orbit of the Galilean satellites are

Fig. 1.3 Photograph of Jupiter taken by Hubble Space Telescope

Table 1.4 Some parameters of the Galilean satellites

Moon	Diameter (km)	Radius of orbit (km)
Io	3630	421,600
Europa	3140	670,900
Ganymede	5260	1,070,000
Callisto	4800	1,883,000

given in Table 1.4. We note that the moon Ganymede is larger than the planet Mercury.

Jupiter has been explored by six flyby spacecraft and two orbiting spacecraft at the time of this writing. The most detailed information about the planet was obtained from the two orbiting spacecraft: Galileo and Juno.

Minimum energy launch windows to send spacecraft to Jupiter occur about every 13 months (399 days). These windows correspond to the time of inferior conjunction when Earth lies between Jupiter and the sun. Each closest approach distance is different. The smallest closest approach distance is 587 million km. The average of closest approach distances between July 2020 and May 2030 will be 628 km. The minimum in that time period was 590 million km in September 2022.

Characteristics of Saturn

Saturn is the second largest planet in the solar system. Its diameter of 116,464 km is about nine times the diameter of Earth. The spectacular rings around the planet have awed astronomers since the introduction of the telescope. A photograph of Saturn taken by the Hubble Space Telescope in 2020 is shown in Fig. 1.4.

Saturn orbits the sun with a period (a Saturn year) of 29.4 Earth years. The time for one rotation of the planet (a Saturn day) is 10.7 h. The rotation axis of the planet is tilted 26.7° from the plane of the orbit around the sun.

Saturn is a gas giant planet composed of mainly hydrogen and helium. Molecular hydrogen makes up 96.3% of the atmosphere by volume, and helium makes up 2.4%. Trace amounts of methane and ammonia are present. Liquid hydrogen is thought to lie under the gaseous structure with liquid metallic hydrogen lower down. The planet may have dense metallic core surrounded by rocky material. A layer of clouds covers the hydrogen-helium layer. The temperature at the top of the clouds is about −178 °C.

Winds up to 500 m/s exist near the equator in the upper atmosphere. The winds blow in the same direction as rotation of the planet. A unique, hexagon-shaped jet stream blows around Saturn's north pole. Wind speeds are about 322 km/h. The six-sided wind pattern is about 30,000 km across.

Saturn has a strong magnetic field and an extensive magnetosphere. The magnetic field strength at the equator near the cloud tops is about 21 μT. The axis of the magnetic field is closely aligned with the spin axis of the planet.

Fig. 1.4 Photograph of Saturn taken by the Hubble Space Telescope

Table 1.5 Some parameters of rings of Saturn

Ring	Radius of inner edge from the center of Saturn (km)	Width of ring (km)	Comments
A ring	122,050	14,600	
B ring	91,980	25,500	Bright ring
C ring	74,490	17,500	
D ring	67,000	7500	Faint
E ring	180,000	300,000	Faint
F ring	140,224	30–500	Faint
G ring	166,000	8000	
Cassini division	117,500	4700	

The rings that encircle Saturn are largely composed of particles of water ice. Most of the particles are very small, but a few are several meters in size. Darker material is present in a few rings. The rings extend from about 67,000 km from the center of Saturn to 280,000 km from the center. There is still conjecture in the scientific community about when and how the rings formed.

The rings are named in alphabetical order in the sequence that they were discovered; for example, the "A" ring was discovered first. Some parameters of the rings as listed by Britannica are given in Table 1.5. The widths given are in the ring plane. A large gap in the rings, called the Cassini Division, is listed in the table. The gap is apparent just outside of the bright B ring in the photograph of Saturn.

Saturn has 63 confirmed and named moons. An additional 20 moons await confirmation. The largest moon, Titan, is 5150 km in diameter, about three times larger than Earth's moon. Titan orbits 1.22 million km from the center of Saturn with a period of 16 Earth days. Other prominent moons in order of size after Titan are Rhea, Iapetus, Dione, Tethys, Enceladus, and Mimas. Mimas has a diameter of 198 km. Many moons are less than 10 km in size.

Some of the moons, referred to as shepherd moons, orbit at the edges of rings and keep the rings well defined. For example, the moon Prometheus orbits just inside of the F ring, and Pandora orbits just outside the F ring.

Characteristics of Uranus

Uranus is a large planet, four times the size of Earth. It orbits 2.87 billion km from the sun, 19 times farther out from the sun than Earth. It has been explored by only one spacecraft, Voyager 2, which made a flyby in 1986. A photograph of Uranus taken by Voyager 2 is shown in Fig. 1.5. The blue-green color of the planet is a result of methane in the atmosphere that reflects blue-green light from the sun while absorbing red light.

Uranus is unique in the sun's planetary system in that its spin axis is nearly perpendicular to the plane of its orbit around the sun. The period of the spin (one

Fig. 1.5 Photograph of Uranus taken by Voyager 2 spacecraft

Uranus day) is about 17 h, and the period of rotation about the sun (one Uranus year) is about 84 Earth years. As a result, the north and south polar regions alternately see sunlight and darkness for long periods during a Uranus year.

Uranus is an ice giant planet. Most of the planet is made up of a dense icy mixture of water, methane, and ammonia. It does not have a firm surface, but it may have a small rocky core. An atmosphere of hydrogen and helium lies above the heavier icy mixture. The composition of the atmosphere by volume is 82.5% molecular hydrogen, 15.2% helium, 2.3% methane, and trace amounts of other gases. The average temperature in the atmosphere is −195 °C. Cloud layers are present in the upper atmosphere. The clouds are carried around the planet by winds that blow up to 900 km/h.

The magnetic field of Uranus is tilted 58.6° from the spin axis. The field strength is about 23 μT at the surface of the planet.

Uranus has 27 known moons. Five of the moons are very large, and that puts them in a separate class. Those large moons orbit between 129,900 and 583,500 km from the center of Uranus. The other moons orbit either closer or further away from the planet. The five large moons range in diameter from 471 to 1578 km. The diameters of the other moons are less than 70 km.

Uranus also has a well-defined ring system with ten distinct rings. The rings lie 41,837–51,149 km from the center of the planet. The farthest ring, Epsilon, has the largest width of about 43 km. The other rings have widths less than 3.9 km.

Characteristics of Neptune

Neptune is the most distant of the outer planets. It orbits 4.515 billion km from the sun. Neptune has been explored by only one spacecraft. Voyager 2 performed a close flyby in 1989. A photograph of Neptune taken by Voyager 2 is shown in Fig. 1.6. The blue-green color of the planet is a result of methane in the atmosphere. The white patches are clouds.

Neptune is an ice giant planet like Uranus. A large portion of the planet is made up of a dense icy mixture of water and other ices. It does not have a firm surface. The planet is thought to have a large solid core. The diameter of Neptune is 49,528 km, about 3.9 times the size of Earth. The rotation period (length of day) is 16.1 h. The period of orbit around the sun (a Neptune year) is about 164 Earth years. The rotation axis is tipped 29.6° towards its orbital plane.

The composition of the atmosphere laying over the surface is 88% hydrogen, 19% helium, 1.5% methane, and trace amounts of other gases. The temperature at the 1 bar pressure level in the atmosphere is −201 °C. The planet has the strongest winds in the solar system, with winds up to 2000 km/h.

Fig. 1.6 Photograph of Neptune taken by Voyager 2 spacecraft

The axis of the magnetic field of Neptune is tilted 46.9° from the spin axis. The magnetic field strength on the surface varies between 10 and 90 μT.

Neptune has 14 known moons. Of these, the moon Triton dominates at 2706 km in diameter with a mass of 214×10^{20} kg. The other moons are less than 350 km in diameter and have a mass less than 0.5×10^{20} kg.

Neptune has several rings around it. The Voyager 2 spacecraft found five continuous rings around the planet and several partial rings. The rings were spaced from 41,900 to 62,930 km from the center of Neptune.

Characteristics of Pluto

Pluto is a dwarf planet that resides in the Kuiper Belt of icy bodies beyond the orbit of Neptune. It is a relatively small body, 2376 km in diameter. It is only about 18% the size of Earth. The only spacecraft that has explored Pluto is New Horizons in 2015. A photograph of Pluto taken by New Horizons is given in Fig. 1.7.

Pluto has an eccentric orbit around the sun. The average distance of the orbit is 5.906 billion km. Perihelion distance is 4.437 billion km, and aphelion is at 7.376 billion km. At perihelion, it passes inside the orbit of Neptune. The period of the orbit (a Pluto year) is 248 Earth years. The rotation period (a Pluto day) is 6.39 Earth days.

Fig. 1.7 Photograph of Pluto taken by New Horizons spacecraft

Pluto is thought to have a solid core surrounded by a mantle of water ice. Frosts of methane and nitrogen coat the surface. The surface has mountains up to 3 km high consisting of huge blocks of water ice. It also has wide areas of cratered plains. The surface temperature is −247 to −233 °C. The composition of the atmosphere is 99% nitrogen, 0.5% methane, 0.05% carbon monoxide, and trace amounts of other gases.

Pluto has five known moons. The largest moon, Chardon, orbits 19,596 km from the planet and has a diameter of 1372 km. The other moons have diameters less than 100 km. No rings have been observed around Pluto.

References

Couper, Heather, et al. SMITHSONIAN, *Planets*, DK Publishing, New York, 2014

Drake, Nadia, *Saturn: What Put the rings on It?*, National Geographic, March 2023

Jet Propulsion Laboratory, NASA, *Planet Compare*, NASA Science Solar System Exploration

Owen, Tobias Chant, *Solar System*, Encyclopedia Britannica, https://www.britannica.com/science/solar-system

Stern, S. Allan, et al., *The Pluto System After New Horizons*, Annual Review of Astronomy and Astrophysics, Vol. 56, 2018

Williams, David R., *Jupiter Fact Sheet*, NASA, https://gsfc.nasa.gov/planetary/factsheet

Williams, David R., *Mercury Fact Sheet*, NASA, https://gsfc.nasa.gov/planetary/factsheet

Williams, David R., *Neptune Fact Sheet*, NASA, https://gsfc.nasa.gov/planetary/factsheet

Williams, David R., *Saturn Fact Sheet*, NASA, https://gsfc.nasa.gov/planetary/factsheet

Williams, David R., *Uranus Fact Sheet*, NASA, https://gsfc.nasa.gov/planetary/factsheet

Williams, David R., *Venus Fact Sheet*, NASA, https://gsfc.nasa.gov/planetary/factsheet

Chapter 2
Spacecraft That Explored the Outer Planets

Eight spacecraft have explored the outer planets in detail up to the year 2024. All were developed and launched by the United States. The names of the spacecraft and the planets explored are listed in Table 2.1. The Pioneers, Voyagers, and New Horizons were flyby spacecraft. The spacecraft Cassini orbited Saturn and deployed the European Space Agency's Huygens probe that landed on Saturn's moon, Titan. Galileo orbited Jupiter and dispatched an instrumented probe through Jupiter's heavy atmosphere. Juno conducted detailed investigation of Jupiter from polar orbits.

All of the spacecraft carried out extensive exploration. That exploration included color imaging from wide- and narrow-angle cameras. Flyby encounters with the planets lasted several weeks as the spacecraft bent partially around a planet in their trajectories.

The Voyager 2 "Grand Tour" mission was remarkable. Rare alignment of the planets Jupiter, Saturn, Uranus, and Neptune allowed taking advantage of gravitational assist from each planet in turn to set up a trajectory to the next planet as the spacecraft traveled outward. The favorable alignment occurs every 176 years.

The Galileo mission overcame a serious setback when the high-gain antenna failed to deploy and the low-gain antenna had to be used for the entire mission. Ingenious workarounds in data handling and transmission timing were performed by scientists to salvage most of the planned science investigations.

A description of each of these spacecraft and of the scientific instruments carried is given in the following chapters. Information obtained during investigation of each planet by scientific instruments is also summarized. It is notable that every one of these spacecraft was successful. It speaks to the maturity of the US space program at the time.

T. Lund, *Spacecraft Exploring Outer Planets*, Springer Praxis Books,
https://doi.org/10.1007/978-3-031-96544-9_2

Table 2.1 Spacecraft that explored the outer planets

Spacecraft	Launch date	Planets visited	Mission
Pioneer 10	24 March 1972	Jupiter	Fly by Jupiter and investigate the planet using 11 scientific instruments
Pioneer 11	6 April 1973	Jupiter, Saturn	Fly by Jupiter and obtain gravitation assist to fly by Saturn. Explore Jupiter and Saturn using 12 scientific instruments
Voyager 1	5 September 1977	Jupiter, Saturn	Fly by Jupiter and Saturn. Return images and scientific data
Voyager 2	20 August 1977	Jupiter, Saturn, Uranus, Neptune	Fly by all outer planets on a "Grand Tour." Return images and scientific data
Cassini/ Huygens	15 October 1997	Jupiter, Saturn	Gravity assist at Jupiter and orbit Saturn. Deploy Huygens probe that landed on Saturn's moon, titan
Galileo	18 October 1989	Jupiter	Orbit Jupiter and deploy a probe through atmosphere. Return extensive images and scientific data
New Horizons	19 January 2006	Jupiter, Pluto, KBO Arrokoth	Gravity assist at Jupiter. Fly by and investigate Pluto and the KBO Arrokoth
Juno	5 August 2011	Jupiter	Polar orbit of Jupiter. Return extensive images and scientific data

Chapter 3
Launch Vehicles for Spacecraft to Outer Planets

As an introduction to launch vehicles, it would be instructive to briefly review early space exploration by the United States.

Early US Space Programs

The United States conducted an ambitious space exploration program beginning in the 1960s with several impressive successes. The first US Satellite, Explorer 1, was launched in January 1958. There was keen interest in learning more about Earth's sister planets through probes lofted by launch vehicles newly available in the early space era. The inner planets, Venus and Mars, were within reach of early launch vehicles, and they were the first to be explored. The outer planets, which are at much greater distance, were explored as launcher technology advanced. Both the United States and the Soviet Union conducted extensive investigation of the inner planets by spacecraft.

Spacecraft by the United States to the planets began with Mariner 2 that reconnoitered Venus by a flyby in August 1962 and Mariner 4 that reconnoitered Mars by a flyby in November 1964. In total, between the years 1962 and 2021, the United States sent three spacecraft to flyby Venus and three spacecraft to orbit Venus. One spacecraft was sent to make multiple flybys of the planet Mercury, and another spacecraft orbited Mercury. In that same time span, the United States sent two spacecraft to fly by Mars and ten spacecraft to orbit Mars. Impressively, missions were conducted to land three stationary spacecraft on Mars: two Viking landers in 1975 and the Insight lander in 2018. Missions to land mobile rovers on Mars included the small rover Sojourner in 1997, the slightly larger Spirit and Opportunity rovers in 2004, the midsize Curiosity rover in 2012, and the large Perseverance

T. Lund, *Spacecraft Exploring Outer Planets*, Springer Praxis Books,
https://doi.org/10.1007/978-3-031-96544-9_3

rover in 2021. The many missions to the inner planets paved the way for the eight missions to the outer planets described in this book.

The United States also conducted a methodical exploration of the moon in the 1960s and 1970s. Those explorations included the Ranger, Lunar Orbiter, Surveyor, and Apollo programs. Ranger was a photography mission as the spacecraft descended towards impact on the moon. The first impact was in July 1964. The Lunar Orbiter program orbited five spacecraft around the moon from August 1966 to August 1967 and made photographic maps of 99% of the lunar surface. The surveyor program soft-landed five spacecraft on the moon between May 1966 and January 1968. The impressive Apollo program soft-landed six manned spacecraft on the moon between July 1969 and December 1972. A crew of two astronauts from each mission walked on and explored the moon. The last three missions carried a dune buggy-type vehicle to carry astronauts on explorations farther from the landing site.

Later US missions to the moon included Lunar Prospector in 1988, Lunar Reconnaissance Orbiter in 2009, Grail A and Grail B in 2011, and Lunar Atmosphere and Dust Environment Explorer in 2013. These were all lunar orbiting spacecraft.

Launch Sites for Planetary Spacecraft

Probes to the planets from the United States were launched from Cape Canaveral in Florida. Construction of the facility began in May 1950. It first served as a launch site for several different military missiles. The first two launches were V-2 rockets.

Cape Canaveral is located on the eastern coast of Florida at a latitude of 28.4°. The Cape's location in a sparsely populated area was an advantage at the time when missile launch accidents were common. The 28.4° latitude allowed taking advantage of about 88% of the maximum velocity imparted by Earth's rotation when launching to the east. Easterly flying spacecraft would pass over the ocean, and stages could be safely dropped in secure areas when they burned out.

Another reason for the selection was that there was an Air Force station named the Joint Range Proving Ground a few miles to the south. That facility was renamed Patrick Air Force Base in 1950. Patrick Air Force Base supports and administers the Cape Canaveral Air Force Station.

New extensive launch facilities were required for the Apollo program that was planned to land astronauts on the moon. To that end, NASA acquired much of the land of Merritt Island that lies just north of Cape Canaveral. Two launch pads, 39A and 39B, were constructed on the island along with the vertical assembly building and other support structures for Apollo. Apollo 11, which landed the first astronauts on the moon, was launched from launch pad 39A in July 1969. NASA's land holding on Merritt Island was named the Kennedy Space Center.

A segment of a map showing Cape Canaveral Air Force Station and the Kennedy Space Center is shown in Fig. 3.1.

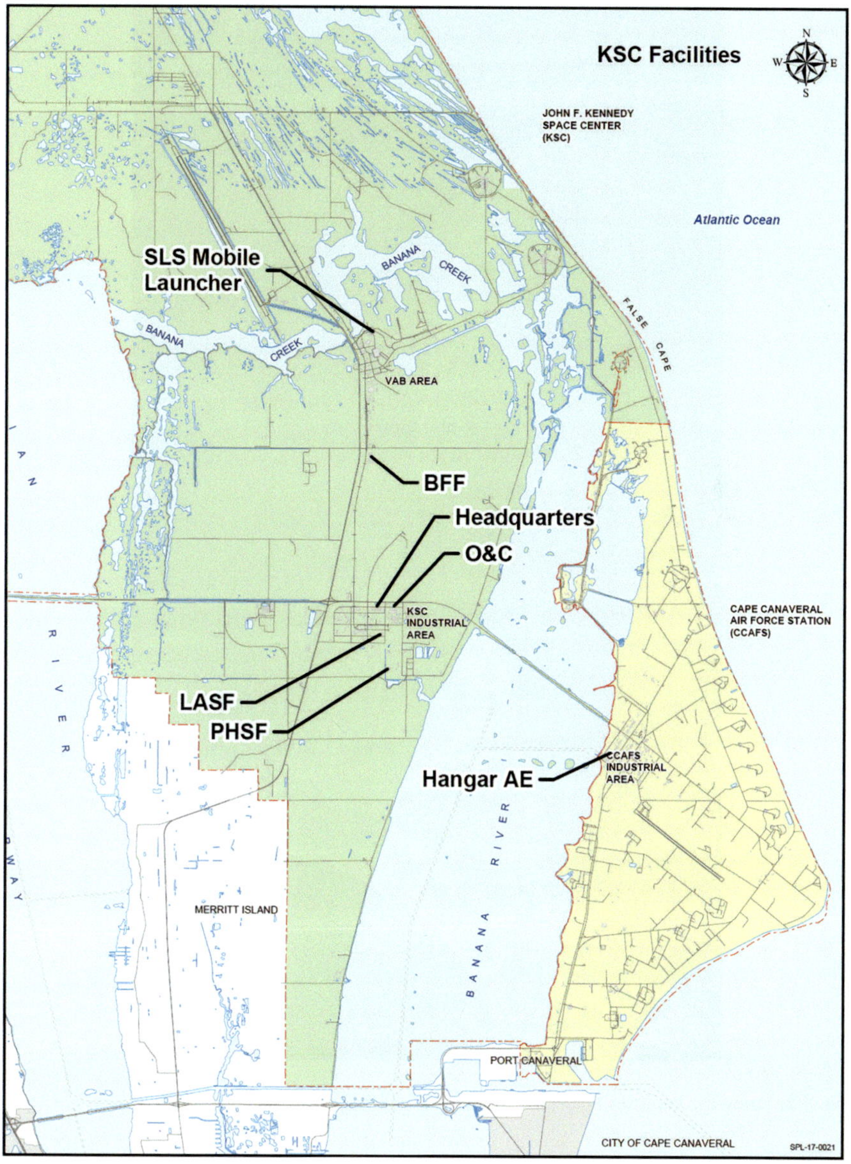

Fig. 3.1 Map showing Cape Canaveral Air Force Station in yellow and Kennedy Space Center in green. (NASA graphic)

A more detailed map of launch complexes at the Cape Canaveral Air Force Station is shown in Fig. 3.2. Names of active launch complexes (LCs) are lettered in red on the detailed map. Refurbished launch complexes were given the name Space Launch Complex (SLC). Launch complexes LC-39A and LC-39B are part of the Kennedy Space Center.

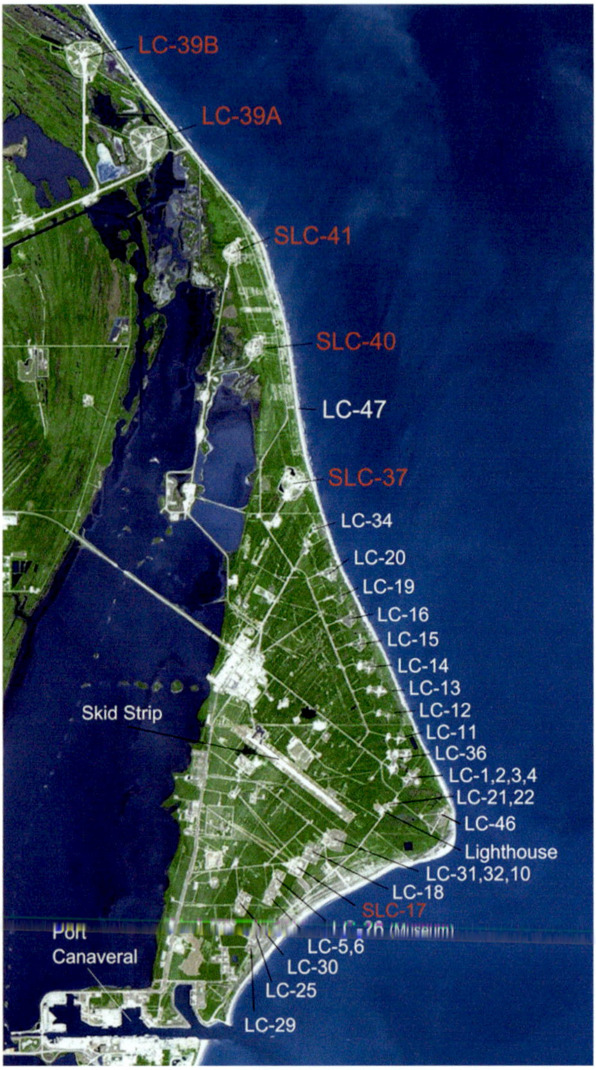

Fig. 3.2 Launch complexes at Cape Canaveral. (NASA graphic)

Presently, LC-39A is leased to SpaceX for launch of their Falcon 9 and Falcon Heavy launch vehicles. SLC-40 is used for the launch of SpaceX Falcon 9 vehicles. SLC-41 is used by the United Launch Alliance for Atlas V launches, and SLC-37B is used by the United Launch Alliance for Delta IV Heavy launches. Launch Complex-13 is used to land spent boosters from SpaceX launch vehicles.

There were a total of 72 launches from the combined Cape Canaveral Air Force Station and Kennedy Space Center in 2023. Of those, 66 were by SpaceX Falcon

9 launchers and 5 were by SpaceX Heavy launchers. Falcon 9 launches included three cargo missions and two crew transport missions to the International Space Station.

Launch Vehicles for Spacecraft That Explored the Outer Planets

Launch vehicles that lofted US spacecraft to the outer planets are listed in Table 3.1.

Launch vehicles for early planetary probes were based on the Atlas, Titan, and Thor intercontinental ballistic missiles (ICBMs). Launch vehicles that originated with the Thor ICBM were given the name Delta. Upper stages were added to the launch vehicles to accelerate spacecraft to trajectories to the planets. Pioneer 10 and 11 spacecraft were launched towards Jupiter by Atlas SLV-3/Centaur launch vehicles. Voyage 1 and 2 spacecraft were launched to the outer planets by Titan III/Centaur launch vehicles. Some planetary spacecraft including Magellan to Venus and Galileo to Jupiter were launched by the Space Shuttle. The Cassini spacecraft that orbited Saturn was launched by a Titan IV/Centaur launch vehicle. The New Horizons spacecraft that explored the dwarf planet Pluto and the Juno spacecraft that orbited Jupiter were launched by Atlas V/Centaur launch vehicles. Europa Clipper was launched by a SpaceX Falcon Heavy launch vehicle.

Table 3.1 Launch vehicles for spacecraft to outer planets

Spacecraft	Mission	Launch date	Launch vehicle
Pioneer 10	Fly by Jupiter	2 March 1972	Atlas SLV-3D/ Centaur D-1A
Pioneer 11	Fly by Jupiter and Saturn	6 April 1973	Atlas SLV-3D/ Centaur D-1A
Voyager 1	Fly by Jupiter and Saturn	5 September 1977	Titan IIIE/Centaur D-1T
Voyager 2	Fly by Jupiter. Saturn, Uranus, and Neptune	20 August 1977	Titan IIIE/Centaur D-1T
Galileo	Equatorial orbit of Jupiter	18 October 1989	Space Shuttle Atlantis
Cassini	Orbit Saturn. Deploy Huygens probe to the moon, Titan	15 October 1997	Titan IVB/Centaur
New Horizons	Flyby Pluto. Flyby Kuiper Belt Object Arrokoth	19 January 2006	Atlas V/Centaur
Juno	Polar orbit of Jupiter	5 August 2011	Atlas V/Centaur
Europa Clipper	Multiple flybys of Europa	14 October 2024	Falcon Heavy

Atlas SLV-3D/Centaur D-1A

Atlas was a successful ICBM, and it was also well suited as a base vehicle for the launch of spacecraft. Atlas D was the first operational ICBM fielded by the United States. That model was first deployed in October 1959, and a total of 30 were deployed, some in silos, early in the cold war with the Soviet Union. The E and F models that followed continued the ICBM vigil from protective silos. Atlas D was adapted to serve as launch vehicle for the Mercury manned space flight program. Later, with an upper stage, Atlas became a workhorse for launching spacecraft to the planets and for launching satellites. Atlas SLV-3D was a standardized version of Atlas D and built for space use. A photograph of the launch of Pioneer 11 to Jupiter by an Atlas SLV-3D/Centaur D-1A is shown in Fig. 3.3.

Atlas was developed and built by General Dynamics. The Atlas launch vehicle was unique in that the fuel and oxidizer tanks that constituted much of the upper structure of the vehicle used thin stainless walls and the shape of the vehicle was maintained by the internal pressure of helium gas. This resulted in significant savings in weight compared to a rigid spacecraft.

The SLV-3D launcher had a diameter of 3.05 m and length of 22.9 m. The engines were located at the bottom end of the launcher. The booster, sustainer, and vernier

Fig. 3.3 Launch of Pioneer 11 by Atlas SLV-3D/Centaur D-1A. (NASA photograph cropped by author)

engines burned rocket propellant-1 (RP-1), which was highly refined kerosene, with liquid oxygen as oxidizer. RP-1 fuel was contained in a cylindrical tank located above the engines at the bottom of the vehicle. Liquid oxygen was contained in a cylindrical cryogenic tank above the fuel tank. The engines were ignited by injecting a charge of hypergolic fluid along with RP-1 fuel into the combustion chamber where it spontaneously ignited upon contact with liquid oxygen. Burning continued on RP-1 fuel and liquid oxygen. Loaded with propellants, the launcher weighed 128,736 kg.

Five engines powered Atlas SLV-3D at launch: two booster engines, one sustainer engine, and two vernier engines. The sustainer engine was mounted at the centerline of the launcher, and the booster engines were mounted on each side of the sustainer. The two booster engines were mounted so that they could be jettisoned at the proper time after launch during the ascent trajectory. The boosters were shut down by command from the ground and jettisoned along with their associated fuel pumps. The burn time could be up to 174 s. The sustainer engine continued to burn until cut off by ground command. It could burn up to 430 s from ignition.

The booster engines were Rocketdyne type LR-89-7 with a combined thrust of two engines of 1670 kN (375,476 pounds) at sea level. The sustainer engine was a Rocketdyne type LR-105-5 with a thrust of 369.4 kN (86,866 pounds) in vacuum and 269 kN (60,473 pounds) at sea level. The vernier engines were Rocketdyne type LR101 that had a thrust of about 5155 N (1154 pounds) in vacuum and 4450 N (1000 pounds) at sea level. The total thrust at sea level of the two booster engines, one sustainer engine, and two vernier engines was 1968 kN (442,400 pounds).

The booster engines were mounted on gimbals that allowed each engine to pivot 5° in pitch and 5° in yaw with respect to the centerline of the Atlas. The pivoted booster engines were used to steer to vehicle to a preprogrammed trajectory after launch. The trajectory followed an arc that gradually tilted from vertical at launch towards the horizontal as the vehicle gained altitude and speed. The sustainer engine was also gimballed, and it could be pivoted 3° in pitch and 3° in yaw about the centerline. The engine thrusted along the centerline while the boosters were firing. Its pivoting ability was used for steering after the booster engines burned out and were jettisoned.

The vernier engines of Atlas could be oriented within a 140° arc in pitch and 50° arc in yaw. This positioning capability allowed the launch vehicle to be rolled to the desired orientation and to be controlled in pitch and yaw. The vernier engines were supplied propellants by the turbopump for the sustainer engine.

Centaur D-1A Upper Stage

The Centaur D-1A upper stage provided thrust to accelerate the spacecraft to escape velocity from Earth and on a trajectory to the planets. Centaur was developed and built by the Convair division of General Dynamics.

The Centaur D-1A upper stage was an advanced, high-performance stage that burned liquid hydrogen with liquid oxygen oxidizer. Liquid hydrogen/liquid oxygen propellants give the highest specific impulse of known propellants. The specific impulse (I_{SP}) of Centaur D-1A was 444 s. For comparison, the sustainer engine of Atlas that burned RP-1 with liquid oxygen as oxidizer had an I_{SP} of 316 s.

The Centaur D-1A stage was cylindrical, 3.05 m in diameter, and about 9.14 m long. The main propulsion came from two Pratt & Whitney RL10A-type engines that provided a total thrust of 133.45 kN (30,000 pounds). Three different variations of the engine, RL10A-3-1, RL10A-3CM-1, and RL10A-3-3, could be used depending on the mission. All three engines had essentially the same thrust. The engines were gimballed over ±4° in a square pattern to allow control of pitch, yaw, and roll of the stage during powered flight. A series of small thrusters powered by steam produced by the reaction of hydrogen peroxide with a catalyst allowed attitude control when the main engines were not firing.

Titan IIIE/Centaur D-1T

The Titan IIIE launch vehicle was developed by Martin Marietta Aerospace. Titan IIIE with a Centaur upper stage provided a significant step-up in payload weight that could be sent to the planets compared to the Atlas SLV-3D/Centaur D-1A launcher. Titan IIIE/Centaur D-1T was used to launch two Voyager spacecraft to the outer planets. It had launched two Viking spacecraft that orbited Mars and dispatched landers to the surface before launching the Voyagers. A photograph of Titan IIIE/Centaur on the launch pad with a Viking spacecraft is shown in Fig. 3.4.

Titan IIIE consisted of a two-stage core vehicle with two solid-fuel booster rockets attached to the sides. The core vehicle was 3.05 m in diameter. The length of the launch vehicle without the Centaur stage was 29.9 m. The two solid-fuel booster rockets, built by the United Technology Corp., were 3.05 m in diameter and 25.9 m long. Propellants for the boosters were powdered aluminum fuel with ammonium perchlorate oxidizer. The two boosters firing together generated a total of 10.60 million N (2.4 million pounds) of thrust at sea level. The burn time was about 117 s. The vehicle was steered during burn of the solid-fuel boosters by a thrust control system that included injectors in four quadrants in the sides of the nozzles of the booster rockets. Nitrogen tetroxide under pressure was applied to selected injectors, and that changed the thrust flow angle of the solid-fuel engines.

The core portion of the launch vehicle consisted of two stages. The two stages, referred to as stage 1 and stage 2, were 3.05 m in diameter, and the total core structure was 29.4 m long. Stage 1 was 19.2 m long, and it was the lower stage. Stage 2, mounted above stage 1, was 7.01 m long. The propellants for the stages consisted of a mixture of equal parts of hydrazine and unsymmetrical dimethyl hydrazine as fuel and nitrogen tetroxide as oxidizer. The fuel mixture was referred to as Aerozine 50. Stage 1 was powered by an Aerojet General Corp. LR-87-AJ-11 rocket engine. The engine had two combustion chambers, and the combined thrust in vacuum was 2413 kN (542,442 pounds).

Fig. 3.4 Titan III/Centaur with Viking spacecraft on launch pad. (NASA photograph)

Stage 1 was ignited 112 s after liftoff, a few seconds before burnout of the booster rockets. It burned for about 150 s. The stage 1 rocket motor was gimballed to allow steering of the launch vehicle. Stage 2 was ignited after burnout and jettisoning of stage 1.

Stage 2 was powered by an Aerojet LR-91-AJ-11 engine with a thrust of 460 kN (103,400 pounds) in vacuum. Stage 2 burned for about 205 s. The engine was gimballed to allow steering, while stage 2 was burning. About 18 s after burnout of stage 2, Titan was separated from Centaur and dropped away. The first burn of Centaur's engine occurred a few seconds after separation.

Centaur D-1T was an energetic upper stage that burned liquid hydrogen with liquid oxygen as oxidizer. Its two Pratt & Whitney RL10A-3 engines yielded an impressive specific impulse (I_{SP}) of 444 s. The thrust of Centaur D-1T was 133,452 N (30,000 pounds). Centaur D-1T was 3.05 m in diameter and 9.6 m long. Centaur contained two cryogenic propellant tanks. One tank held 11,290 kg of liquid oxygen, and the other tank held 2232 kg of liquid hydrogen.

Titan IVB/Centaur Launch Vehicle

The Cassini spacecraft, which orbited Saturn, was launched by a Titan IVB/Centaur launch vehicle. A photograph of Titan IVB/Centaur on the launch pad with Cassini within the payload fairing is shown in Fig. 3.5. Titan IVB incorporated a new and improved set of solid-fuel boosters. The two new boosters were equipped with LR-87-AJ-11A engines that provided a combined thrust of 12.84 million N (2.88 million pounds). The length of the new boosters was 34.1 m, which was 8.2 m longer than those used on Titan IIIE. The burn time of the boosters was 136 s.

The core portion of Titan IVB was essentially the same as that in Titan IIIE. A different version of Centaur was used with Titan IV. The diameter of the liquid hydrogen tank of Centaur was increased from 3.05 to 4.3 m to shorten the length of the tank. The overall length of Centaur for Titan IV was 9 m. The diameter of the liquid oxygen tank remained at 3.05 m. The Titan IV Centaur used two RL10A-3A engines with a total thrust of 146.8 kN (33,000 pounds).

Space Shuttle Launch Vehicle

The Space Shuttle evolved from the desire for a more economical access to space than by the massive, one-time-use rockets then in use. The configuration of the new launch system, referred to as the Space Shuttle, was established in 1972. It was a three-element system consisting of an orbiter, an expendable external tank carrying liquid hydrogen and liquid oxygen for the orbiter's engines, and two recoverable solid rocket boosters. The payload was carried in a large bay with hinged covers in the orbiter.

A photograph of a launch of Space Shuttle Columbia is shown in Fig. 3.6. The large red structure in the photograph is the external fuel tank. The solid rocket boosters are attached to either side of the fuel tank, and the orbiter is the airplane-like object in front in the photograph. At the end of a mission, the orbiter reentered into the Earth's atmosphere and glided to a landing at either the Kennedy Space Center in Florida or Edwards Air Force Base in California.

Contracts to develop and build the Space Shuttle were awarded in 1972. Rockwell International was selected to develop the Space Shuttle orbiter, Martin Marietta developed the external tank, and Morton Thiokol developed the solid rocket boosters. Rocketdyne, a division of Rockwell, developed the orbiter's main engines.

Fig. 3.5 Titan IVB/Centaur on the launch pad with Cassini spacecraft within the payload fairing. (NASA photograph)

Six orbiters were built: Columbia, Challenger, Discovery, Atlantis, Endeavour, and Enterprise. Tragically, Challenger and Columbia and their crew were lost due to accidents. Endeavour was built to replace Challenger. The orbiter Enterprise was only used for testing in the atmosphere. The first manned flight of the Space Shuttles occurred in April 1981. The last flight occurred in August 2011. Space Shuttles flew 135 missions during their 30 years of service. The Magellan spacecraft to Venus and the Galileo spacecraft to Jupiter were launched by the Space Shuttle.

Fig. 3.6 Launch of Space Shuttle Columbia. (NASA photograph)

Orbiter

The orbiter was about the size of a regional passenger jet airplane. The orbiter was 37.2 m long, and the width at mid-fuselage was about 5.2 m. The wingspan was 23.8 m. Major structural sections of the orbiter were the forward fuselage, which contained the pressurized crew compartment; the mid-fuselage, which contained the 4.6 m wide and 18.3 m long payload bay and associated doors; the aft fuselage; and the vertical tail. The crew compartment had a flight deck on top of a middeck. The usual crew was seven people. The payload bay was designed to carry cargoes weighing up to 29,000 kg (32 tons) to Earth's orbit.

The orbiter needed to be protected from the searing heat due to friction when reentering the Earth's atmosphere. The thermal protection system kept the temperature of the aluminum skin below 177 °C (350 °F), while temperatures on the leading edge of the wings could rise to 1510 °C (2750 °F) during reentry. The nose cap and leading edges of the wings were protected with an all-carbon composite consisting of layers of graphite cloth in a carbon matrix. Areas subjected to lower heat were shielded with high-temperature ceramic tiles.

The orbiter contained three main rocket engines that used liquid hydrogen for fuel and liquid oxygen for oxidizer. The thrust of each engine was 1.67 million N (375,000 pounds) at sea level. The thrust increased to 2.09 million N (470,000 pounds) in a vacuum. The engines could be throttled over a range of 56–104% of maximum rated thrust.

External Tank

The external tank, which was 47 m long and 8.38 m in diameter, carried 143,000 gal of liquid oxygen and 383,000 gal of liquid hydrogen. As well as carrying fuel, the external tank formed the structure to which the orbiter and the two solid rocket boosters were attached. The orbiter was attached to the tank at the forward end by a wishbone attachment beneath the crew compartment. The aft connection to the orbiter was through tripods at the base of the external tank.

Solid Rocket Boosters

The two solid rocket boosters each generated 11.79 million N (2.65 million pounds) of thrust at sea level. The total thrust of the orbiter's three engines and two booster rockets at liftoff was 28.59 million N (6.43 million pounds).

The solid rocket boosters were 45.5 m long and 3.7 m in diameter. The weight of the propellants in each booster was 500,000 kg (1.1 million pounds). The burn time was about 127 s. After the boosters exhausted their fuel at about 45 km altitude, they were jettisoned from the external tank. The boosters were slowed by parachutes and fell into the ocean where they were recovered, refueled, and reused.

Atlas 5 Launch Vehicle

Atlas 5 was the latest and most energetic of the Atlas family of launch vehicles. Developed by Lockheed Martin, the first flight of Atlas 5 was in 2002. Atlas 5 differed from previous Atlas versions in that its fuel and oxidizer tanks were rigid structures instead of the thin-walled tanks supported by internal pressure as in previous models. All versions of the Atlas 5 used a Common Core Booster (CCB) first stage. The CCB was 3.8 m in diameter and 32.5 m long.

The CCB was powered by one RD-180 rocket engine. The RD-180 engine was developed by NPO Energomash in Russia. It was built in the United States by Pratt & Whitney/NPO Energomash. The RD-180 engine provided substantial increase in performance over existing US engines at the time. The engine had two combustion chambers with associated exhaust nozzles. The propellants were RP-1 (kerosene) with liquid oxygen as oxidizer. Thrust of the engine at sea level was 3830 kN (860,000 pounds). Thrust in a vacuum was 4150 kN.

Up to five strap-on solid-fuel rockets could be mounted around the Common Core Booster to increase liftoff thrust. The solid-fuel boosters were 1.55 m in diameter and 19.5 m long. The thrust of each solid-fuel booster was 1686 kN (379,550 pounds). The boosters burned for 90 s.

Models of the Atlas V included the Atlas V 400 (401) and the Atlas V 500 (501, 511, 521, 531, 541, and 551) series of launchers. The first digit of the numbering system indicates the diameter in meters of the payload shroud. The second digit indicates the number of solid rocket boosters attached to the Common Core Booster, and the third digit indicates the number of engines in the Centaur upper stage. The New Horizons and Juno spacecraft were launched by Atlas V type 551 launch vehicles. A photograph of the New Horizons spacecraft about to be launched by an Atlas V 551 is shown in Fig. 3.7.

Both the Atlas V 400 series and 500 series incorporated what was referred to as the Common Centaur upper stage. Common Centaur was 3.05 m in diameter and 12.7 m long. It could be fitted with one or two Pratt & Whitney RL10A-4-2 engines. Launch vehicles for the New Horizons and Juno spacecraft used one Centaur engine. The engines could be stopped and restarted. The thrust from one engine was 99.2 kN (22,300 pounds). Fuel for the engine was liquid hydrogen with liquid oxygen as oxidizer. A total of 20,830 kg (45,922 pounds) of propellant was carried.

Falcon Heavy (Expendable)

Europa Clipper was launched by a Falcon Heavy (Expendable) launch vehicle. Falcon Heavy is a heavy-lift launch vehicle produced by Space Exploration Technologies Corporation. The corporation, founded by Elon Musk, is commonly known as SpaceX. Falcon Heavy is an extension of the very successful Falcon 9 launch vehicle that has launched several hundred payloads to space.

Falcon Heavy consists of the central core of a Falcon 9 with two Falcon 9 first stages attached to the sides. The Falcon 9 stages are reinforced for Falcon Heavy, and the two extra booster stages have nose cones at the front ends. A picture of Falcon Heavy prepared to launch Europa Clipper is shown in Fig. 3.8.

The three boosters that constitute the first stage contain nine Merlin engines each. The Merlin engine, developed by SpaceX, burns kerosene (RP-1) in the presence of liquid oxygen oxidizer. The thrust of the engine at sea level is 845,000 N (190,000 pounds). The 27 engines firing together generate 22.815 million N (5.13 million pounds) of thrust at liftoff. The core second stage generates a thrust of

Fig. 3.7 Atlas V launcher with New Horizons spacecraft on launch pad. (NASA photograph cropped by author)

981,000 N. The height of Falcon Heavy is 70 m, and the diameter of the central core stage is 3.66 m.

Normally, the three booster sections are recovered after their work is finished. The boosters fly back to a landing site, legs are deployed, and boosters are guided to a soft vertical landing. The Falcon Heavy (Expendable) launch vehicle used for the launch of Europa Clipper was a maximum performance version that used all of its fuel to loft the spacecraft without saving some for recovery of the boosters. Landing legs were omitted from the boosters for the expendable model.

Fig. 3.8 Falcon Heavy on launch pad with Europa Clipper. (NASA image cropped by author)

References

Convair Aerospace and Martin Marietta Aerospace, *Titan IIIE/Centaur D-1T Systems Summary,* September 1973

Ezell, Edward Clinton, Exell, Linda Neuman, *On Mars, Exploration of the Red Planet 1958-1978*, NASA SP-4212, 1984 (Appendix E presents details of Atlas/Centaur, and Titan IIIE/Centaur launch vehicles)

Glenn Research Center, *NASA Glenn Efforts Launch Cassini Toward Saturn*, May 2008

http://b14643.eu/spacerockets_2/United_States_6/Titan_iii/Propulsion/engines.htm

Lockheed Martin, *Atlas Launch System Mission Planner's Guide*, December 1998

Lockheed Martin, *Atlas V Launch Services Users Guide,* March 2010

NASA, *Space Shuttle Fact Sheet*, https://www.nasa.gov/centers/johnson

NASA, *The Space Shuttle and its Operations*, https://www.nasa.gov/centers/johnson

Odom, James B., *Space Shuttle Propulsion Master's Forum Presentation Mitigating risk through Testing,* May 2009

SpaceX, *Falcon User's Guide*, 2021

Chapter 4
Pioneer 10 and 11 Spacecraft

Two successful spacecraft, Pioneer 10 and 11, provided the first detailed scientific information about the outer planets Jupiter and Saturn. The well-instrumented spacecraft explored Jupiter during flybys by Pioneer 10 in December 1973 and by Pioneer 11 in December 1974. After flyby, Pioneer 10 altered its path to explore Jupiter's massive moon, Ganymede, and it continued travel out beyond the orbits of the outer planets and into interstellar space. Pioneer 11 used gravity assist from Jupiter to travel another 4.7 years out to Saturn for exploration of that planet during a flyby in September 1979. It then journeyed out beyond the orbits of the outer planets and into interstellar space.

Both spacecraft maintained contact with Earth for over 20 years and at distances well past the orbits of the outer planets. A wealth of scientific information was received along with hundreds of images of Jupiter and Saturn systems.

Pioneer 10 and Pioneer 11 spacecraft were essentially the same. An artist's conception of the appearance of Pioneer 10 at Jupiter is shown in Fig. 4.1.

Background of Pioneer Jupiter Program

The Pioneer Jupiter program had its genesis in a series of studies and papers by NASA and industry that showed the unique utility of a spacecraft sent to explore Jupiter. The spacecraft could be used to explore interplanetary space, explore the environment of Jupiter, and use gravity assist from Jupiter to travel to Saturn and beyond to explore interstellar space.

NASA approved the mission to Jupiter in February 1969. Management of the mission was assigned to the Pioneer Project Office of NASA AMES Research Center located in Mountain View, California. The Pioneer Project Office at AMES

T. Lund, *Spacecraft Exploring Outer Planets*, Springer Praxis Books,
https://doi.org/10.1007/978-3-031-96544-9_4

Fig. 4.1 Artist's concept of Pioneer 10 at Jupiter. (NASA graphic, cropped by author)

had previously managed a series of spacecraft missions involving Pioneers 6, 7, 8, and 9 that were sent into orbit around the sun to measure the solar wind, solar magnetic field, and cosmic rays. Charles Hill became Pioneer Project Manager for the Pioneer Jupiter program at NASA AMES.

Two identical spacecraft were planned that were initially referred to as Pioneer F and Pioneer G, but later became known as Pioneer 10 and Pioneer 11. The TRW Systems Group in Redondo Beach, California, was contracted to build the two spacecraft. NASA Lewis Research Center in Cleveland, Ohio, was responsible for the launch vehicle, and NASA Jet Propulsion Laboratory (JPL) provided tracking and data system support. NASA Goddard Space Flight Center was responsible for worldwide communications with the spacecraft.

Minimum energy launch windows to Jupiter occur about every 13 months. The launch of Pioneer 10 was planned for the 1972 launch opportunity, and the launch of Pioneer 11 was planned for the 1973 opportunity. Pioneer 10 was launched on 2 March 1972 and made its closest approach to Jupiter during a flyby on 4 December 1973. Pioneer 11 was launched on 5 April 1973 and made its closest approach to Jupiter on 3 December 1974. Using gravity assist from Jupiter, Pioneer 11 traveled on to Saturn and made the closest approach to Saturn on 1 September 1979.

Both spacecraft were successful, and both traveled out past the orbit of Neptune towards interstellar space. Last contact with Pioneer 10 occurred 30 years after launch when it was about 12 billion km from Earth. Last contact with Pioneer 11 occurred 22 years after launch when it was about 6.6 billion km from Earth.

A descriptive plaque was mounted on each spacecraft to give information about Earth in case the spacecraft was intercepted by an interstellar civilization. The plaque was promoted by Carl Sagan, a leading US scientist who was esteemed for his direct talk about space physics to the general public. An illustration of the plaque

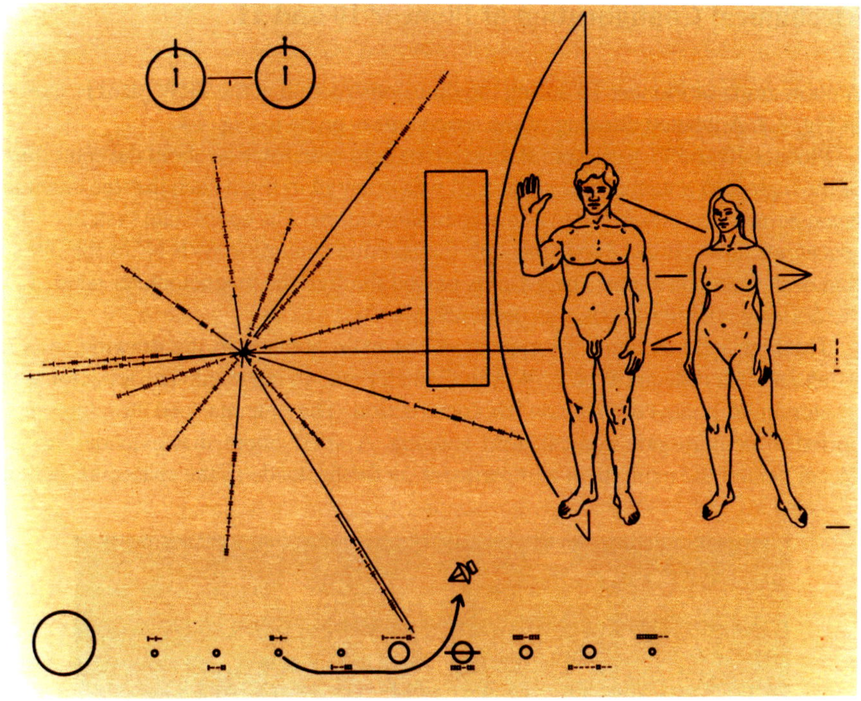

Fig. 4.2 Plaque carried by Pioneer 10 and 11. (NASA graphic)

from NASA Report SP-446 is given in Fig. 4.2. The plaque was 15 cm by 23 cm in size.

The graphic in the upper left of the plaque represents the two energy states of the neutral hydrogen atom. The vertical lines in the center and perimeter represent the spin axis of the neutron and electron, respectively. It is difficult to see at the scale of the graphic, but the left atom has a dot at the top of the neutron spin axis symbol and a dot on the bottom end of the electron spin axis symbol indicating that the spin axes are in different directions. The right atom has dots on the upper end of both the neutron and electron spin axes indicating that the neutron and electron spin axes are aligned. The two states have slightly different energy levels, and energy in the form of radiation is released when an atom transitions from the high-energy to the low-energy state. The wavelength of the radiation is about 21.1 cm. This signature radiation is observed in clouds of hydrogen in deep space and would be familiar to advanced civilizations.

A bar between the two atoms was provided for scaling purposes in the plaque. The length represents the wavelength of the radiation of 21.1 cm. There are two tick marks defining the height of the female Earth inhabitant, and the binary equivalent of decimal 8 is located in the space between ticks. Eight 21.1 cm bars would put the height of the female inhabitant at 1.69 m or about 5 ft 6 in.

Mechanical Configuration of Pioneer 10 and 11

A photograph of Pioneer 10 during the final construction at the TRW plant is shown in Fig. 4.3. A labeled drawing of the spacecraft is given in Fig. 4.4.

Pioneer 10 and 11 were spin stabilized in flight with rotation about the centerline of the high-gain antenna. The spacecraft was oriented in flight such that the antenna beam pointed towards Earth to allow communications. The coordinate system of the spacecraft directed the Z-axis through the antenna centerline with $+Z$ in the direction of the antenna beam. The $+X$-axis passed through the magnetometer labeled in the drawing, and the Y-axis was perpendicular to it.

The high-gain antenna, referred to as the main antenna in the drawing, was a parabolic dish reflector type 2.74 m in diameter and 46 cm deep. The reflector was of aluminum honeycomb sandwich construction. The antenna feed structure supported a medium-gain conical horn antenna. A short boom that extended about 76 cm behind the equipment compartment on the backside of the spacecraft held a

Fig. 4.3 Pioneer 10 spacecraft during the final construction at TRW. (NASA photograph)

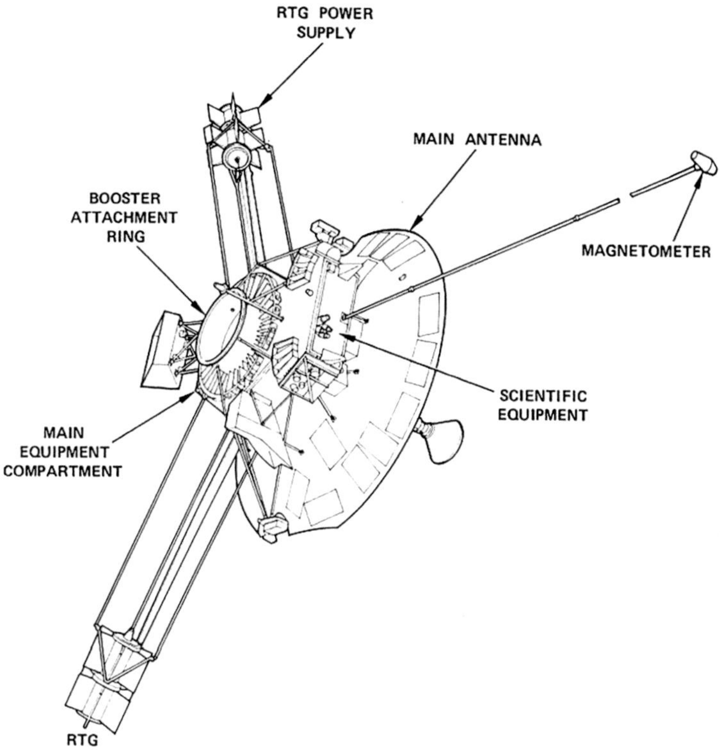

Fig. 4.4 Drawing of Pioneer 10 and 11 spacecraft. (NASA graphic)

low-gain, omnidirectional antenna. The boom and antenna are shown in Fig. 4.4, but they are not labeled.

Appendages on the spacecraft included two booms that held radioisotope thermoelectric generators (RTGs) that supplied power to the spacecraft. The deployable booms spaced the RTGs about 3 m from the centerline of the spacecraft to reduce the effects of radiation on the electronics. Another boom, which extended 6.6 m from the center of the spacecraft, held a magnetometer. The spacing reduced the influence of the residual field of the spacecraft.

A hexagonal equipment compartment was mounted on structure behind the high-gain antenna. It contained electronics for spacecraft systems and a tank for hydrazine propellant for the thrusters. Most of the electronic equipment was mounted to the sides of the hexagon. The sides unfolded for access. As per the scaling from drawings, the equipment compartment was about 1.1 m across the sides of the hexagonal structure and 42 cm high. A scientific equipment compartment containing electronics for the experiments was mounted next to the main equipment compartment on the back of the antenna. That compartment had a complex shape with extents of 1.1 m long, 42 cm wide, and 35 cm high.

The spacecraft was controlled in space by small thrusters that reacted hydrazine with a catalyst in a thrust chamber. The resulting hot gases from decomposition of the hydrazine exited the thrust chamber through a nozzle to provide a thrust of 4.5 N (1 pound). There were six thrusters grouped in three pairs. The thrusters were located in the cutouts on the periphery of the high-gain antenna reflector.

Structure on the backside of the spacecraft held a booster attachment ring that held the spacecraft firmly to a solid-state booster rocket during launch. The length of the spacecraft from the booster attachment ring to the end of the medium-gain antenna was about 2.4 m. At launch, the spacecraft weighed 258 kg.

Pioneer 10 and 11 Systems

Electrical Power

Electrical power for Pioneer 10 and 11 was generated by four SNAP 19 radioisotope thermoelectric generators (RTGs). The spacecraft was intended to travel far from the sun into interstellar space. Energy from the sun would be too low to generate sufficient power from solar arrays at those large distances. RTGs generate electrical power by applying heat from radioactive decay of plutonium-238 material to a series of thermoelectric couples. The half-life of the decay process of plutonium-238 is 87.7 years, which far exceeded the mission requirements.

The SNAP-19 RTGs were cylindrical in shape with six narrow radial fins running the length of the device on the periphery of the cylinder. The fins radiated excess heat to space. The length of the RTG was about 28 cm, and the span across the tips of opposing fins was about 40 cm. An RTG weighed about 14.4 kg.

Each RTG held a total of 90 thermoelectric couples arranged in six groups of 15 mounted around the heat source. A group of 45 thermoelectric couples were connected in series, and the series was connected in parallel with a second series of 45 thermoelectric couples. The nominal output voltage of the RTG under load was 4.2 V DC. The nominal output power available was about 40 W at launch and 33 W minimum during encounter with Jupiter. The four RTGs generated a total of 160 W at launch and about 132 W at Jupiter.

The RTGs were mounted at the ends of booms to space them away from the electronics to lessen radiation levels. The spacing was about 3 m from the spin axis. Two RTGs were mounted in tandem at the ends of each of the two booms as shown in the spacecraft drawing in Fig. 4.4. The output of each RTG was applied to an inverter in the electrical power subsystem. The inverters converted 4.2-V DC power to an AC waveform at a voltage of 61 V rms and a frequency of 2.5 kHz. The four inverters were synchronized so that the outputs of the four inverters could be summed. The summed output of the inverters was applied to a power control unit (PCU) and a central transformer rectifier filter (CTRF).

The PCU transformed the AC waveform to a lower voltage and rectified it to form a regulated 28-V DC power bus. The PCU also charged the battery and accepted power from the battery when spacecraft power load was more than could be supported by the RTGs. The CTRF contained separate rectifiers and filters to provide regulated DC voltages of ±12, ±16, and +5 V to power spacecraft electronic circuits.

Attitude Control

Pioneer 10 and 11 were spin stabilized by rotation around the *Z*-axis of the spacecraft. The spin rate of Pioneer 10 was 4.8 rpm as intended, but the spin rate of Pioneer 11 was 7.8 rpm because of a thruster failure. The *Z*-axis corresponded to the centerline of the beam of the high-gain antenna. The antenna beam was directed closely to Earth to maximize signal level for communications. The attitude control subsystem of the spacecraft used a sun sensor assembly and a stellar reference assembly as reference for controlling thrusters to maintain the desired attitude of the spacecraft. A conical scan procedure using the high-gain antenna was used periodically to fine-tune the pointing direction to Earth.

The sun sensor had three segments giving a total angular coverage in the *Y–Z* plane of 169° and narrow angle coverage in the perpendicular plane. As the spacecraft rotated about its *Z*-axis, the fanlike field of view swept through space. Detectors in the instrument, operating differentially, generated a pulse when the field of view was swept by the sun. The accuracy of the measurement of roll angle to the sun was better than 0.2° with the elevation of the sun relative to the *Y*-axis between ±40°. The accuracy decreased to about 0.5° when the elevation angle was ±80° from the *Y*-axis.

The stellar reference assembly had a fanlike field of view that was also swept through space as the spacecraft rotated. The field of view was 38° in a direction parallel to the spin axis and 0.5° in the perpendicular direction. The centerline of the instrument was located in the *X–Y* plane, displaced 25° from the *Y*-axis. The instrument was normally used to detect the star Canopus. The accuracy of measurement of roll angle to the star was about 0.16°.

When conical scan was ordered, the high-gain antenna was put into a conical scan mode by offsetting the feed horn. A signal was sent up from Earth, and the signal level at the output of the receiver was monitored. Rotation of the spacecraft about the *Z*-axis caused a sinusoidal error signal to appear at the receiver output if the line of sight to Earth and the angle of the antenna centerline were different. The magnitude of the signal was proportional to the angle between the line of sight to Earth and the centerline of the antenna, and the phase of the signal gave the direction of the error.

A control electronics assembly processed inputs from the sun and star sensors and inputs from the receiver during conical scan. Spacecraft attitude was primarily determined from the knowledge of the direction to Earth and the direction to the sun. Attitude error information was used to drive a series of six thrusters to change the attitude of the spacecraft.

The thrusters were arranged in three pairs on the spacecraft. Two thrusters in a pair thrusted in opposite directions. One pair, located on the periphery of the high-gain antenna at the −*Y*-axis, thrusted parallel to the *X*-axis and was used for spin control. Another pair at the same location thrusted parallel to the *Z*-axis and was used for precession control and to induce velocity parallel to the *Z*-axis. A similar pair located at the antenna periphery at the +*Y* axis was also used for precession control and velocity increments.

The thrusters generated a thrust of 4.5 N (1 pound) each. Thrust was obtained by feeding hydrazine into a combustion chamber where it reacted with a catalyst. The hot gases resulting from decomposition of the hydrazine exited a nozzle on the combustion chamber and generated thrust in the opposite direction.

Hydrazine was contained in a spherical tank about 42 cm in diameter. The tank, which held 27.3 kg (60 pounds) of hydrazine, contained a diaphragm within it. Nitrogen gas was present on one side of the diaphragm, and hydrazine was on the other side. The nitrogen, pressurized to about 3.5 MPa (500 psi) at launch, forced hydrazine out of the tank and into combustion chambers of the thrusters when valves to the thrusters were opened.

Telecommunications Subsystem

The telecommunications subsystem received and demodulated command data uplinked from the very large antennas of the Deep Space Network (DSN) on Earth, and it transmitted scientific and engineering data on the downlink to the DSN. Impressively, communication was maintained with Pioneer 10 out to a distance of 12 billion km from Earth. Communication with the DSN was carried out in the S-band frequency range. The uplink frequency was about 2110 MHz, and the downlink frequency was about 2290 MHz.

Antennas

Three different antennas were available on the spacecraft for uplink and downlink functions. The primary antenna was a high-gain parabolic reflector type 2.74 m in diameter. It had a gain of about 33 dB (factor of 2000 above an isotropic radiator) and half-power beamwidth of about 3.4° at the downlink frequency. The high-gain antenna beam was directed along the spacecraft +*Z*-axis. The high-gain antenna could be used in a conical scan mode by displacing the feed. That action moved the antenna beam about 1° from the normal centered position, and rotation of the spacecraft resulted in conical scan operation. The conical scan operation, which was described in the last section, allowed close alignment of the centerline of the high-gain antenna beam with Earth.

The medium-gain antenna was a conical horn mounted to the support structure holding the feed for the high-gain antenna. The beamwidth of the medium-gain antenna was about 32°. The antenna was canted 9.3° away from the *Z*-axis of the spacecraft to allow coarse conical scanning operation as the spacecraft rotated. An omnidirectional antenna was mounted on a short boom on the opposite side of the spacecraft from the high-gain antenna. The outputs of the medium-gain antenna and the omni antenna were combined to give antenna coverage over most of the region around the spacecraft.

Either the high-gain antenna or the combined medium-gain/omni antennas could be selected for downlink or uplink. The antenna ports were connected to a diplexer that allowed transmitting and receiving by the same antenna. The spacecraft contained two transmitters and two receivers for redundancy. Transfer switches allowed either transmitter or either receiver to be connected to either the high-gain or the medium-gain/omni antenna ports.

Receiver

The spacecraft carried two receivers for redundancy. One receiver was connected to the high-gain antenna port and the other to the medium-gain/omni antenna port. Either receiver could be connected to either antenna port. The two receivers operated at different frequencies, and a particular receiver was selected by transmitting a particular frequency from Earth.

Commands from Earth were transmitted to the spacecraft by very large antennas of the Deep Space Network. The data rate was rather slow at 1 bit/s. The low data rate allowed commands to be processed in the spacecraft at very long distances planned for the mission. The command words were 22 bits in length. It took 22 s to transmit each word. A total of 222 different commands were available. Of those, 149 were used to control spacecraft systems and 73 were used to control the experiments.

The receiver detected and processed the uplink signal from Earth and interpreted the command words. The command word output from the receiver was applied to one of the two digital decoder units (DDUs). The output of either receiver could be applied to either DDU. Each DDU in turn drove a command distribution unit, which routed commands to various functions in the spacecraft.

The receiver also generated a signal at a frequency of about 114 MHz, which was coherent with the received signal from Earth at a ratio of 12/221. This signal was applied to the transmitter driver which amplified and multiplied to a frequency of about 2292 MHz. The ratio of received signal to transmitted signal was 240/221. The coherent transmit signal allowed two-way Doppler to be measured, which gave very accurate measure of velocity of the spacecraft relative to Earth.

Transmitter

The spacecraft contained two redundant transmitters with associated transmitter drivers. The transmitter driver accepted a coherent signal from the receiver at a frequency of about 114 MHz. In case the signal from the receiver was not available, a precision oscillator within the driver generated a signal that provided a signal at the output of the driver at 114 MHz. The driver amplified the signal and multiplied its frequency by a factor of 20. The transmitter driver accepted a subcarrier at a frequency of 32.768 from the digital telemetry unit. The subcarrier was biphase modulated by telemetry data. The data bit rate varied from 2028 bits/s when the spacecraft was still near Earth to 1024 bits/s when near Jupiter and as low as 16 bits/s when at great distances. The output of the transmitter driver was at a power level of about 50 mW and a frequency of about 2292 MHz and was biphase modulated by telemetry data.

The output of the transmitter driver was applied to the signal input of the transmitter. The transmitter was a traveling wave tube amplifier with nominal output power of 8 W at a frequency of about 2292 MHz. Only one of the two transmitters was powered at a time in response to commands from Earth. The output of the selected transmitter could be applied to either the high-gain antenna or the medium-gain/omni antenna, again by commands from Earth.

Data Handling Subsystem

The data handling subsystem gathered data from spacecraft systems and experiments, converted the data to digital form when required, and organized the data into a digital data stream representing telemetry data. The data was encoded by biphase modulation on a subcarrier, which was applied to the transmitter driver. The resulting transmitter output was biphase modulated by telemetry data.

The data handling system included a digital telemetry unit (DTU) and a digital storage unit (DSU). The DTU included a series of multiplexers to select particular engineering or science experiment data. It could select 55 serial digital data streams, 92 bi-level signals, and 111 different analog signals. Selected analog signals were converted to digital form by a 6-bit analog-to-digital converter.

Downlink data was contained in a series of 64-bit words. There were 13 different arrangements of data contained in the words that could be selected by commands from Earth. The various arrangements were optimized for particular phases of the mission such as interplanetary cruise, encounter with planets, and interstellar cruise.

The digital storage unit, which provided 49,152 bits of storage, was a ferrite core type. It was used to store telemetry data when the data rate exceeded the available real-time data rate or when communications was lost as when the spacecraft was occluded by a planet. It could also be used to store up to five uplink commands when required.

Flight of Pioneer 10 and 11

Initial Flight of Pioneer 10 and 11

Pioneer 10 was launched on 3 March 1972 from Pad 36A at Cape Canaveral in Florida. The launch vehicle was an Atlas SVL-3D/Centaur D-1A with a Thiokol TE36-4 solid rocket third stage added. The initial flight trajectory entailed direct ascent to a heliocentric orbit to intercept Jupiter. Pioneer 11 was launched on 6 April 1993 from Pad 36B at Cape Canaveral in Florida. The launch vehicle was the same type as used as for Pioneer 10.

The two booster engines, sustainer engine, and two vernier engines of Atlas generated a total thrust of 1.914 million N (431,300 pounds) at liftoff. After liftoff, the vehicle rose vertically for about 2 s, and then it was rolled and pitched to the proper attitude. The booster engines were cut off at about 139 s after liftoff by a signal from the Centaur guidance system after it sensed an acceleration of 5.7 g. The boosters were jettisoned 142 s after liftoff, and the sustainer engine continued to fire. The sustainer engine continued to fire until fuel was depleted about 248 s after liftoff. Atlas was separated from the spacecraft at 250 s after launch, and it dropped away.

Centaur was fired at 260 s after liftoff, and the nose fairing was jettisoned 20 s after that. Centaur fired for 455 s and was cut off at 715 s after liftoff. The spacecraft coasted for 70 s after cutoff of Centaur. Small rockets were then fired on a spin plate that mounted the Thiokol TE-M-364-4 solid-fuel rocket to cause the spacecraft to rotate at 30 rpm. Centaur was separated from the spacecraft at 787 s after liftoff, and it dropped away. The solid-fuel rocket was fired at 800 s after liftoff, and it fired until fuel was expended at 844 s after liftoff. The solid-fuel rocket was separated from the spacecraft at 945 s (15.75 min) after liftoff.

Firing of the solid-fuel rocket increased the velocity of the spacecraft to 51,592 km/h. That was well above the escape velocity of Earth of 40,270 km/h and allowed a direct flight to intercept Jupiter.

The booms holding the RTGs and the magnetometer were deployed about 20 min after liftoff. Deployment of the booms reduced the spin rate, and thrusters were fired to set the spin rate to the intended value of 4.8 rpm.

The spacecraft was maneuvered to align the high-gain antenna with Earth. Radiation monitoring instruments were turned on for checkout while the spacecraft was passing through the Van Allen radiation belts of Earth. The scientific instruments were turned on sequentially between the second and tenth days after launch.

The trajectory of the spacecraft was accurately determined after a few days of flight by tracking stations of the Deep Space Network (DSN). The DSN included three major tracking stations: Goldstone near Bakersfield, California; Tidbinbilla near Canberra, Australia; and Madrid near Madrid, Spain. The stations were spaced around the rotating Earth so that at least one station would be able to communicate with a spacecraft at all times. Each station contained a 64-m-diameter parabolic antenna along with smaller antennas. The stations provided angular and range tracking, sent commands to the spacecraft, and received telemetry data from the spacecraft.

Flight of Pioneer 10

A course correction was made 4 days after launch to correct dispersion of the trajectory caused by imperfections in the launch vehicle. A second course correction was made on 26 March 1972 to refine the trajectory. The last correction was made on 26 November 1973, just before the spacecraft entered the magnetosphere of Jupiter.

The spacecraft passed through the asteroid belt on its journey to Jupiter. The asteroid belt is a belt lying between the orbits of Mars and Jupiter that contains over a million asteroids greater than 1 km in diameter and millions of smaller asteroids. The largest known asteroid is Ceres, which is about 470 km in diameter. The belt is about 170 km wide and extends from 380 million km to 550 km from the sun. Fortunately, for traversing spacecraft, the average distance between asteroids is about 1 million km.

Pioneer 10 first encountered the asteroid belt on 15 July 1972 and emerged in February 1973. It passed through the asteroid belt with only a few minor asteroid strikes. A NASA graphic illustrating the trajectory of Pioneer 10 to Jupiter and location of the asteroid belt is given in Fig. 4.5.

Pioneer 10 began imaging Jupiter with its photopolarimeter on 6 November 1973 when still about 29 million km away. Other scientific instruments began providing data shortly thereafter. The spacecraft reached the boundary of magnetosphere of Jupiter, which corresponded to the location of the bow shock, at a distance of about 10 million km, on 26 November 1973.

The trajectory of Pioneer 10 was designed to have the spacecraft pass about three planet radii (209,733 km) from the center of Jupiter. The spacing was chosen to survive high radiation levels from Jupiter while still providing acceptable resolution from the scientific instruments. Pioneer 10 made its closest approach to Jupiter on 4

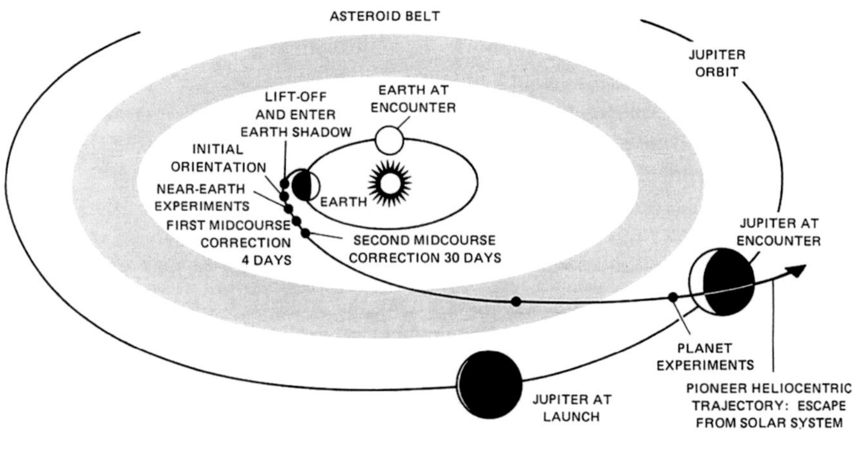

Fig. 4.5 Trajectory of Pioneer 10. (NASA graphic)

December 1973 at about 130,354 km above the cloud tops. Spacecraft velocity at the time of closest approach was 126,000 km/h.

Gravity of Jupiter bent the trajectory so that the spacecraft passed behind Jupiter about an hour after the closest approach. This passage allowed an occultation experiment to be conducted that provided information on Jupiter's atmosphere. The spacecraft emerged from behind the planet, and probing by the scientific instruments continued. Data from the scientific experiments and images from the photopolarimeter continued to be gathered during the encounter and until 2 January 1974 when the encounter period was declared over.

Pioneer 10 took about 500 images of Jupiter from the first image on 6 November to the end of encounter period. In addition, it imaged the Jupiter moons Callisto, Ganymede, and Europa. An image of Jupiter taken by the imaging photopolarimeter at a range of 2.5 million km is given in Fig. 4.6. The shadow of the moon Io is near the center.

Pioneer 10 continued outbound through interplanetary space. It became the first spacecraft to enter interstellar space when it crossed the heliopause on 25 August 2012. Regular contact with Pioneer 10 was stopped in March 1997 due to budgetary considerations. The last communication with Pioneer 10 occurred in January 2003 when it was about 12 billion km from Earth.

Fig. 4.6 Image of Jupiter taken by the photopolarimeter of Pioneer 10 at a range of 2.5 million km. (NASA image)

Flight of Pioneer 11

Pioneer 11 was launched on 6 April 1993 from Pad 36B at Cape Canaveral in Florida. The launch vehicle and launch procedure were the same as for Pioneer 10 described previously.

The process of configuring Pioneer 11 after launch was also the same as for Pioneer 10. One RTG boom did not deploy completely initially, but it deployed fully after firing thrusters to shake it loose. The spin rate of Pioneer 11 was intended to be 4.8 rpm, but because of an inoperative thruster, it settled at 7.8 rpm.

Pioneer 11 embarked on its long cruise towards Jupiter, and it had successfully passed through the asteroid belt by March 1974. Midcourse corrections were made on 20 April and again on 26 April 1974. The corrections, made by thrusters, increased the velocity of the spacecraft by 230 km/h and set it on a course for closest approach of about 43,000 km over the cloud tops of Jupiter on the far side of the planet.

The trajectory of Pioneer 11 was designed to go under the south pole of Jupiter to avoid Jupiter's high equatorial radiation and pass closer to the planet than Pioneer 10 to take advantage of gravitational assist of Jupiter to pick up additional speed necessary to travel on to Saturn. A sketch of the trajectories of Pioneer 10 and 11 looking down at the north pole of Jupiter is shown in Fig. 4.7. A sketch of the trajectories of Pioneer 10 and 11 as seen from Earth is shown in Fig. 4.8.

Pioneer 11 encountered the bow shock of Jupiter on 25 November 1974 when about 10 million km out from the planet. Imaging and polarimetry of Jupiter were conducted nearly full time from this time until several days after the closest passage. Imaging and polarimetry of the moons Ganymede, Callisto, and Io were also conducted during approach and departure from Jupiter. The best images of Jupiter were taken from 1 day before the closest approach to 1 day after the closest approach when Pioneer 11 was within a million miles of Jupiter. Vivid images of the Great Red Spot were obtained along with good images of the north and south polar areas. Several hundred images were made of Jupiter and its moons.

Pioneer 11 approached Jupiter below the south pole and was deflected by the immense gravity of Jupiter around the bottom of the planet and upwards. About 20 min after the spacecraft became occluded by Jupiter (blackout of the radio signal), it reached the point of closest approach, 43,000 km above the cloud tops, on the backside of the planet. The speed of the spacecraft was about 173,000 km/h at the time of closest approach. The spacecraft emerged from blackout 42 min after entering.

The trajectory of Pioneer 11 carried it above the plane of the ecliptic and on a path towards Saturn. The travel time from Jupiter to Saturn was about 5.7 years. Course corrections were made on 26 May 1976 and on 13 July 1978 to refine the trajectory to intercept Saturn in its orbit. An image taken of Saturn by the photopolarimeter while still about 2.5 million km out is given in Fig. 4.9. The spacecraft encountered the bow shock of Saturn about 1.5 million km from the planet on 31 August 1979. It passed through the ring plane of Saturn from above and beyond the

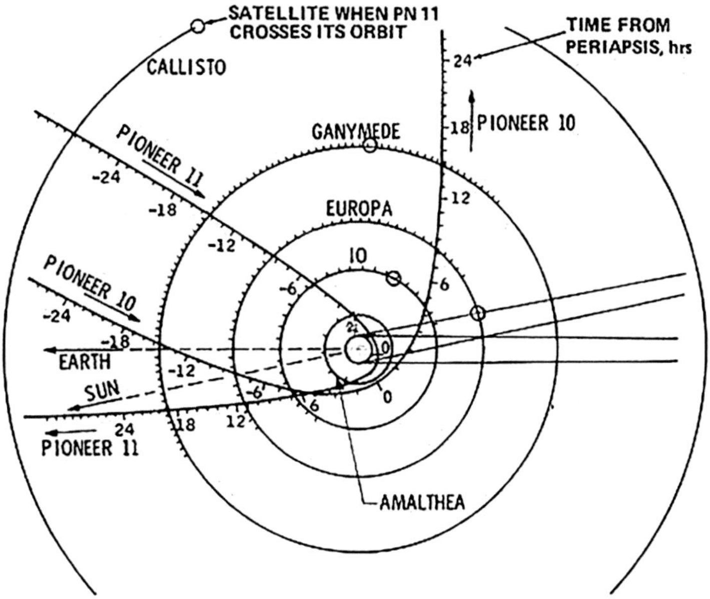

Fig. 4.7 Trajectories of Pioneer 10 and 11 looking down from north pole of Jupiter. (NASA graphic)

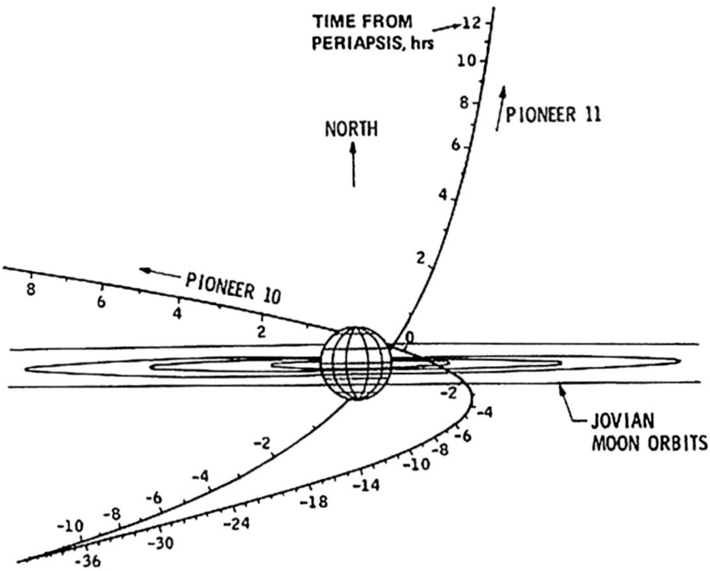

Fig. 4.8 Trajectories of Pioneer 10 and 11 at Jupiter as seen from Earth. (NASA graphic)

Fig. 4.9 Image of Saturn taken by photopolarimeter of Pioneer 11 at a range of 2.5 million km. (NASA image)

outermost ring on 1 September 1979. Pioneer 11 made its closest approach to Saturn at a distance of 20,900 km above the cloud tops later that day.

Pioneer 11 took about 440 images of Saturn and its rings during the encounter. It discovered a new ring just outside of the A ring that was subsequently named the F ring. Temperature measurements by the radiometer indicated overall temperature of the planet to be about −180 °C.

Pioneer 11 continued past Saturn out through interplanetary space. It traveled past the orbit of Neptune in February 1990 and into interstellar space. Last contact with Pioneer 11 occurred in November 1995 when it was about 6.6 billion km from Earth.

Scientific Experiments on Pioneer 10 and 11

Pioneer 10 and 11 carried a total of 11 scientific instruments on their historic journeys. Pioneer 11 was the only one of the pair to explore Saturn, and it carried an extra instrument: a flux gate magnetometer. A list of experiments, principal investigator, and a summary of purpose of each are given in Table 4.1.

A drawing of Pioneer 10 and 11 showing locations of the scientific instruments on the spacecraft is given in Fig. 4.10.

Table 4.1 Experiments carried out by Pioneer 10 and 11

Experiment	Principal investigator	Purpose
Imaging photopolarimeter (IPP)	Thomas Gehrels, University of Arizona	Rotation of the spacecraft was used to scan a narrow field of view of a telescope across a planet in azimuth. The telescope was incremented in elevation at each scan to generate an image. Photometry and polarimetry data were also gathered
Helium vector magnetometer (HVM)	Edward Smith, JPL	Measure weak magnetic fields in interplanetary space and strong magnetic fields of planets
Flux gate magnetometer	Mario Acuna, NASA GSFC	Measure strong magnetic fields of Jupiter and Saturn
Infrared radiometer (IRR)	Andrew Ingersoll, California Institute of Technology	Measure the temperature of clouds of Jupiter and the temperature of some of its moons. Pioneer 11 measured the temperature of clouds of Saturn and of its rings
Ultraviolet photometer (UV)	Darrell Judge, University of Southern California	Measure the density of interplanetary neutral hydrogen and measure the density and temperature of hydrogen and helium gas in interplanetary space and in the atmosphere of Jupiter
Trapped radiation detector (TRD)	K. Walker Fillius, University of California San Diego	Explore radiation belts of Jupiter and Saturn as well as background radiation in interplanetary space
Cosmic ray telescope (CRT)	Frank McDonald, NASA GFSC	Detect galactic and solar cosmic rays
Geiger tube telescope (GTT)	James Van Allen, University of Iowa	Measure the flux of protons and electrons in interplanetary space and in the vicinity of Jupiter and Saturn
Plasma analyzer (PA)	Aaron Barnes, NASA Ames Research Center	Detect and measured particles in the solar wind and investigated solar wind interaction with magnetic fields of Jupiter and Saturn
Charged particle instrument (CPI)	John Simpson, University of Chicago	Measure the properties of low-intensity particles in interplanetary and interstellar space and measure the properties of particles that made up intense radiation belts of Jupiter
Meteoroid detector	William Kinard, NASA Langley	Record the number of meteoroid particles that penetrated a series of gas-filled cells
Asteroid meteoroid detector (AMD)	Robert Soberman, General Electric	Detect particles by sensing reflected sunlight from the particle

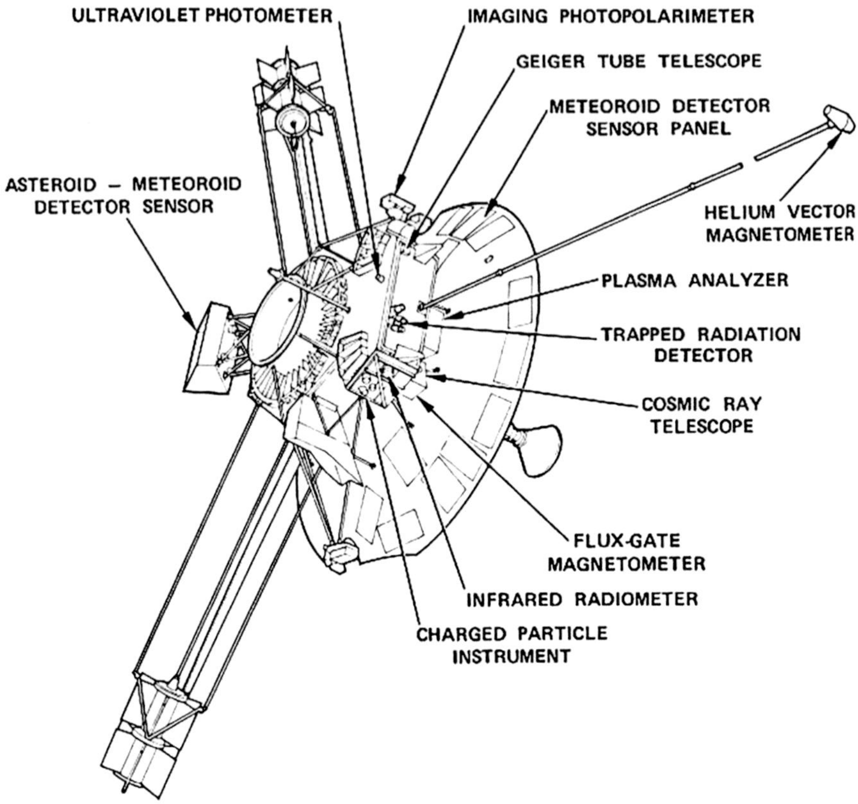

Fig. 4.10 Locations of scientific instruments on Pioneer 10 and 11. (NASA drawing)

Imaging Photopolarimeter (IPP)

The imaging photopolarimeter used rotation of the spacecraft to scan the narrow field of view of a telescope in azimuth across an object. It incremented elevation angle at each scan to generate an image. The field of view for imaging, as set by a selectable aperture, was 0.028° by 0.028°. The field of view for polarimetric and photometric operation was 0.46° by 0.46°. The input aperture of the telescope was 2.54 cm in diameter.

Light at the output of the telescope was applied to a Wollaston prism that separated the light into two linearly polarized beams with orthogonal polarization. The two beams were dispersed in angle. Both beams were directed at a dichromatic mirror set at a 45° angle. Light having wavelengths more than 3500 Å (red light) passed through the mirror, while shorter wavelengths (blue light) were reflected. Red light passing through the mirror consisted of separate orthogonally polarized beams displaced slightly in angle. The two beams were applied to a lens having a coating tuned to pass red light. The two red beams (at different angles) at the output of the lens were directed to a detector device with two photocathodes. One photocathode intercepted each beam. The analog signal from each photocathode, which represented intensity of the reflected light from an object, was amplified and digitized by a 6-bit analog-to-digital converter.

Blue light reflected by the dichromatic mirror was handled in the same manner as the red-light channel. The output of the imaging photopolarimeter in the imaging mode was amplitude of two orthogonal polarizations of red light and amplitude of two orthogonal polarizations of blue light reflected from the object.

The outputs of the detectors were read out at each 0.15° of spacecraft roll angle, alternating with red and blue channels. The digital data was stored in a 6144-bit buffer. The buffer was read out at a rate of 512 bits/s for data assembly and transfer to Earth.

Color images require red, blue, and green image information. Green information was synthesized during data processing on Earth from the red and green image data plus color content of photographs taken of Jupiter from Earth. An image of Jupiter made from data from Pioneer 10 at a distance of 2.5 million km from the planet was shown previously in Fig. 4.6. The famous Great Red Spot is visible on the left side of the planet, and the moon Io cast its shadow near the center.

New information about Saturn and its rings was learned from the imaging photopolarimeter on Pioneer 11. A new ring, named the F ring, was discovered at 2.33 Saturn radii. It is separated from the A ring by the newly named Pioneer division. A new moon was found at 5.53 Saturn radii. The equatorial radius of Saturn was found to be 60,000 ± 500 km, and the ratio of polar to equatorial radius was 0.912. An image of Saturn made from data from Pioneer 11 at a distance of 2.5 million km from the planet was shown previously in Fig. 4.9.

Helium Vector Magnetometer (HVM)

The purpose of the magnetometer on Pioneer 10 and 11 was to measure the magnetic field of the planets and fine structure of the magnetic field in interplanetary space. The magnetometer was mounted at the end of a deployable boom 6.6 m long to space it away from residual magnetic field of the spacecraft.

The magnetometer was a helium vapor type that made use of the magnetic properties of two electrons in a helium atom. That type of magnetometer was sensitive to weak magnetic fields, and it had been successfully used on the Mariner 4 spacecraft to Mars in 1964. The magnetometer had eight operating ranges that could be selected by command from the ground or automatically selected by the instrument itself. Automatic selection was based on the strength of the magnetic field encountered. The ranges were ±4 nT, ±14 nT, ±42 nT, ±640 nT, ±4 μT, ±22 μT, and ±140 μT where T denotes Tesla. The automatic range selecting mode was entered upon turn-on. For reference, the magnetic field at the Earth's surface at the equator at the prime meridian is about 32 μT.

The operation of the magnetometer was based on the transparency of a plasma of helium gas to radiation at the resonant frequency of 1.083 μm. The pumping frequency of 1.083 μm, which is in the infrared wavelength region, ionized helium creating a plasma. The helium plasma was contained in an enclosure referred to as the absorption cell. The transparency of the plasma was dependent on magnetic fields enveloping the plasma.

A triaxial set of Helmholtz coils was used to generate rotating magnetic field vectors in the plasma. A beam of infrared radiation at 1.083 μm was transmitted through the absorption cell, and an infrared detector was used to sense modulation of the infrared beam by the magnetic fields. In the absence of an ambient magnetic field, the modulation of the infrared beam was at the frequency of the rotating magnetic field. The presence of an external magnetic field generated a frequency component of the infrared beam at the second harmonic of the fundamental frequency of the rotating field.

The presence of a second harmonic frequency component was synchronously detected. The resulting DC signal was amplified and applied to a servo amplifier that applied a DC current to a particular Helmholtz coil set to null out the fundamental frequency component. The amount of current was proportional to the strength of the ambient magnetic field. Three Helmholtz coils were used to allow measurement of three axes of the ambient magnetic field. The nulling current in each coil was converted to a DC analog voltage and filtered. The resulting three DC signals were proportional to the three components of the ambient magnetic field. The three signals were applied to a multiplexer that fed a 9-bit analog-to-digital converter. The range scale used for a particular measurement was encoded as a 3-bit word and added to the three 9-bit amplitude words to form a 30-bit output word. The 30-bit word was applied to a shift register and the data clocked out into the data stream telemetered to Earth.

Magnetic field strength of Jupiter measured by Pioneer 10 ranged from 300 to 550 microtesla (μT) in the equatorial region, 1150 μT at the north pole, and 900 μT at the south pole. Measurements by Pioneer 11 were 1400 μT at the north pole and 1100 μT at the south pole. The direction of the magnetic field of Jupiter was found to be opposite that of Earth. The dipole axis of the field was tilted away from the axis of rotation by about 10°.

The equatorial magnetic field on the surface of Saturn, measured by the helium magnetometer on Pioneer 11, was 20 μT. The tilt between the magnetic dipole axis and the rotation axis was found to be less than 1°.

Flux Gate Magnetometer (FGM)

Pioneer 11 carried a flux gate magnetometer as well as a helium magnetometer. The flux gate instrument was added to extend the measurement range beyond that of the helium magnetometer. The magnetic field of Jupiter was expected to be relatively strong. The flux gate magnetometer had a single measurement range of ±1000 μT. The resolution of the instrument was 1.0 μT. Since it was intended to measure relatively strong magnetic fields, it was mounted directly to the spacecraft rather than on a boom.

Flux gate magnetometers were first used by aircraft in World War II to detect submarines by measuring slight perturbations of the Earth's magnetic field as the aircraft flew over the submerged submarine. They have since been widely used in terrestrial and space applications to measure magnetic fields.

In basic form, flux gate magnetometers employ a highly permeable toroidal core that is driven in and out of saturation by an alternating current in a drive winding around the core. Taking an arbitrary line through the lane of the core, the half of the core on one side of the line is magnetized in one direction and the other half is magnetized in the other direction. A sense coil, wound around the core structure, senses any difference in induced magnetism in the two halves of the core. In the absence of an external magnetic field, the transitions in and out of saturation are the same for both halves of the core, and there is no net current in the sense winding. The presence of an external magnetic field along the arbitrary line causes one half of the core to come out of saturation early and the other half to come out of saturation later. A current pulse is induced in the sense coil in response to the unbalance at the time of transition. The magnitude of the current pulse is proportional to the strength of the external field.

The main magnetic field of Jupiter measured by the flux gate magnetometer indicated a dipole term with a strength of 428,000 nT. The magnetic field intensity of Saturn was measured by the flux gate magnetometer to be 20,000 nT at the cloud tops in the equatorial region.

Infrared Radiometer (IRR)

An important goal of Pioneer 10 and 11 was to measure the temperature of clouds of Jupiter and the temperature of some of its moons. Pioneer 11 traveled on to Saturn and measured the temperature of clouds of Saturn and its rings. Remote measurement of temperature from substantial flyby distance was performed by a two-channel infrared radiometer. A radiometer measures the natural electromagnetic radiation from a surface. Any object or surface above a temperature of absolute zero (0 K or −273 °C) radiates with intensity that is a function of the temperature of the body and wavelength of the radiation.

A special state of an object called a black body, which is a perfect radiator and perfect absorber of energy, is the basis for radiometric work. The radiation from an ordinary surface is lower than from a black body by a factor known as the emissivity. Black body characteristics can be used as a first approximation for an unknown body. Black body radiation was investigated by German physicist Max Planck who derived what is now known as Planck's law in 1900. Planck's law for the spectral radiance per steradian emitted from 1 m^2 of the surface of a black body involves the wavelength of radiation, temperature, velocity of light, Planck's constant, and Boltzmann's constant. Planck's law can be used to compute temperature from radiant energy measured at a given wavelength.

The infrared radiometer used a telescope having a Cassegrain optical system with an aperture diameter of 7.6 cm. The field of view was 1.0° by 0.3°. The optical beam was directed at a 75° angle from the spin axis of the spacecraft. Temperature measurements were made as the beam scanned across a planet as the spacecraft spun on its axis. The instrument operated in two wavelength bands established by filters: 30–56 and 14–25 μm. The shorter wavelength band was sensitive to temperatures in the range of 200–700 K, and the longer wavelength band was sensitive to temperatures in the range of 80–300 K. The incoming infrared radiation in each band was applied to an 88-element thin-film thermopile detector. The detector output was amplified and converted to a digital data sequence and included in the data stream telemetered to Earth.

Pioneer 10 measured temperatures of Jupiter cloud tops of −147 °C in the 14–25 μm band and −128 °C in the 30–56 unit band. Pioneer 11 measured an effective temperature of Jupiter of −148 °C. Pioneer 11 went on to Saturn and measured the temperature of the tops of clouds of Saturn of −179 °C. The temperature of the rings was measured at −213 to −203 °C on the sunlit side and <−213 to −206 °C on the side of rings in Saturn's shadow.

Ultraviolet Photometer (UV)

The purpose of the ultraviolet photometer was to measure the density of interplanetary neutral hydrogen and determine the extent of the heliosphere from hydrogen distribution. It was also used to measure the density and temperature of hydrogen and helium gas in interplanetary space and in the atmosphere of Jupiter.

The ultraviolet photometer measured the intensity of reflected ultraviolet light in the wavelength band of 200–1400 Å (20–140 nm). The field of view of the instrument was 1.15° by 9.3°. The photometer was mounted such that the optical axis of the instrument was pointed 20° from the spacecraft spin axis. As a result, the field of view swept an annular ring as the spacecraft rotated.

The ultraviolet photometer used separate detector types for the hydrogen line at a wavelength of 1216 Å and the helium line at 584 Å. The detector outputs were amplified, converted to digital form, and included in the digital data stream telemetered to Earth.

The ultraviolet photometer detected an ultraviolet glow around Jupiter. The ultraviolet glow due to hydrogen had brightness of about 1000 rayleighs, and the ultraviolet glow of helium had brightness of 20–40 rayleighs. The term "rayleigh" is a measure of photon flux. For reference, the night sky of Earth has a flux of about 250 rayleighs. The background brightness near the transition of interplanetary and interstellar space was measured at about 40 rayleighs. The interstellar density of hydrogen in that region was computed to be about 0.04 per cm^3. The density of helium was computed to be 0.01 per cm^3.

Trapped Radiation Detector (TRD)

Charged particles in the solar wind are trapped by the magnetosphere (magnetic field) of planets and form radiation belts. An example is the well-known Van Allen radiation belts located around Earth. Jupiter has a very strong magnetosphere and extremely intense radiation belts. The charged radiation detector on Pioneer 10 and 11 was designed to explore these radiation belts as well as the background radiation in interplanetary space. The Pioneer 10 mission found that Jupiter releases bursts of electrons and protons that contribute to the strong radiation belts.

The trapped radiation detector was mounted on the side of the scientific instrument compartment facing the magnetometer boom. The instrument had five separate sensors serviced by five different apertures that protruded from the instrument enclosure. The sensor types and sensitivities are given in Table 4.2.

The SP and SE scintillator sensors were only present on the Pioneer 11 spacecraft.

The Cherenkov counter sensors detected flashes of light as high-velocity electrons passed through a water-methanol liquid medium. Cherenkov radiation occurs when the velocity of a charged particle exceeds the speed of light in the medium. The water-methanol solution was contained in a glass bottle 1.4 cm in diameter and 5.5 cm long. A photodetector that responded to flashes of light was placed against one flat surface of the bottle. The amplitude of the flash of light was proportional to the length of the path of the particle before its velocity dropped below the Cherenkov limit. The photodetector output was amplified, shaped, and applied to a linear commutator that preserved pulse height information. The commutator serviced pulse outputs from the electron scatter sensor and the minimum ionization sensor as well.

Table 4.2 Sensors in trapped radiation detector

Sensor type	Channel	Particle sensitivity
Cherenkov counter	C1	>6 MeV electrons
Cherenkov counter	C2	>9 MeV electrons
Cherenkov counter	C3	>13 MeV electrons
Electron scatter	E1	>0.16 MeV electrons
Electron scatter	E2	>0.25 MeV electrons
Electron scatter	E3	>0.46 MeV electrons
Minimum ionizing	M1	>35 MeV
Minimum ionizing	M2	Background
Minimum ionizing	M3	>80 MeV protons
SP scintillator	SPDC	>10 keV electrons, >100 keV protons
SE scintillator	SEDC	>10 keV electrons, >100 keV protons

The electron scatter sensor used a crooked path for electrons such that only electrons scattered from the surface entered the solid-state detector. The detector generated a voltage pulse proportional to the energy of the scattered electron that impinged on the detector.

The minimum ionizing sensor responded to particles with sufficient energy to penetrate a shield 1 cm thick. One channel detected particles with energies less than 3 MeV, and the other two channels detected protons with energies in the range of 50–350 MeV. The minimum ionizing sensor used a solid-state detector that generated a voltage pulse with amplitude proportional to the energy of the penetrating particle.

In the case of the Cherenkov counter and the electron scatter sensors, the commutator output fed three pulse height discriminators. The discriminators provided counts of electrons above three different energy values. The commutator also fed the output of the minimum ionization sensor to pulse height discriminators for both electron and proton energies. The outputs of the discriminators were counts of number of particles above certain energy levels. The count was accumulated for 1.5 s for each discriminator output.

The fields of view of the sensors were perpendicular to the spin axis and swept around the azimuth plane of the spacecraft as the spacecraft rotated. The field of view of the Cherenkov counter sensors was dependent on the energy level of the electrons. At an energy level of 23 MeV, the half-power width was 120° for C1, 90° for C2, and 65° for C3. The field of view of the scattered electron sensor was 55° for all channels.

Data giving fluxes of electrons and protons measured by Pioneer 11 when near its closest approach to Jupiter of 43,000 km was reproduced in a paper by Bolton et al. (2004). The data indicates flux maximums of about 2×10^7 electrons/cm^2/s for electron energies greater than 5 MeV, 1×10^7 electrons/cm^2/s for electrons with energy greater than 35 MeV, and 7×10^5 electrons/cm^2/s for protons with energy above 80 MeV.

Measurement of flux of electrons near Saturn was given in a paper by Fillius and McIlwain (1980). The data shows flux of electrons with energy above 0.46 MeV of about 1×10^6 electrons/cm^2/s in the area just inside the inner (A) ring. The flux was essentially zero within the ring region and picked up again outside of the ring region. The data indicated that charged particles are swept up by the rings.

Cosmic Ray Telescope (CRT)

The purpose of the cosmic ray telescope was to detect galactic and solar cosmic rays. The instrument used three telescopes with solid-state detectors to measure cosmic rays. The telescopes were referred to as the high-energy telescope, medium-energy telescope, and low-energy telescope. The high-energy telescope measured flux of protons between 56 and 800 MeV. The medium-energy telescope measured flux of protons between 3 and 22 MeV. The low-energy telescope measured flux of electrons between 0.05 and 1 MeV and the flux of protons between 0.05 and 20 MeV. The telescopes all pointed in a direction normal to the spin axis.

The instrument also measured ions of the ten lightest elements (hydrogen to neon).

A paper by Lopate (1990) presented cosmic ray count rates from 1983 to 1989 for Pioneer 10. Curves were presented for the number of counts for energies above 70 MeV and for energies above 109 MeV. The curves had a broad peak about the year 1987 with an average of 0.5 counts/s for energies above 109 MeV and 0.95 counts/s for energies above 70 MeV. Pioneer 10 was at a distance of about 40 AU at the time.

Geiger Tube Telescope (GTT)

The purpose of the Geiger tube telescope was to measure the flux of protons and electrons in interplanetary space and in the vicinity of Jupiter and Saturn. The instrument incorporated seven miniature Geiger-Mueller tubes arranged in three groups. Geiger-Mueller tubes have long been used to detect radiation levels on Earth, and they have been used in satellites to investigate the radiation belts around the Earth.

Geiger-Mueller tubes detect ionizing radiation in the form of electrons, protons, or gamma rays. The Geiger-Mueller tubes are usually cylindrical with a wire stretching along the axis of the cylinder and a window at one end. The window is commonly made of mica. The wire is connected through a resistor to a positive voltage source of several hundred volts with respect to the case of the cylinder. The tube is filled with a gas, usually argon. When an energetic particle enters the tube, it produces ions in the gas that are swept towards the cylinder walls and the wire by the electric field. The ions produce other ions by collision, and an avalanche of ions occurs in the tube with resulting current flow. The current causes a voltage drop across the series resistor, and the voltage across the tube falls below the value required to sustain a discharge, and the current stops. The voltage again rises to full potential, and the cycle repeats. The pulses of current produce voltage pulses in a load resistor. Each pulse is referred to as a count. The number of counts per unit time is proportional to the energy of the radiation impinging on the tube.

One group of Geiger-Mueller tubes in the GTT consisted of three EON Corporation-type 6213 tubes stacked one above the other in a magnesium frame with lead shielding all around. The tubes were referred to as A, B, and C with tube C located in the middle of the stack. Cutouts were provided in the lead shielding in front of the active end of tubes A and B. Tube C had lead shielding on all sides and was used for background readings in conjunction with tubes A and B. Tubes A and B had different collimators formed from magnesium in front of the cutouts in the lead shielding. The field of view angle of tubes A and B was 60° with the centerline directed towards the +*X*-axis of the spacecraft.

NSSDCA 1972-012A-11 reported that the combination of tubes A and C responded to electrons in the range of 5–21 MeV and protons in the range of 30–78 MeV. The combination of tubes B and C responded to electrons in the range of 0.55–21 MeV and protons above 78 MeV.

A second group of three Geiger-Mueller tubes were arranged in a triangular array with lead shielding on all sides. Those tubes were EON type 5107, which were similar to the 6213 type but with a larger dynamic range. Each tube responded to electrons with energies above 31 MeV and protons with energies above 80 MeV.

A third group consisted of a single EON 6213 tube, which was located at the end of a tube with a 90° bend. The bent tube was gold plated on the inside, and only electrons that scattered off the surface found their way to the Geiger-Mueller tube. The arrangement strongly discriminated against protons. Electrons above 0.06 MeV were detected. The field of view of the arrangement was towards the +*X*-axis of the spacecraft.

Counts for each detector were accumulated for periods of time related to the telemetry data rate of the spacecraft. A data rate of 1048 bits/s was used at Jupiter, and that corresponded to an accumulation time of 0.1875 s. The counts were alternately accumulated in one or the other of two 24-bit memory devices. One memory was read out, while the other was accumulating. The 24-bit data was compressed to 12 bits by a logarithmic-like procedure. The 12-bit Geiger tube telescope data was part of the telemetry data stream.

Analysis of GTT data from Pioneer 10 was reported by Baker and Van Allen (1976). The counting rates were primarily due to electrons with energies above 0.06 MeV. That held for a thin magnetosphere disk extending from about 20 to 100 Jupiter radii (R_J) as well as for the central magnetosphere at distances less than 20 R_J. The magnetosphere disk was oriented along the magnetic equatorial plane. Electron densities were a maximum at about 50 R_J during the inbound flight of the spacecraft and at about 70 R_J during the outbound flight.

Plasma Analyzer (PA)

The purpose of the plasma analyzer was to detect and measure particles in the solar wind and investigate solar wind interaction with magnetic fields of Jupiter and Saturn. The instrument was mounted to structure on the back of the high-gain antenna. It viewed the sun through a wide slit in the parabolic reflector. The slit presented a field of view of about 146° by 2.0° to the instrument.

The plasma analyzer measured properties and direction of arrival of particles by means of two pairs of quadraspherical plates. The plates constituted a high-resolution analyzer and a medium-resolution analyzer. The high-resolution analyzer used plates with radius of curvature of 9 cm and plate separation of 0.5 cm. The medium-resolution analyzer used plates with radius of curvature of 12 cm and plate spacing of 1.0 cm.

A voltage was applied between the plates of each pair. The voltage was varied in 64 steps, one step each revolution of the spacecraft. The electric field caused charged particles to deflect. When the electric field strength was matched to the particle charge and velocity, the particle made it all the way through the gap in the curved plates and struck a detector. In the case of the high-resolution analyzer, there were 26 channeltron (electron multiplier type) detectors mounted in a semicircle at the exit of the plates. Angle of arrival was determined by which channeltron detected the particle. The channeltrons detected angle of arrivals between ±51°.

In the case of the medium-resolution analyzer, five current collector detectors were followed by associated electrometer amplifiers. The three center detectors each covered an angular range of 15°, and the three of them covered an angular range of ±22.5°. The two outer detectors each covered an angular range of 47.5°, and they were placed ±46° from center.

The high-resolution analyzer was used to measure ions only. It measured the number of ions per second for energy levels from 100 to 8000 eV. The medium-resolution analyzer measured count rates of ions of energies between 100 and 18,000 eV and electrons with energies from 1 to 500 eV.

A paper by Wolfe et al. (1974) indicates that the solar wind was abruptly influenced by Jupiter at a distance of 109 Jupiter radii while Pioneer 10 was inbound. The bulk speed of the solar wind, in the spacecraft frame of reference, decreased from 420 to 250 km/s, and the proton density changed from about 0.03 protons/cm^3 to about 0.1 protons/cm^3. Also, the direction of the flow changed from the sunline to at least 40° away from the sunline. Another paper by Wolfe et al. (1975) describes measurements made by Pioneer 11. At a distance of 109.7 Jupiter radii inbound, the bulk speed of the solar wind, in the spacecraft frame of reference, decreased from 480 to 328 km/s. The direction of flow shifted about 50° from the sunline.

Charged Particle Instrument (CPI)

The charged particle instrument contained two types of sensors to measure low-intensity particles in interplanetary and interstellar space. It also contained two different types of sensors to measure properties of particles that made up intense radiation belts of Jupiter.

One of the low-energy particle measuring sensors was called the main telescope (MT). It incorporated seven elements to identify light elements in particles with atomic numbers from 1 through 8 (hydrogen to oxygen). Other elements measured energy spectra of protons from 3 to 68 Mev and electrons from 6 to 30 MeV. One element resolved the isotopes of hydrogen and helium to allow distinguishing particles coming from the sun or from the galaxy. The telescope had a field of view of 60° with the axis about perpendicular to the spin axis of the spacecraft. The other low-energy sensor was the low-energy telescope (LET). It measured protons in the energy range of 0.3–9 MeV. The field of view of the LET was 70°.

Two types of sensors were used to measure the high energies of particles within the radiation belt of Jupiter. One was an electron current detector consisting of silicon detector shielded by beryllium. It measured high fluxes of electrons with energies above 3 MeV. The other was a fission cell detector that consisted of a thin slab of thorium 232 sandwiched between two silicon detectors. High-energy nucleons striking the thorium caused fission in the thorium, and the fission fragments were sensed by the detectors. It was used to measure the flux of protons with energy above 30 MeV. It was insensitive to electrons.

The charged particle instrument was developed by the NASA Goddard Spaceflight Center. A paper by J. H. Trainor et al. gives plots of flux of high-energy protons and electrons observed as the spacecraft crossed the bow shock and magnetosphere of Jupiter. At the time of crossing the magnetopause at about 95 R_J, the flux of high-energy protons was about 10^3 protons/cm^2/s, and the flux of high-energy electrons was about 3×10^4 electrons/cm^2/s. The term R_J is the radius of Jupiter. At a distance of 10 R_J, high-energy proton flux was about 3×10^6 protons/cm^2/s, and high-energy electron flux was above 10^5 electrons/cm^2/s.

Meteoroid Detector (MD)

Meteoroid detectors were carried by the Pioneer 10 and 11 spacecraft to measure the cumulative number of meteoroid impacts as the spacecraft traveled from Earth to Jupiter and into interstellar space. Each spacecraft carried a total of 13 meteoroid detector assemblies that had a plan view dimension of 32 cm by 18 cm. The 13 assemblies were mounted on the back of the reflector of the high-gain antenna near the periphery.

The active area of each meteoroid detector had dimensions of 29.2 cm by 15.5 cm. Each assembly was divided up into 18 pressurized cells. The cells, which were about 29 cm long and 0.86 cm wide, were pressurized at 150.6 kPa with an argon-nitrogen gas mixture. The exposed side of the cell was made from stainless steel 25 μm thick for Pioneer 10 and 50 μm thick for Pioneer 11. There were a total of 234 pressurized cells on each spacecraft.

Each pressurized cell contained a pressure switch with a wire cathode and cylinder anode. A DC voltage between the anode and cathode was set at 525 V. With full pressurization of the gas, conduction did not occur between anode and cathode. If the pressure decreased to about 11% of the initial pressure, a glow discharge between electrodes occurred with attendant flow of current. If the pressure decreased further to about 0.17% of the initial value, the glow discharge ceased and current became zero again. If a meteoroid struck and caused a hole in the pressurized cell, the loss in pressure was sensed by a flow of current across the pressure switch. As gas pressure continued to diminish due to the hole, a point was reached where current flow stopped. The time between switch events was used to determine leak rate and hence the size of the hole. The size of the hole in turn was used to estimate properties of the meteoroid.

In the case of Pioneer 11, a total of 40 pressurized cells had been penetrated after 2 years of mission time, 65 cells after 4 years, and 105 cells after 8 years. The particle flux increased by three orders of magnitude during the encounter with Saturn. Researchers thought that the increase was due to particles in the rings.

Asteroid Meteoroid Detector (AMD)

The purpose of the asteroid meteoroid detector was to detect small particles at distances up to 1 km from the spacecraft by measuring reflected sunlight from the particle. Larger objects were detected at a greater distance. The instrument incorporated four telescopes, each with a field of view of 7.5°. Each telescope was fitted with a photomultiplier-type detector to sense sunlight reflected from a particle. The fields of view of the telescopes overlapped slightly, and a signal was required to be present simultaneously in at least three telescopes before detection was declared.

The instrument was mounted to the appendage shown at the forefront on the lower right side of the photograph of the spacecraft shown previously in Fig. 4.3. The center of the overlaps of the fields of view of the four telescopes was directed 135° from the +*Z*-axis. The *Z*-axis was the spin axis of the spacecraft. The +*Z*-direction was along the centerline of the beam of the high-gain antenna that was pointed towards Earth. The instrument viewed an annular region of space as the spacecraft rotated.

During the first 10 months of flight, which took the spacecraft from 1.0 to 3.3 astronomical units (AU) from the sun, over 200 meteoroid and asteroid events were detected. The spacecraft entered the asteroid belt (between the orbits of mars and Jupiter) after about 4.5 months of travel, and it cleared the asteroid belt after about 11 months of travel. The asteroid belt stretches from about 2.2 to 3.2 AU from the sun. The sizes of particles detected during travel between 2.0 and 3.5 AU from the sun ranged from about 35 μm to 10 cm.

References

Acuna, M. H. and Ness, N. F. *Summary of Initial results from the GSFC Fluxgate Magnetometer on Pioneer 11*, NASA-TM-X-70899, 1975

Baker D. A. and Van Allen, J. A., *Energetic electrons in the Jovian magnetosphere*, Journal of Geophysical Research, Vol. 81, Issue 4, 1976

Bolton, Scott J., et al., *Jupiter's Inner Radiation Belts*, Jupiter: The Planet, Satellites and Atmosphere, Edited by Fran Bagenal, et al., Cambridge University Press, 2004

Dyal, Palmer, *Pioneers 10 and 11 Deep Space Missions*, NASA Technical Memorandum 102269, 1990

Fillius, W., et al., *Trapped Radiation Belts of Saturn: First Look*, Science, Vol. 207, Issue 4429, 1980

Fillius, Walker, Mcllwain, Carl, *Very Energetic Protons in Saturn's Radiation Belt*, Journal of Geophysical Research, Vol 85, Issue A11, 1980

Gehrels, T., et al., *The Imaging Photopolarimeter experiment on Pioneer 10*, Science, Vol. 183, Issue 4122, 1974

Lopate, C., et al., *Pioneer 10 and 11 Gradients of Galactic Cosmic Ray Nuclei and Anomalous Components Through the Period of Solar Minimum and to 45 AU from the Sun*, NASA Astrophysics Data System, 1990

NASA, *Geiger Tube Telescope (GTT)*, NSSDCA/COSPAR ID 1972-012A-11

NASA, *Imaging Photopolarimeter (IPP).* NSSDCA/COSPAR ID 1972-012A-07

NASA, *Pioneer 10*, NSSDCA/COSPAR ID 1972-012A

NASA, *Pioneer 11*, NSSDCA/COSPAR ID 1973-019A

Opp, Albert G., *Pioneer 10 Mission: Summary of Scientific Results from the Encounter with Jupiter*, Science, Vol. 183, No. 4122, 1974

Trainor, J. H., et al., *Jovian Protons and Electrons: Pioneer 11*, NASA-TM-X-70845, 1975

Wolfe, John H., et al., *Preliminary Pioneer 10 Encounter results from the Ames Research Center Plasma Analyzer Experiment*, Science, Vol. 183, January 1974

Wolfe, John H., et al., *Pioneer 11 Encounter: Preliminary Results from the Ames Research Center Plasma Analyzer Experiment*, Science, Vol. 188, 1975

Uri, John, *45 Years Ago, Pioneer 11 Explores Jupiter*, NASA "The Solar System," 2019

Chapter 5
Voyager 1 and 2 Spacecraft

The twin Voyager spacecraft, Voyager 1 and Voyager 2, were developed to explore the outer planets. They were impressive spacecraft, still operating 46 years after launch. Senior editor, Michael Bakich, writing in Astronomy magazine in 2018, stated: "The twin probes explored more planets, discovered more moons and offered more breaking news than any other spacecraft." Voyager 2 was launched in August 1977, and Voyager 1 was launched in September 1977.

Voyager 1 conducted flybys of Jupiter, Saturn, and Saturn's very large moon, Titan. It then proceeded through interplanetary space and out into interstellar space. Voyager 2 conducted a "Grand Tour" by using gravitational assist from each planet in turn and conducted flybys of Jupiter, Saturn, Uranus, and Neptune before traveling out into interstellar space. Photographs and detailed scientific observations were made of each planet encountered on the long voyages. Both spacecraft functioned well and returned excellent photographs and scientific data from examination of the planets.

Presently in 2024, Voyager 1 is farther from Earth than any other man-made object at about 24.342 billion km. Voyager 2 is 20.386 billion km from Earth. An artist's rendering of Voyager 1 entering interstellar space is shown in Fig. 5.1.

Background of Voyager Program

A fortuitous rare positioning of the outer planets Jupiter, Saturn, Uranus, and Neptune occurred in the period late 1970s and early 1980s. The alignment allowed taking advantage of gravitational assist from each planet in turn to set up a trajectory to the next planet as the spacecraft traveled outward. The event, which occurs about

T. Lund, *Spacecraft Exploring Outer Planets*, Springer Praxis Books,
https://doi.org/10.1007/978-3-031-96544-9_5

Fig. 5.1 Artist's rendering of Voyager 1 in interplanetary space. (NASA graphic)

every 176 years, was discovered by a graduate student named Gary Flandro at the California Institute of Technology (Caltech) in 1965.

NASA formulated an ambitious plan to take advantage of the rare alignment and send several spacecraft on a "Grand Tour" of the four outer planets plus Pluto. The program proposed was deemed too expensive by Congress, and it was not funded. A more modest program was substituted that entailed only flybys of Jupiter and Saturn. Funding for the scaled-down program was approved by Congress. The program would include two identical spacecraft and a mission time of about 4 years. The two spacecraft were called Voyager 1 and Voyager 2.

The two spacecraft were developed by the Jet Propulsion Laboratory (JPL) of the California Institute of Technology. The Jet Propulsion Laboratory was, and still is, managed by the California Institute of Technology (Caltech) under contract from NASA. The original manager for the Voyager program at JPL was Bud Schumerier.

The spacecraft were built with reliable, long-life electronic components and circuits available at the time. Electrical power was furnished by a radioisotope-based electrical power generator that had a half-life of about 87 years. In reading accounts of spacecraft development, one detects an undercurrent of hope by the engineers that the spacecraft would provide excellent photographic and scientific data and funding for an extended mission to Uranus and Neptune would be forthcoming after the flyby of Saturn.

Mechanical Configuration of Voyagers 1 and 2

The overall Voyager vehicle included a mission module and a propulsion module. The mission module was usually referred to as the Voyager spacecraft, and it was shown in Fig. 5.1. The propulsion module contained a solid-propellant rocket motor that fired after shutdown of the Centaur upper stage of the Titan IIIE launcher. A launch vehicle adapter connected the propulsion module to the Centaur upper stage.

The solid-propellant rocket engine provided 68,000 kN (15,287 pounds) of thrust. The burn time was 42 s. The propulsion module increased the velocity of the spacecraft by about 2 km/s (7200 km/h), which was sufficient to send the spacecraft on a trajectory to encounter Jupiter. The propulsion module was jettisoned after the rocket burned out, and the mission module (Voyager spacecraft) continued on the long cruise to Jupiter.

A photograph of the Voyager spacecraft being tested at JPL in 1976 is shown in Fig. 5.2. The extendable booms were folded in the picture. The launch vehicle adapter that mated with the Centaur upper stage on the launch vehicle is at the bottom of the picture. It is partially covered by an orange thermal blanket. The launch vehicle adapter remained with the Centaur upper stage after separation of Centaur

Fig. 5.2 Voyager vehicle undergoing testing at JPL. (NASA photograph)

from the Voyager. The next item up from the bottom is the propulsion module that is covered with a silver-colored thermal blanket. The Voyager spacecraft is the upper portion of the spacecraft covered with black thermal blankets and includes the white high-gain parabolic antenna.

The launch weight of the spacecraft, including the propulsion module, was 2066 kg. After jettisoning the burned-out propulsion module, the spacecraft weighed 825 kg. The spacecraft weight included 117 kg of scientific instruments.

A photograph of the Voyager 2 spacecraft separated from the propulsion module and being tested at JPL in 1977 is shown in Fig. 5.3. A boom that contained several scientific instruments is deployed at the left side of the photograph.

A labeled drawing of the Voyager spacecraft is given in Fig. 5.4. The central structure of the spacecraft was a ten-sided hollow polygon that was 1.78 m across the flats and 0.47 m high. Each segment of the polygon formed a narrow compartment that held electronics for spacecraft systems and scientific experiments. As per the scaling from drawings, the compartments were about 47 cm high, 52 cm wide, and 12 cm thick. The high-gain communications antenna, which used a 3.66-m-diameter parabolic reflector, was mounted to the central structure by several struts. A spherical tank holing hydrazine fuel for thrusters was mounted in the open space of the polygon central structure. The tank was 61 cm in diameter and held 104 kg of hydrazine.

Voyager's coordinate system placed the *Z*-axis (roll axis) along the boresight of the high-gain antenna with the boresight pointing in the $-Z$ direction. The *Y*-axis of the spacecraft (yaw axis) was near the vertical in the drawing of Fig. 5.4. The

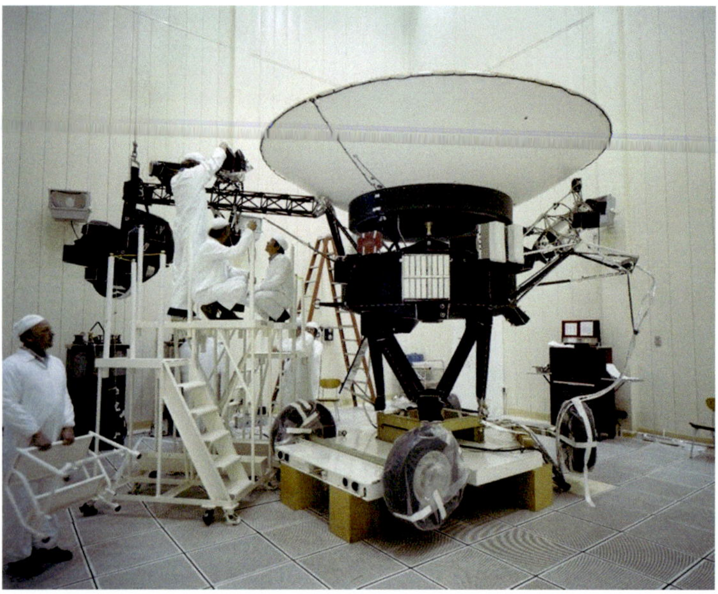

Fig. 5.3 Voyager 2 undergoing testing at JPL in 1977. (NASA photo)

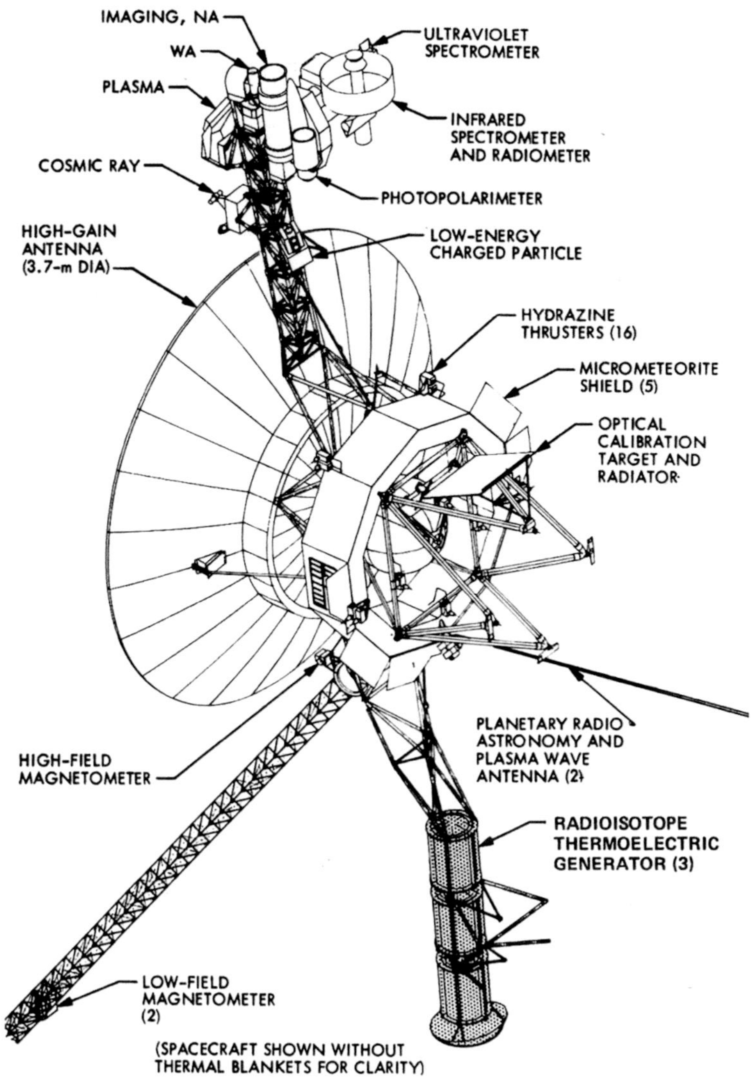

Fig. 5.4 Drawing of Voyager spacecraft. (NASA graphic)

radioisotope thermoelectric generator (RTG) was supported in the $-Y$ direction below the ten-sided spacecraft structure. A boom supporting the magnetometers extended along the $-X$ axis. The magnetometer boom was 13 m long to distance the magnetometers from stray magnetic fields of the spacecraft. A boom, referred to as the science boom, extended in the positive Y-axis direction.

The science boom, which was 2.3 m long, is pointed upwards in Fig. 5.4. It contained three science instruments mounted along the boom. A steerable scan platform that held five optical instruments was mounted at the end of the boom. The science boom, with the scan platform at the end, is extended on the left side in the picture of Voyager in Fig. 5.3 and upwards in Fig. 5.4.

The scan platform could be rotated about two orthogonal axes under the control of the attitude and articulation control system. The two axes were referred to as the azimuth axis and the elevation axis. The azimuth axis was perpendicular to the centerline of the science boom, which put it 7° from the *Z*-axis of the spacecraft. The elevation axis was perpendicular to the azimuth axis. The angular range of the platform was 360° in azimuth and 210° in elevation. Four different slew rates from 0.052° to 1.0°/s could be commanded. Pointing accuracy of the platform was within 0.14°. Five science instruments, including the cameras, were mounted to the scan platform.

The spacecraft had 12 hydrazine-powered thrusters arranged on the spacecraft for attitude control and 4 additional thrusters for trajectory correction maneuvers (TCMs). The thrusters were all the same configuration, and each generated 0.89 N (0.2 pounds) of thrust. A few of the thrusters can be seen mounted on the spacecraft in the drawing (Fig. 5.4). The attitude control thrusters were grouped in two redundant pairs of six thrusters each. Each pair controlled rotation at about a given coordinate axis. The four TCM thrusters thrusted in the direction of the +*Z*-axis.

Temperature control of the spacecraft was accomplished by means of louvers on the sides of high-heat-dissipating electronic compartments on the ten-sided structure. The louvers were operated automatically by bimetallic springs. Some sections of the spacecraft required heating to maintain acceptable operating temperatures. Small radioisotope heating units, which provided about 1 W of thermal energy, were located near the magnetometers and near the sun sensors. Other portions of the spacecraft were heated as needed by electric heaters.

Voyager Systems

Electrical Power

Electrical power for the spacecraft was produced by three radioisotope thermoelectric generators (RTGs). The spacecraft were intended to travel far from the sun into outer interplanetary space where energy from the sun was too low to generate sufficient power from solar arrays. The RTGs, which were type MHW-RTG, were cylindrical in shape, 40.6 cm in diameter, and 50.8 cm long. The three RTGs were mounted in tandem on the end of a deployable boom as shown in Fig. 5.4. Placing the RTGs at the end of a boom lessened radiation levels from the RTGs on spacecraft electronics and science instruments.

RTGs generate electrical power by applying heat from radioactive decay of a material such as plutonium-238 to a series of thermoelectric couples. Plutonium-238 decays into uranium-234 by emitting alpha particles. The half-life of the decay process is 87.7 years. The kinetic energy of the alpha particles is converted to heat when the particles strike the surrounding material.

The MHW-RTG used pellets of plutonium-238 oxide pressed into spherical shape. There were 24 such spheres in a cylindrical region at the center of the RTG. A graphite cylindrical shell enclosed the spheres. A total of 312 silicon-germanium thermoelectric couples were mounted around the graphite shell. The devices were mounted such that one end was in contact with the hot surface and the other end was in contact with the much cooler outer case of the RTG.

Each of the RTGs had initial thermal output of about 2400 W, which resulted in initial electrical power output from the field of thermocouples of 157 W. The electrical outputs of the three RTGs were connected in parallel giving an initial total power output of about 471 W. The power output of the RTGs decreased with time. The power available was estimated to decrease to about 430 W after 38 months into the mission when the spacecraft was expected to encounter Saturn. JPL reported that the power available from both spacecraft was 249 W in the year 2020.

The voltage at the parallel output of the three RTGs was somewhat over 30 V. A regulated 30-V power bus was formed by using a shunt regulator set at 30 V. Excess power from the regulator was dissipated in resistors attached to a radiator plate that radiated excess heat into space. The radiator plate, which was also used as a calibration object for optical sensors, is shown in the drawing of the spacecraft, Fig. 5.4. The plate was about 1.2 m^2.

The 30-V power bus supplied power to the telecommunication system, some heaters, gyros, and electric motors. The bus also supplied power to an inverter that generated AC square wave output power at a frequency of 2.4 kHz and voltage of 50 V RMS. The 2.4 kHz square wave power form was used by most spacecraft systems and science instruments. The spacecraft had two redundant inverters. In case of failure of the main inverter, power circuits were switched to the backup inverter. Distribution of power to various subsystems in the spacecraft was controlled by a series of command-activated relays.

Attitude Control

The spacecraft was attitude stabilized by using a sun sensor, a Canopus star tracker, and a three-gyro inertial subsystem. Signals from the sensors were applied to a computer that managed firing of attitude control thrusters to set and maintain a given attitude. The attitude was set to point the boresight of the high-gain antenna towards Earth during long periods of cruise to the planets. There were two redundant sun sensors and two redundant Canopus trackers.

The computer was referred to as attitude and articulation control system (AACS). It contained two redundant 4096-word memory units, two redundant processors, and redundant input/output circuits. The input/output circuits included driver circuits to activate thrusters. The memory units had a fixed memory section and a reprogrammable section.

The spacecraft had two redundant sets of six thrusters for attitude control. Each redundant set had three pairs of thrusters arranged to adjust the attitude of the spacecraft around the pitch, yaw, and roll axes. The thrusters were MR-103 type built by Aerojet. Each thruster, which was about 34 mm in diameter and 15 cm long, had a nominal thrust of 0.89 N (0.2 pounds). Thrust was generated by the flow of gases out of a nozzle. The gases resulted from decomposition of hydrazine fuel in the presence of a catalyst in the reaction chamber of the thruster. The catalyst was granular aluminum oxide coated with indium.

Voyager had two redundant sun sensors that were co-aligned and looked through a cutout in the reflector of the parabolic antenna. Although not labeled in Fig. 5.4, the back of the sun sensor is the small rectangular object on the left side of the reflector across from the louvered segment of the ten-sided structure. The sun sensors could be biased in the pitch and yaw axes up to ±20°. The biasing effectively put the zero-error point of the sensors at the angles selected. The biasing was done by the computer command subsystem, which had knowledge of angles to Earth. That allowed a null to be obtained from the sun sensor at angles away from the center and away from the boresight of the high-gain antenna. The pitch and yaw error outputs of the sun sensor were applied to the AACS computer. The computer used those error signals to fire thrusters to point the high-gain antenna close to Earth.

After setting the pitch and yaw angles to the necessary values, the spacecraft was rotated around the roll axis until one of the two redundant Canopus trackers acquired the star Canopus. The Canopus tracker fed signals to the AACS computer to maintain that roll angle. In brief, outputs from the sun sensor stabilized the spacecraft in pitch and yaw to point the high-gain antenna towards Earth, and the Canopus tracker stabilized the spacecraft in roll.

The spacecraft could also be stabilized in three axes by signals from three gyros located in the inertial reference unit. Each gyro provided attitude information in two orthogonal axes. Gyro A provided pitch and yaw information, Gyro B provided roll and pitch, and Gyro C provided roll and yaw. The three gyros provided a measure of redundancy. Normally, only two gyros were active at a time.

Gyro information was needed during trajectory correction maneuvers (TCMs). TCMs were under the control of the computer command subsystem (CCS). Spacecraft attitude angles were noted and put into memory at the beginning of a TCM. The CCM determined spacecraft attitude required for the TCM thrusters to make the necessary trajectory correction. Data from the gyros was used to set the attitude of the spacecraft to that needed for the TCM. After the TCM, the attitude of the spacecraft was returned to values before the TCM.

Command and Control

The Voyager spacecraft was controlled by three computer systems: the computer command system (CCS), flight data subsystem (FDS), and attitude and articulation control system (AACS). Two computer subsystems of each type were carried for redundancy. The CCS and AACS computers were based on the CCS computer developed previously for the Viking orbiter. The Viking mission soft-landed two instrumented spacecraft on the surface of Mars in 1976.

The very high input/output requirements for the flight data subsystem of Voyager required that a new computer be developed for it by JPL. After development, the computer was built by the Aerospace Systems Department of General Electric. In fact, all of the computers used by the Voyager spacecraft were built by General Electric to JPL specifications.

The CCS was the brains of the spacecraft. It managed operation of the spacecraft, controlled all major spacecraft systems, handled commands sent up from Earth, communicated with the other computer systems, and monitored health of the spacecraft and of its science instruments. The CCS did not contain a microprocessor that benefitted later spacecraft computers. Instead, it used a series of TTL logic chips that were common at the time. The computer used an 18-bit word where the least significant 6 bits were used for operating codes and the most significant 12 bits for addressing. The 6-bit operating codes resulted in 64 available instructions, and the 12-bit address section allowed 4096 direct addresses. The average instruction cycle time was 88 μs.

Each CCS had a 4096-word memory. The memory was plated wire type where magnetic plating on the wire provided read/write capability. The memory contained 2800 words of fixed memory and 1290 words of reprogrammable variable memory. The reprogrammable memory allowed updates from Earth to enable functions required for the many different phases of the mission. The reprogrammable feature was of great benefit in reducing the overall size of memory required for the complex mission. Programming was performed in the form of assembly language.

The AACS computer was essentially the same as the CCS computer. The CCS issued commands to the AACS for spacecraft attitude and maneuvers and for positioning of the scan platform that held scientific instruments.

The flight data subsystem (FDS) computer was developed by JPL to provide fast processing required to gather, format, and transmit large amounts of data to Earth. The FDS computer was designed a few years after the design of the CCS, and it was able to take advantage of CMOS chips then available. The computer was structured as a 16-bit machine. It used CMOS random access memory. While fast and efficient, CMOS memory required a constant source of power or the information would be lost. This was provided by a direct power line from the RTGs to the memory section. The size of the CMOS memory of each of the redundant FDS computers was 8192 words.

Telemetry Modulator

Engineering data, data from science instruments, and image data were gathered and assembled in a form to be transmitted by the flight data subsystem (FDS) computer. The data was sent in serial form to the telemetry modulator. General science and engineering data was assembled and transferred by the FDS at a fixed information data rate of 3.6 kilobits per second (kbps). The data was encrypted by a Golay encoder to reduce bit error rate, which increased the needed data rate to 7.2 kbps.

Imaging data from the cameras consisted of 800 lines at 800 pixels/line. The amplitude of each pixel was digitized as an 8-bit word. As a result, transmission of one picture required 5.12 million bits. The maximum data rate of the system was 115.2 kbps. The science and engineering data was transmitted along with camera data at each transmission interval. The net result was that one image could be transmitted in about 48 s. The signal-to-noise ratio was adequate for an acceptable bit error rate at a data rate of 115.2 kbps when Voyager was in the vicinity of Jupiter. Farther out, in the vicinity of Saturn, the data rate had to be reduced to 29.9 kbps for acceptable bit error rate, and the time required to transmit one image was about 226 s.

The extended mission to Uranus and Neptune involved much longer ranges to Earth, and image compression was used to reduce the file size of images. In addition, error correcting coding was employed to obtain acceptable bit error rates.

The data was sent to the telemetry modulation unit (TCU) in serial digital form. The TCU received two data channels. One was referred to as the low-rate channel and the other as the high-rate channel. The low-rate data rate was 40 bps. The high-rate channel transferred data at rates from 10 bps to 115.2 kbps. The TCU biphase modulated either channel on either a 22.5 kHz subcarrier or a 360 kHz subcarrier. The 360 kHz subcarrier was used for data rates of 7.2 kbps and higher. Separate modulators were provided for the S-band downlink channel and the X-band downlink channel. The modulated subcarrier for S-band was sent to the S-band exciter, and the modulated carrier for X-band was sent to the X-band exciter.

Downlink data rates varied from 40 bits per second (bps) of engineering data during cruise to 115,200 bps while imaging during encounter with a planet. If the distance to Earth was too far to achieve adequate signal-to-noise ratio on the communication downlink, data was sent to a tape recorder and played back later at a slower rate. The recorder could store 650 million bits of data. Playback from the tape recorder could be selected to be 57.6, 33.6, 21.6, or 7.2 kbps. The lower rates were used at greater distances from Earth.

Telecommunications System

The Voyager spacecraft communicated with Earth through the very capable Deep Space Stations (DSSs) of the Deep Space Network (DSN) Three DSNs were used to communicate with Voyagers. Locations of the DSNs were Canberra, Australia; Madrid, Spain; and Goldstone, California. Uplink communication from the DSN to Voyager was conducted in S-band at a data rate of 16 bps. Downlink from Voyager to the DSN could be at either X-band or S-band. X-band was primary, and S-band was secondary for downlink.

Antenna Subsystem

Voyager contained two types of antennas. One antenna, referred to as the high-gain antenna (HGA), was a large parabolic reflector type 3.66 m in diameter. The antenna used a Cassegrain feed structure. The HGA had feed provisions for both S-band and X-band. The feed for S-band produced right-hand polarized operation. The feed for X-band was dual-polarized with right-hand or left-hand circular polarization depending on which of the two ports was fed. The gain of the HGA was 48 dB, and the beamwidth was 0.5° at X-band. The gain was 36 dB, and beamwidth was 2.3° at S-band. The boresight of the antenna was along the $-Z$-axis (longitudinal) of the spacecraft.

The other antenna was referred to as the low-gain antenna (LGA). It only operated at S-band. It was a right-hand circularized polarized radiator mounted on the backside of the subreflector for the high-gain antenna. The gain of the LGA was 7.0 dB, and the beamwidth was 60°. The boresight of the LGA was also along the $-Z$-axis of the spacecraft.

The low-gain antenna, with its broad beamwidth, was used when Voyager was at short to moderate distances from Earth. The high-gain antenna was used when the spacecraft was more distant. The sun sensor could be biased by 20° in both pitch and yaw, which allowed the spacecraft attitude to be adjusted so that the high-gain antenna beam covered the orbit of Earth with the sun sensor nulled at the angles to the sun. The high-gain antenna was used exclusively after about 80 days of travel from Earth.

Downlink of engineering data and data from scientific instruments could be transmitted in X-band or in S-band through the high-gain antenna or through the low-gain antenna at S-band. X-band was the primary downlink communication band using the high-gain antenna, and S-band was secondary. The transmit frequency was normally made coherent with the uplink carrier frequency for a particular spacecraft. The ratio was 880/221 for X-band and 240/221 for S-band.

An uplink carrier frequency of 2114.7 MHz was transmitted from the DSN to Voyager 1, and a carrier frequency of 2113.3 MHz was transmitted to Voyager 2. The carrier was modulated with command information and a ranging signal. Based on the uplink carrier frequencies, the transmit frequency at X-band was 8420.5 MHz for Voyager 1 and 8414.9 MHz for Voyager 2. The transmit frequency in S-band was 2296.5 MHz for Voyager 1 and 2295 MHz for Voyager 2.

Receiver and Demodulator

The S-band signals received through the high-gain antenna or the low gain antenna were applied through RF switches to one or the other of two redundant S-band receivers. The receiver was a narrowband type with a phase-lock loop that generated a signal coherent with the received carrier but displaced in frequency from it. That signal was applied to an exciter unit that drove the downlink transmitter. The coherent downlink carrier allowed two-way Doppler to be extracted, which allowed very accurate relative velocity of the spacecraft to be determined. Command and ranging modulation on the S-band carrier was sent from Earth to Voyager in the form of biphase modulation on a 512 Hz subcarrier. Commands were in the form of 16-bit words with Manchester format. Uplink data rate was 16 bps.

The receiver used a signal developed by its phase-lock loop to extract the 512 Hz subcarrier. The subcarrier was sent to the modulation demodulation subsystem (MDS) where the command bits were detected and sent on to the CCS computer. The range modulation, which was in the form of a sequence of pseudorandom coders, was also decoded. The received range modulation data was included in the downlink signal structure. The range code, when received at the DSS on Earth, allowed accurate measure of range to the spacecraft.

Exciter and Transmitter

Separate exciter units were used for S-band and X-band downlink transmit channels. Two redundant exciters were provided for each band. The exciters used the coherent signals from the S-band receivers to generate low-powered signals at the transmitter frequency for the S-band downlink. One of the two redundant receivers was paired with one of the two redundant exciter units, and together they formed a redundant pair. There were two such redundant receiver-exciter pairs, and only one pair was powered at a time.

The transmit frequency was normally made coherent with the uplink carrier frequency for a particular spacecraft. The ratio of transmit to receive frequencies was 240/221 for S-band. The transmit frequency in S-band was 2296.5 MHz for Voyager 1 and 2295 MHz for Voyager 2. In case the uplink signal was not available, an auxiliary oscillator in the exciter unit was used to generate a reference for the transmitted frequency. A more stable frequency reference was available from an ultra-stable oscillator and could be selected by command. The ultra-stable oscillator allowed quite accurate one-way Doppler tracking of relative velocity of the spacecraft.

The exciter applied telemetry data and range modulation data to the downlink signal in the form of biphase modulation. The telemetry data signal was generated and supplied by the modulation demodulation subsystem, and range modulation was detected and supplied by the receiver.

The S-band downlink signal from the exciter was applied to one of the two S-band amplifier units. One of the amplifiers was a solid-state type with a power output of 9.4 W, and the other was a traveling wave-type amplifier with a power output of 28.3 W. Only one of the amplifiers was powered at a time. Neither amplifier was powered if downlink was being conducted at X-band. The output of the selected S-band power amplifier was fed to RF switches that allowed the transmitter signal to be applied to either the low-gain antenna or the high-gain antenna for transmission.

Transmission of engineering and science data was made in X-band for all but the early portion of the mission. A reference carrier signal from the active S-band exciter was applied to one of the two redundant X-band exciters. The selected

X-band exciter upshifted the reference carrier to X-band. The exciter output was amplified by a traveling wave tube amplifier to an output power of 23 W. The transmit frequency was 8420.5 MHz for Voyager 1 and 8414.9 MHz for Voyager 2.

The exciter phase modulated the X-band carrier by telemetry data and by the range modulation signal. Telemetry data was generated and supplied by the modulation demodulation subsystem, and range modulation was detected and supplied by the receiver.

Flight of the Voyagers

Flight of Voyager 1

Voyager 1 was launched on 5 September 1977 from Launch Complex 41 at Cape Canaveral in Florida by a Titan III/Centaur D-1TR launch vehicle. Titan IIIE consisted of a two-stage core vehicle with two solid-fuel booster rockets attached to the sides. Centaur D-1T was an energetic upper stage that burned liquid hydrogen with liquid oxygen as oxidizer.

At launch, the solid rocket boosters fired first. The combined thrust of the two boosters was 10.68 million N (2.4 million pounds). At 6.3 s after launch, Titan began a roll to the proper orientation. Stage 1 of the core vehicle ignited about 112 s after liftoff. Stage 1 burned out about 258 s after liftoff, and stage 2 ignited. Stage 1 was separated from stage 2, and it was dropped away. The shroud over the spacecraft was jettisoned at about 10 s after separation of stage 1. Stage 2 burned out at about 468 s after liftoff. Titan was separated from Centaur 486 s after liftoff, and Titan was dropped away.

Centaur was fired at about 496 s after launch to place the joined Centaur and Voyager spacecraft into low Earth orbit. It was shut down at about 92 s later, and Centaur/Voyager coasted in orbit for proper positioning. After coasting for about 42.5 min, Centaur was fired for the second time, this time for about 356 s. Centaur was separated from Voyager at about 61 min after launch, and the propulsion module was ignited 15 s later.

The propulsion module, which was supported from the Voyager framework, developed about 68 kN (15,300 pounds) of thrust. It was fired for about 45 s and increased velocity of Voyager by about 2 km/s (7200 km/h). The propulsion module was then jettisoned. The resulting velocity of Voyager was sufficient to send it on a trajectory to Jupiter. The travel time to Jupiter would be about 20 months.

The sun was acquired about 5.5 h after launch, and the science boom, magnetometer boom, and radioisotope thermoelectric generator boom were deployed. The star Canopus was acquired on 24 August, which allowed the spacecraft to be stabilized in three axes for the long cruise to Jupiter. Science instruments were turned on to verify operation and perform calibration.

Communication with the Voyagers and precise tracking of the two spacecraft were performed by the Deep Space Network (DSN). The DSN stations were located at Goldstone, California; Madrid, Spain; and Canberra, Australia. Each station included one 64-m-diameter parabolic antenna and two 26-m-diameter antennas. Spacing the DSN on three continents allowed nearly continuous coverage of the Voyager flights.

Key events in the flight of Voyager 1 are given in Table 5.1. In March 2024, Voyager 1 had been traveling through interstellar space. It is at a distance of 24.389 billion km from the sun and traveling at a speed of 61,184 km/h. Four of the ten original scientific instruments are still operating, and the spacecraft is still sending data back to Earth periodically, 46 years after launch.

A sketch of the trajectories of Voyager 1 and Voyager 2 to the outer planets is given in Fig. 5.5. Voyager 1 visited Jupiter and Saturn and traveled out of the ecliptic to visit Saturn's large moon, Titan. The trajectory of Voyager 2 stayed in the ecliptic, and the spacecraft visited Jupiter, Saturn, Uranus, and Neptune.

Flight of Voyager 2

Voyager 2 was launched 2 weeks before Voyager 1. The launch and early flight sequence were essentially the same as described for Voyager 1. The first trajectory correction maneuver for Voyager 2 was made on 11 October 1977. The resulting correction was within 1% of the desired value, and Voyager 2 continued cruise along its new trajectory to Jupiter. On 10 April 1978, the spacecraft was 465 million km from Earth and 360 million km from Jupiter. Voyager 2 performed a second trajectory correction maneuver on 3 May 1978. The maneuver involved a burn of 203 s and resulted in a velocity change of 0.615 m/s.

Key events in the flight of Voyager 2 are given in Table 5.2. In March 2024, Voyager 2 had been traveling through interstellar space in the direction of the constellation Pavo. It is at a distance of 20.363 billion km from the sun and traveling at a speed of 55,355 km/h. Five of the ten original scientific instruments are still operating, and the spacecraft is still sending data back to Earth, 46 years after launch.

Table 5.1 Key events in the travel of Voyager 1

Event	Date	Distance from sun (AU)
Launch from Earth	5 September 1977	1.0
Closest approach to Jupiter	5 March 1979	5.2
Closest approach to Saturn	12 November 1980	9.6
Traverse heliopause and enter interstellar space	25 August 2012	122
Cruising in interstellar space	30 March 2024	163.1

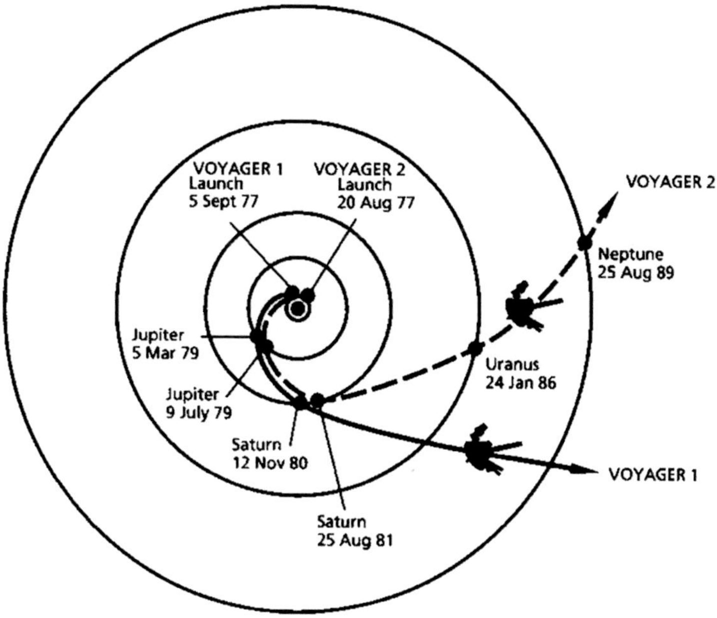

Fig. 5.5 Trajectories of Voyager 1 and Voyager 2 to the outer planets. (NASA graphic)

Table 5.2 Key events in the travel of Voyager 2

Event	Date	Distance from sun (AU)
Launch from Earth	20 August 1977	1.0
Closest approach to Jupiter	9 July 1979	5.2
Closest approach to Saturn	25 August 1981	9.6
Closest approach to Uranus	24 January 1986	19.2
Closest approach to Neptune	25 August 1989	30.0
Traverse termination shock	30 August 2007	83.6
Traverse heliopause and enter interstellar space	5 November 2018	119.7
Cruising in interstellar space	30 March 2024	136.1

Flyby of Jupiter

Voyager 2 began observing Jupiter at a range of 54 million km in late April 1979. Images were taken every 2 h to generate a time-lapse sequence of the Great Red Spot. A final trajectory correction maneuver before encounter with Jupiter was made on 27 June 1979. The first crossing of the bow shock of Jupiter was noted on 2 July at a distance of about 7 million km from the planet. Several of Jupiter's moons were observed while inbound to the closest approach to Jupiter, and several others were observed while outbound. A tabulation of closest approach of Voyager 2 to some of Jupiter's moons is shown in Table 5.3. Images were made of many of the moons.

Closest approach to Jupiter by Voyager 2 occurred on 9 July 1979 at a distance of 650,180 km from the cloud tops. The spacecraft passed slightly below the planet. Observations and photographs were made of the rings of Jupiter on 8 July and on 10 July. In total, Voyager 2 took about 14,000 pictures of Jupiter and its moons on its mission. A photograph of a section of Jupiter containing the Great Red Spot taken from a distance of 9 million km is shown in Fig. 5.6.

A photograph of the small moon Europa from a distance of 241,000 km is shown in Fig. 5.7. The fractured surface was thought to be a form of ice.

Shortly after passing the point of closest approach to Jupiter, a trajectory correction maneuver was made to take advantage of gravity assist from Jupiter to sling the spacecraft in a direction to fly by Saturn and Uranus. A trajectory correction was made on 23 July 1979 to refine the trajectory, and the spacecraft continued a 25.5-month-long cruise to Saturn.

Flyby of Saturn

Observance of Saturn began on 5 June 1981, about 12 weeks before the closest approach. Photographs were taken by the narrow-angle camera at each 72° of rotation of Saturn as the spacecraft closed on the planet. Other science instruments were also active. A photograph of Saturn taken on 4 July 1981 from a distance of 21 million km is shown in Fig. 5.8. A photograph of the ring structure of Saturn taken from a distance of 2.5 million km is shown in Fig. 5.9. The hundreds of rings around Saturn were found to extend from near the cloud tops to 75 km above the cloud tops.

A trajectory correction maneuver was made on 18 August in preparation for a gravity assist from Saturn to change the direction of travel towards Uranus. Closest approach to Saturn occurred on 25 August 1981 at a distance of 100,830 km from the cloud tops.

The trajectory of Voyager 2 passed above Saturn. The trajectory descended and crossed the ring plane about 2 h after the time of the closest approach to the planet. Photographs of the rings were taken above the rings, in the ring plane, and from below the rings.

Table 5.3 Closest approach of Voyager 2 to Jupiter's moons

Moon	Closest approach distance (km)	Date of closest approach
Callisto	214,900	8 July 1979
Ganymede	62,297	9 July 1979
Europa	205,848	9 July 1979
Amalthea	558,565	9 July 1979
Io	1,129,850	9 July 1979

Fig. 5.6 Photograph of Jupiter by Voyager 2 showing Great Red Spot

Fig. 5.7 Photograph of fractured surface of the moon Europa taken by Voyager 2

Fig. 5.8 Photograph of Saturn taken by Voyager 2

Several of Saturn's moons were observed while inbound to the closest approach to Saturn, and several others were observed while outbound. A tabulation of the closest approach of Voyager 2 to some of Saturn's moons is shown in Table 5.4. Images were made of the moons listed. Eight small moons also orbit Saturn. Several of them appear to be shepherding various rings.

The trajectory of Voyager 2 was bent sharply by the encounter with Saturn to set the spacecraft on a trajectory to intercept Uranus. A trajectory correction maneuver was made on 29 September to refine the trajectory. The spacecraft then continued its 53-month (4 years and 5 months) cruise to Uranus.

Fig. 5.9 Photograph of ring structure of Saturn taken by Voyager 2

Table 5.4 Closest approach of Voyager 2 to some of Saturn's moons

Moon	Closest approach distance (km)	Date of closest approach
Iapetus	910,000	22 August 1981
Hyperion	470,000	24 August 1981
Titan	665,000	25 August 1981
Dione	502,000	25 August 1981
Mimas	310,000	25 August 1981
Enceladus	87,000	25 August 1981
Tethys	93,000	26 August 1981
Rhea	645,000	26 August 1981
Phoebe	2,080,000	4 September 1981

Flyby of Uranus

Observations of Uranus began on 4 November 1985, while Voyager 2 was about 103 million km from the planet and 81 days from the closest approach. A picture of Uranus taken on 18 January 1986 by the narrow-angle camera during the approach

Fig. 5.10 Photograph of Uranus taken by Voyager 2

is shown in Fig. 5.10. The blue-green color of the planet is a result of methane in the atmosphere that reflects blue-green light from the sun and absorbs red light.

Uranus is unique in the sun's planetary system in that its spin axis is nearly perpendicular to the plane of its orbit around the sun. The spin axis was directed closely towards the sun and Earth at the time of Voyager's approach.

A correction maneuver was performed on 23 December 1985 to refine the spacecraft's trajectory towards Uranus to enable a flyby of Neptune. The spacecraft passed to the right side and below the Uranus. Closest approach occurred on 24 January 1986 at a distance of 81,440 km from the cloud tops.

There were five known moons of Uranus before the flight of Voyager 2. Voyager 2 discovered ten additional moons. The smallest moon and closest to Uranus is Cordelia, which has a diameter of about 40 km and orbits 49,752 km from the center of Uranus. The largest moon is Titania, which has a dimeter of 1560 km and orbits 436,340 km from the center of the planet. The moons all orbit near the plane of Uranus's equator.

Uranus was known to have nine rings before the flight of Voyager 2. Voyager 2 discovered two additional rings. The inner ring, named epsilon, is flanked by two shepherding moons, Cordelia and Ophelia. Ophelia orbits 53,764 km from the center of Uranus. The thickness of the rings was found to be much thinner than the rings

Fig. 5.11 Rings of Uranus photographed by Voyager 2

of Saturn, which were about 60,000 km thick. The width of the rings of Uranus varied from 2 to 100 km. A photograph of the rings taken by Voyager 2 from a distance of 236,000 km is given in Fig. 5.11.

The magnetic field of Uranus was found to be tilted about 60° from the rotation axis and offset from the center by about one-third of the planet's radius. Magnetic field strength measured at the closest approach distance was 413 nT.

The atmosphere of Uranus was found to consist of about 83% molecular hydrogen and 16% helium with a trace of other gases including methane. The average temperature of the atmosphere was about −213 °C. Wind speeds varied from 15 to 220 m/s.

Voyager 2 used gravity assist from Uranus to send it on a 43-month (3 years 7 months) long trajectory to Neptune. A trajectory correction maneuver was performed on 14 February 1986 to adjust the trajectory.

Flyby of Neptune

Voyager 2 began close observation of Neptune on 5 June 1989. A small trajectory correction maneuver was made on 1 August to fine-tune the trajectory. Neptune is the smallest of the outer planets, but it is still about 3.9 times the size of Earth. It orbits about 30 AU from the sun. An AU is the average distance from Earth to the sun (149.6 million km).

A photograph of the planet taken from a distance of 7.1 million km and a few days before the closest approach is shown in Fig. 5.12.

The dark area near the center of the photograph is called the Great Dark Spot. It was a large storm with its largest dimension about the size of Earth. The white cloud just below the dark area remained with the dark spot but changed in appearance over time. The Great Dark Spot was found to circle Neptune with a period of about 18 h. The associated wind speed was calculated to be about 1000 km/h. The small triangular white cloud, which appears lower and to the left side of the photograph, was called "Scooter" because it moved rapidly around the planet.

Voyager 2 passed above and nearly over the north pole of Neptune. Closest approach occurred on 25 August 1989 at a distance of 4850 km above the cloud tops. All of the scientific experiments were still operative at Neptune encounter,

Fig. 5.12 Photograph of Neptune taken by Voyager 2

12 years after launch from Earth. Its cameras recorded over 9000 images of Neptune and its moons and rings.

Prior to the Voyager missions, Neptune was known to have two distinct moons, Triton and Nereid, and a possible third moon. Voyager 2 discovered six additional moons. The largest moon, Triton, has a diameter of 2707 km, about 78% of the size of Earth's moon. It is unique in the solar system in that it is the only moon that orbits a planet in the opposite direction to the planet's rotation. It orbits 354,759 km above Neptune. Two geyser plumes rising about 8 km above the icy surface of Triton were observed by Voyager 2.

Neptune was observed to have several rings circling the equator. Images taken by Voyager 2 from a distance of 280,000 km showed two bright rings and a faint inner ring. The rings were at distances from the center of Neptune of 41,900, 53,200, and 62,900 km. Close examination of the photographs indicated another broad ring extending from about 53,200 to 59,000 km.

The atmosphere of Neptune was found to consist of 85% molecular hydrogen, 13% helium, and 2% methane. The abundance of methane gives the planet its blue color. The temperature near the cloud tops was measured to be −218 °C.

The magnetic field of Neptune was measured to be about 10,000 nT near the closest approach at 4850 km above the cloud tops. The magnetic field was aligned about 47° from the spin axis.

After the closest approach to Neptune, Voyager 2 passed by the large moon Triton and continued out into interplanetary space. Its trajectory took it at an angle of 48° below the plane of the ecliptic at a velocity of about 55,057 km/h (482 million km/year). The spacecraft continued travel within the heliosphere amid the solar wind. Data from the plasma spectrometer instrument on Voyager 2 indicated the velocity of ions (solar wind) to vary from 400 to 600 km/s during most of the flight.

On 4 November 2011, commands were sent to Voyager 2 to switch from the primary set of three pairs of thrusters to the backup pair of three thrusters. Thrusters were used to adjust the attitude of the spacecraft to point the high-gain antenna towards Earth. The primary thrusters, with associated on/off controls, had been used for 34 years.

After 18 years of travel from encounter with Neptune, Voyager 2 approached the termination shock where the solar wind was slowed by interaction with gas and fields of interstellar space. The first crossing of the termination shock was detected by data from the plasma spectrometer on 30 August 2007 at a distance of 83.6 AU from the sun.

Beyond the termination shock, the solar wind continues to slow, and finally at the heliopause, which is the boundary between the heliosphere and interstellar space, the flow is nearly completely stopped by gas and fields of interstellar space. Voyager 2 reached the heliopause, as detected by the plasma spectrometer, and passed into interstellar space on 5 November 2018. The heliopause is about 123 AU from the sun.

In March 2024, Voyager 2 had been traveling through interstellar space in the direction of the constellation Pavo. It is at a distance of 138.147 AU (20.363 billion km) from the sun and traveling at a speed of 55,355 km/h. Five of the ten original scientific instruments are still operating, and the spacecraft is still sending data back to Earth, 46 years after launch.

Table 5.5 Experiments carried by Voyagers 1 and 2

Experiment	Principal investigator	Purpose
Imaging science subsystem (ISS)	Bradford Smith, University of Arizona	A narrow field of view camera and a wide field of view camera were used to produce multispectral images of cloud patterns, surface features, and ring structures of planets and moons
Photopolarimeter subsystem (PPS)	Authur Lane, JPL	Determine the properties of particulate matter in the atmosphere of the outer plants and of their rings
Infrared radiometer infrared spectrometer (IRIS)	Rudolf Hanel, Goddard SPC	Determine the thermal structure of atmosphere of the planets and determine the balance of energy radiated by that absorbed from the sun. Measure the abundance of hydrogen and helium
Ultraviolet spectrometer (UVS)	A. Broadfoot, University of Southern California	Determine the constituents of the atmosphere of planets by measuring scattering and absorption of light from the sun at ultraviolet wavelengths
Plasma spectrometer (PLS)	John Richardson, MIT	Measure low-energy ions and electrons in the solar wind and planetary magnetospheres. Characterize interaction of the solar wind with each of the outer planets
Low-energy charged particles (LECPs)	S.M. Krimigis, JHU/APL	Investigate the energy spectra, species, and spatial structures of hot plasmas and energetic particles in the vicinity of the outer planets
Cosmic-ray subsystem (CRS)	Edward Stone, Caltech	Detect very energetic particles from outer space and in the radiation fields of the planets. Measure the energy spectrum of electrons and of cosmic ray nuclei from hydrogen through iron
Magnetic fields (MAG)	Adam Szabo, NASA Goddard SPFC	Measure the magnetic fields and magnetospheres of outer planets and investigate interaction of the fields with the solar wind
Plasma wave subsystem (PWS)	Donald Gurnett, University of Iowa	Measure and develop spectrums of the plasma wave and low-frequency radio wave emissions in the magnetospheres of the outer planets
Planetary radio astronomy (PRA)	James Warwick, University of Colorado	Measure radio waves emitted by the planets
Radio science subsystem (RSS)	G. Tyler, Stanford University	Use the S-band and X-band downlink signals received from Voyager to probe the atmospheres of planets and moons during occultation. Use Doppler shift on signals to measure changes in gravity as spacecraft travels past planets

Scientific Experiments on Voyagers 1 and 2

Voyagers 1 and 2 carried a total of ten scientific instruments plus a radio science experiment on their historic journeys. A list of experiments and a summary of purpose of each are given in Table 5.5.

A drawing of Voyager showing locations of the scientific instruments on the spacecraft is given in Fig. 5.13.

Imaging Science Subsystem (ISS)

The imaging science subsystem (ISS) included a wide-angle camera and a narrow-angle camera. Both cameras were television-like with a raster of 800 lines and 800 pixels/line. The cameras were controlled by the flight data subsystem computer. They were mounted on a scan platform, which could be pointed over a wide range of angles by the attitude and articulation control system computer.

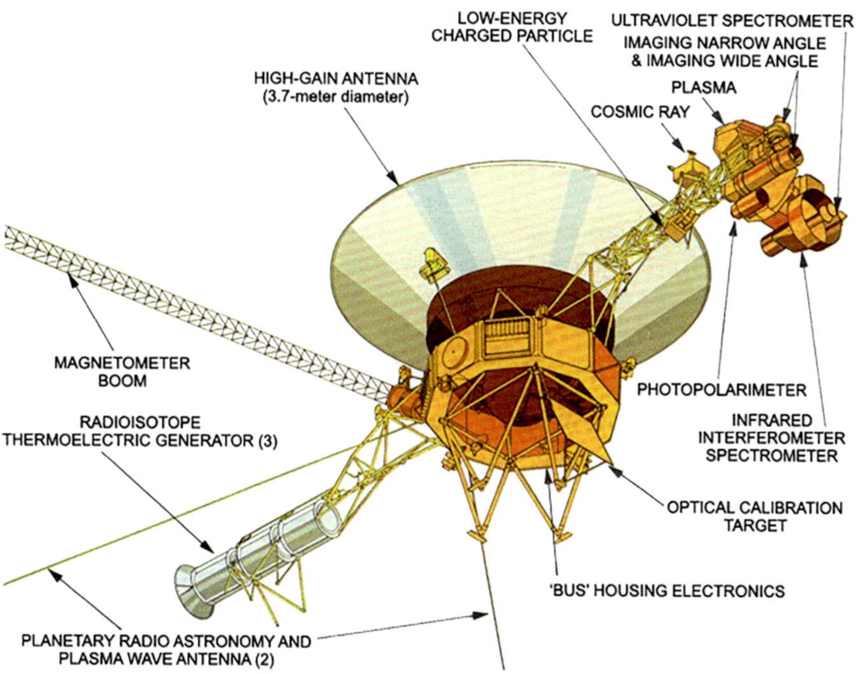

Fig. 5.13 Locations of science instruments on Voyager spacecraft. (NASA/JPL image)

Narrow-Angle Camera

The optics of the narrow-angle camera used a Cassegrain-type telescope with an aperture of 176 mm in diameter. The sensor following the telescope was a vidicon camera tube with a faceplate 25 mm in diameter. The image on the faceplate was read out by an electron beam that was scanned in the horizontal and vertical directions by horizontal and vertical deflection coils. The active imaging area on the faceplate was 11.14 mm^2 resulting in a field of view of the camera of 0.42 deg^2.

The optics of the wide-angle camera used a five-element Petzval-type telescope with an input aperture of 57 mm in diameter and *f*-number of 3.5. The detector following the telescope was the same vidicon tube as used in the narrow-angle camera. The field of view of the camera was 3.17 degrees square.

The electronics associated with the cameras drove horizontal and vertical deflection coils and the focus and alignment coils of the vidicon. The electronics generated a full image frame of 800 lines with 800 pixels/line. The analog video output from the vidicon tube was amplified, filtered, and applied to an eight-bit analog-to-digital converter. The serial digital signal, representing the amplitude of each pixel, was sent to the flight data subsystem for processing.

The normal readout time of one frame was 48 s. Longer frame times were used when imaging the more distant planets where light levels from the sun were lower. Fewer pixels and smaller raster scans could be used to lower image data rates to achieve adequate signal-to-noise ratios for the downlink data.

There were 14 different image modes that could be selected by the flight data subsystem. One of these modes, Mode IM3, was real-time readout of the full 800-line frame with 800 pixels/line. Mode IM2 was a full-frame image of 800 lines and 800 pixels/line that was playback of recorded image data. Some other modes used fewer pixels/line and/or fewer lines. Examples are Mode IM6 with 440 pixels/line and Mode IM14 with 80 pixels/line. Mode IM15 read out the top half of the image in one frame and the bottom half in the next frame.

The vidicon was a black-and-white camera device. It was sensitive to wavelengths from 280 to 640 nm. Color images were synthesized on the ground by combining images taken by the cameras through various filters. A filter wheel with eight positions was located between the output of the telescope and the faceplate of the vidicon. The narrow-angle camera filter wheel contained two clear filters, two green filters, and one each of violet, blue, orange, and ultraviolet filters. The orange filter passed wavelengths from 590 to 640 nm, and the ultraviolet filter passed wavelengths from 280 to 370 nm.

The wide-angle camera filter wheel contained one each of clear, blue, violet, green, and orange filters along with a sodium D filter centered at 589 nm, a methane-JST filter centered at 619 nm, and a methane-U filter centered at 540 nm.

An image of Jupiter taken with the narrow-angle camera at a distance of about 40 million km is shown in Fig. 5.14. Color was obtained by combining images taken through three different filters. An image of the Great Red Spot of Jupiter taken at closer range by the narrow-angle camera is shown in Fig. 5.15. One can imagine the awe of observers when such images were being read out.

Fig. 5.14 Image of Jupiter taken by narrow-angle camera of Voyager 1 at 40 million km distance. (NASA image)

High-quality images were made of Saturn and of its rings by both spacecraft, and high-quality images of Uranus and Neptune were taken by Voyager 2. Images were also made of the moons of each planet. Most of the pictures contained in the NASA photographic gallery of the Voyager missions were taken by the narrow-angle camera with only a few taken by the wide-angle camera.

Photopolarimeter Subsystem (PPS)

The photopolarimeter was used to determine the properties of particulate matter in the atmosphere and rings of the planets. The instrument measured the intensity and polarization of scattered sunlight by particles at eight different wavelengths.

The photopolarimeter included a telescope to focus light onto the overlapping elements on three wheels in series. The telescope was a Cassegrain type with an aperture of 15 cm in diameter and *f*-number of 1.4. The wheels were rotated to

Fig. 5.15 Image of Great Red Spot on Jupiter by narrow-angle camera (NASA image)

expose new elements for each step of the measurement process. The wheels were arranged so that light from the telescope passed through a selected element on each wheel in series.

One of the wheels, referred to as the aperture wheel, included four circular openings that set the field of view of the instrument at 0.12°, 0.25°, 0.8°, or 3.5°. The second wheel, referred to as the analyzer wheel, had positions of open, dark, calibration, and five polarization analyzer positions at 0°, 45°, 60°, 120°, and 135° rotation. The third wheel was a filter wheel with eight different filters. The center wavelengths were 235, 263, 310, 390, 490, 590, 727, and 750 nm.

A typical sequence of operation of the PPS was to step through a programmed 40-analyzer/filter combination for a given aperture setting. The programmed settings included analyzer wheel settings of open, 0°, 60°, 120°, and dark. Eight filter positions were stepped through for each of the five aperture wheel settings for a total of 40 different readings. The dwell time on each combination was 400 ms followed by a 200-ms period to switch to the next wheel position. The total sequence took 24 s.

Photons of light at the output of the three overlapping wheel elements were detected by a photomultiplier tube. The pulses out of the photomultiplier were counted and formatted for output data from the PPS. The output data was in the form of a 30-bit word where 20 bits represented the count over 400 ms integration time and 10 bits were added for instrument status. The 20-bit count data was reduced to 14 bits in the flight data computer by converting the data form to a 10-bit mantissa and 4-bit exponent.

The photopolarimeter subsystem failed on Voyager 1, and no useful data was obtained during its flight. Most functions on the photopolarimeter on Voyager 2 operated well, although some did not. Data was obtained during flyby of each planet. An example is observations made of the atmosphere of the distant planet Neptune where data indicated that the mean radius of atmospheric haze particles was about 0.25 μm.

Infrared Radiometer Infrared Spectrometer (IRIS)

The infrared radiometer infrared spectrometer (IRIS) unit contained an infrared radiometer and an infrared spectrometer. Both instruments were fed infrared light from a telescope with a field of view of 0.25°. The radiometer examined broadband thermal emission in the wavelength range of 0.33–2.0 μm (microns). The spectrometer measured emission lines in a spectrum of wavelengths from 4 to 55 μm.

The purpose of the IRIS was to determine the thermal structure of atmosphere of the outer planets and determine the balance of energy radiated by the planets and that absorbed from the sun. It also measured the abundance of certain molecules in the atmosphere of the planets.

The instrument probed the planets by means of a telescope that had an aperture 500 mm in diameter and field of view of 0.25°. The light out of the telescope was applied to a dichroic mirror that reflected light in the wavelength range of 0.33–2.0 μm (visible to near infrared) to the radiometer channel. Light with wavelengths longer than 2.5 μm was reflected into a Michelson-type infrared interferometer.

Infrared Spectrometer

The spectrometer was a Michelson type that split the incoming infrared light beam into two paths. One path went to a fixed mirror, and the other went to a motor-driven moveable mirror. The moveable mirror was moved at a constant rate over its length of travel and then returned to its original position at the beginning of each IRIS frame. The reflected light from the two mirrors was combined to generate an interference pattern. The interference pattern was detected, and the analog output of the detector was sampled, digitized, and applied to the computer. The computer performed a Fourier transform of the amplitude vs. time data to develop a spectrum of the radiation. The spectral range of the interferometer was 4–55 μm. The resolution of the measurement was 1.3 μm.

The detector for the interferometer was a Schwartz-type four-junction thermopile. The thermopile was a voltage-generating device that used four thermoelectric pairs connected in series. One junction of each pair was exposed to infrared energy from the telescope, and the other junction was shielded from the infrared light. The voltage output of each junction pair was proportional to the temperature difference of the two junctions.

The analog output of the detector was sampled, and each sample was digitized and applied to the computer. The time to record an interferogram was 45.6 s. The wavelengths of emission lines from constituents in the atmosphere of the planets were perceived in the spectrum and related to particular molecules. The atmospheres of the outer planets are dominated by molecular hydrogen and molecular helium.

Data from the spectrometer allowed the ratio of hydrogen to helium to be determined in the vicinity of the planets and in open space. The spectrometer detected other modules in the vicinity of the planets including methane, ethane, ethylene, phosphine, and ammonia.

Infrared Radiometer

The spectral range of the radiometer was restricted to wavelengths between 0.33 and 2.0 μm by the response of the dichroic mirror and coatings on the radiometer side of the mirror. Incoming infrared light from the dichroic mirror was focused onto a thermopile-type detector. The detector was similar to that described for the spectrometer, but with 18 junctions rather than 4.

The analog signal at the output of the detector was amplified and filtered with a time constant of 2.7 s. The amplified, filtered signal was applied to three different output channels. The first channel integrated the signal over a 45.6-s interval. This was equal to the time required by the spectrometer to acquire an interferogram. The second channel provided the analog signal out of the amplifier following the detector. The third channel further amplified that signal by a factor of eight. The outputs of channels two and three were sampled every 6 s, and the amplitudes were digitized and applied to the computer.

Verification of the calibration of the radiometer was performed before and after each encounter with a planet. The verification involved viewing a diffusely scattering target plate mounted on the spacecraft. The plate was shown in the drawing in Fig. 5.13 given previously.

Temperature measurements at Jupiter and Saturn were of the cloud layers at various depths. As the Voyagers approached Jupiter, for example, the IRIS was scanned along the meridian from north to south. The field of view of 0.25° of the instrument resulted in a scan spot diameter of about 13,000 km on the clouds at a distance of 3 million km and smaller spot size at shorter ranges. The diameter of Jupiter is 142,984 km. Temperatures were −113 to −123 °C at a pressure of about 10 mbar and about −163 °C lower in the clouds at pressures greater than 100 mbar.

The temperature of Saturn was found to be about −203 °C in the clouds at a pressure of 70 mbar and −130 °C at a pressure of 1.2 bars. The average atmospheric temperature of Uranus was found to be about −213 °C. The effective temperature of Neptune, averaged from pole to pole, was −214 °C. The temperature of Neptune's moon Triton was −235 °C.

Ultraviolet Spectrometer (UVS)

The ultraviolet spectrometer (UVS) determined constituents of the atmosphere of the outer planets by measuring scattering and absorption of light from the sun at ultraviolet wavelengths. It measured amplitudes in a series of narrow wavelength bands to determine the responding molecules. The instrument was mounted on the scan platform and boresighted in the same direction as the cameras.

The UVS could be operated in either an airglow or an occultation mode. In the airglow mode, it measured radiation due to resonant scattering of the solar flux from molecular constituents in the atmosphere. Resonance scattering from molecular hydrogen, for example, generated radiation at a wavelength of 122 nanometers (nm). In the occultation mode, the instrument measured selective absorption of solar or stellar radiation by molecular constituents in the atmosphere as the spacecraft passed in and out of the planet's occultation zone. It identified a particular molecule by the wavelength of the absorption and its abundance by the amount of absorption.

The spectrometer used a diffraction grating to disperse incoming light onto a linear array of 128 detectors. The grating covered a wavelength range of 53–170 nm. The diffraction grating was concave in shape with a radius of 400 mm. The substrate was 40 by 60 cm in size. The ruling on the grating was 540 lines/mm.

The field of view of the UVS in the dispersion direction of the grating was 0.1°, as set by a series of 13 collimator plates placed between the input aperture of the instrument and the diffraction grating. The field of view in the cross-dispersion direction was 0.87°, as set by the width of the detector in the image plane.

The occultation mode used a mirror to direct sunlight into a separate region of the collimator. The mirror was offset 19° from the boresight of the airglow spectrometer to avoid having to point other instruments on the scan platform towards the sun.

The dispersed light at the output of the diffraction grating was applied to a dual-microchannel plate located in front of the array of detectors. The microchannel plate was made up of a large number of very fine hollow optical fibers. The fibers were short, and the microchannel plates were quite thin. The hollow fibers were coated on the inside with a material that liberated electrons when struck by another electron. When a photon of ultraviolet light struck the entrance to the plate, it liberated an electron. That electron in turn struck the coating on the inside of the fiber and liberated additional electrons. The microchannel plate acted as an electron multiplier. There were two microchannel plates in series, and the photoelectron was amplified by a factor of about a million by the two plates.

Electrons at the output of the microchannel plates struck the anode of a particular detector located behind the plates. The anode accumulated the charge from the electrons until the detector was read out. The array of 128 detectors was read out at a rate of 3125 scans/s. Each individual detector could record photon event rates of about 300/s.

The detector anodes were read out, and the amplitude of the charge pulse was placed into a memory unit with 128 locations of 16-bit words. The memory output was transferred to the flight data system upon command from the FDS. The FDS controls the integration time between readouts to accommodate mission requirements.

Data from the ultraviolet spectrometer indicate that the atmospheres of all of the outer planets are dominated by emissions from the hydrogen Lyman alpha line and emission from molecular hydrogen.

Plasma Spectrometer (PLS)

The plasma spectrometer was used to study the solar wind and its interaction with the outer planets. It measured low-energy ions and electrons in the solar wind and planetary magnetospheres. It was also instrumental in detecting when the spacecraft crossed the heliopause, the boundary between the heliosphere and interstellar space. Voyager 2 was 119 AU from Earth when it crossed the heliopause in November of 2018, 41 years and 3 months after the launch from Earth.

The PLS instrument measured speed, density, and temperature of protons and electrons in the plasma outflowing from the sun (solar wind). The instrument contained four Faraday cups aligned at different angles to conduct the measurements. A photograph of the instrument is shown in Fig. 5.16. The PLS was mounted on the science boom with the centerline of the unit in the direction of the boresight of the high-gain antenna. The centerline is coming out of the page of the photo and in the

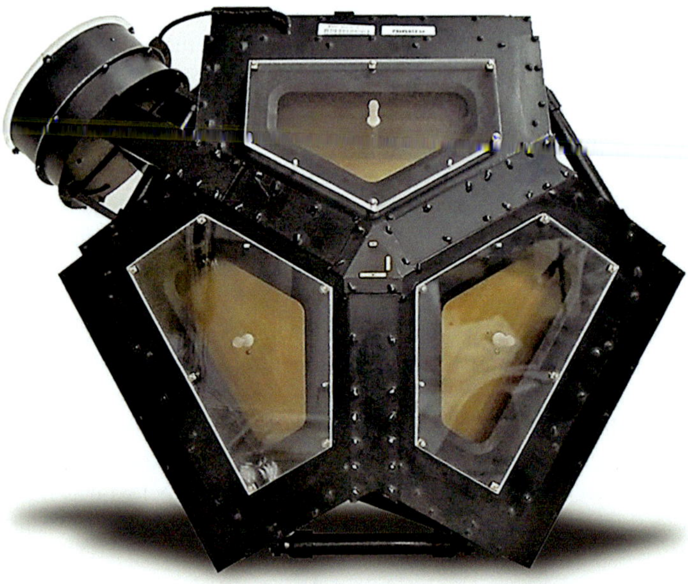

Fig. 5.16 Photograph of plasma spectrometer for Voyager. (NASA photograph)

center of the cluster of three cups. The centerline was pointed in the *Z*-axis of the spacecraft and hence towards Earth when the high-gain antenna was directed at Earth. This placed the centerline of the instrument in the general direction of the sun once the spacecraft was a substantial distance from Earth. The fourth cup was pointed at right angles to the center of the group of three cups. The transparent covers shown over the apertures of the cups in the photograph were removed before flight.

The field of view of each cup was 45°. The center of the field of view was displaced 20° from the centerline, as is apparent in the photograph. The aperture of the Earth facing cups was a five-sided opening with a width of 15.8 cm, vertical sides 8.4 cm long, and distance from the peak to the base of 9.0 cm. The irregular shaped cups were 6.5 cm deep.

The cups contained a collector element near the bottom and a series of nine grids between the aperture and collector. The grids used wires much thinner than the space between wires to allow ions to pass through. The first grid was stretched across the aperture, and it was at ground potential (the same potential as the case). Three grids, referred to as modulation grids, were located below the aperture grid. The modulator grids were connected to a variable positive voltage that repelled positive ions. The variable voltage allowed selection of energies of ions that were able to pass through to the collector element.

Two grids at ground potential were located below the modulation grids. The next grid down was a suppressor grid that was maintained at a potential of −100 V to shield the collector from electrons in the solar wind plasma. Two additional grids at ground potential were located below the suppressor grid. The purpose of the several grids at ground potential was to shield the collector from the electric field of the modulation grids.

The positive voltage applied to the modulation grids consisted of a DC pedestal with a 400 Hz square wave superimposed. The DC pedestal was increased in steps logarithmically from 60 V. A different mode, where a bias of −50 V was subtracted from the modulation signal, was used to detect low-energy ions. The initial value of the bias voltage was 10 V in that mode. The amplitude of the 400 Hz square wave increased as the DC pedestal voltage increased.

The instrument had four operating modes. One mode, referred to as M, was a high-resolution ion measurement mode that used 128 contiguous channels to measure energy/charge ratios in the range of 10–5950 V. A second ion measurement mode, referred to as L, used 16 contiguous channels to cover the same range. There were two electron measurement modes. E1 was a low-energy mode with a measurement range of 10–140 V with 16 contiguous channels. The other electron mode, E2, measured the range of 10–5950 V with 16 contiguous channels.

Ion velocity data from the flight of Voyager 2 indicated velocity fluctuations between about 400 and 600 km/s from the start of the mission. The velocity had dropped to 150 km/s at a distance of about 85 AU. Later in its travel, the spacecraft crossed the heliopause, and the velocity of the plasma from the sun dropped to near zero. Temperature measurements of the plasma made over the first 19 years of the mission indicated large fluctuations between about 6000 and 25,000 K with an average of about 12,000 K.

Low-Energy Charged Particles (LECPs)

The low-energy charged particle instrument investigated the energy spectra, species, and spatial structures of hot plasmas and energetic particles in the vicinity of the outer planets. It measured electrons, protons, alpha particles, and some heavier elements in interplanetary and interstellar spaces. The instrument was developed by the Applied Physics Laboratory of Johns Hopkins University (APL/JHU). The instrument was mounted under the science boom and about midway in the boom's length.

The LECP instrument contained two measurement systems. One, called the low-energy magnetospheric particle analyzer (LEMPA), was designed to measure interplanetary and interstellar particles. The other measurement system was a low-energy particle telescope (LEPT) designed to measure particles during encounters with planets. The two sensors were mounted together as a unit.

The LECP unit contained a rotating platform that allowed scanning of the fields of view of the instrument over 360° in a plane parallel to the *Z*-axis (roll axis) of the spacecraft. The scan normally stopped at eight different look sectors. One of the sectors, sector 8, was a calibration position. The calibration position contained a sun shield to block particles from the sun and a radioactive source with known radiation for each of the sensors. Because the sensors were connected to the electronics unit by cables, the scan did not go continuously around but reversed in direction at the beginning of each scan sequence. The time for a complete scan in the cruise mode was usually 48 min. When the spacecraft approached a planet and was within 60 days of the closest approach, the scan time was reduced to 48 s.

Low-Energy Magnetospheric Particle Analyzer (LEMPA)

Both the LEMPA and the LEPT had several detectors to measure ions, protons, and electrons. The LEMPA used a conical baffle with an opening of about seven cm diameter to establish a field of view of 45° for the primary detector. The baffle fed a cylindrical section that contained a tungsten absorber in the path of another detector.

Detectors used in the LEMPA were labeled by Greek letters α, β, γ, and δ and as A and B. Detector α was the primary detector for measurement of protons and ions with energies less than 15 keV. It was a surface barrier-type detector with an area of 25 mm^2 and located at the centerline and near the small end of the conical baffle. Magnets were used to deflect electrons away from the α detector. The electrons were deflected into two detectors, β and γ, for detection and measurement. The β and γ detector channels were designed for wide dynamic range measurement of electrons with energies less than 15 keV.

Detector δ was intended for measurement in high-intensity environments as encountered during spacecraft travel near the minimum approach distance to a planet. It was used to measure high-intensity alpha particles, protons, and heavier ions. The measurement range was 0.25–2.0 MeV/nucleon. The detector was located on the side of the conical baffle across from the narrowest point. It viewed slightly to the side of the main boresight of the instrument with a field of view of about 29°. Detector B was located farther back in the cylinder that was attached to the conical baffle and behind a 3 mm thick tungsten absorber. It would only respond to highest intensity ions and protons. The detection range was 15–150 MeV/nucleon.

The LEMPA had provisions to also view from the end of the instrument opposite from the conical baffle. The field of view from that end of the instrument was 54°. Detector A was located at that end and behind a magnesium absorber 2 mm thick so that it only responded to high-intensity ions and protons. The detection range was 15–150 MeV/nucleon.

A hemispheric appendage on top of the instrument contained two detectors labeled β′ and δ′ that had their fields of view at an angle of 90° from the other detectors of LEMPA. Additionally, the fields of view of the β′ and δ′ detectors were 90° from each other. The assembly holding the two detectors was rotated by the scan platform of the main sensor. The detection range of the δ′ detector was the same as the δ detector. The β′ detector responded to electrons with energies less than 15 keV.

Low-Energy Particle Telescope (LEPT)

The LEPT was intended to measure low-intensity ions and electrons in interplanetary space. The telescope, which was about 7.1 cm in diameter and 16 cm long, had view ports at each end. One end used a conical baffle that set a field of view of 60°. The other end had a field of view of 51°.

The sensor with the baffle included three detectors labeled D1a, D1b, and D1c located near the small end of the conical baffle displaced from the centerline and 120° apart. Two of the detectors had response from 0.093 to 0.18 MeV/nucleon, and one had response from 0.48 to 1.4 MeV/nucleon. Detector D2 was located behind the plane of the D1 detectors and responded to ions that passed through the D1 detectors. It had response from 3 to 31 MeV/nucleon.

The end of the telescope opposite from the conical baffle had nickel foil over the aperture, and the first detector, D5, was located just behind the foil. That detector was only 90 μm thick, and ions passed through it to strike detector D4. Energetic ions passed through detector D4, which was 2450 μm thick, and struck detector D3. Detectors D3 and D4 had response to particles with energies greater than 200 MeV/nucleon.

The command and data system of the LECP controlled the instrument and processed the large amount of data from the many detectors. It organized and stored the data and handled the interface with the flight data subsystem of the spacecraft. The CDS managed the calibration sequence, and it responded to direction from the FDS to configure the LECP for the various operating modes of the spacecraft. Those operating modes included the following:

Cruise mode: Here, all data except that from the A, B, and δ detectors were processed. The look angle scan rate was set at 1/min.

Far encounter mode: This mode started 60 days before the closest approach. All detectors were processed, and the look angle scan rate was increased to 1/48 s.

Near-encounter mode: All the detectors of the LMPA sensor were activated. Detectors of the LEPT were switched off.

The University of Maryland Space Physics Group generated plots of much of the data from the LECP instrument on Voyager 1 and Voyager 2. One set of data graphs the flux vs. the energy of the ions at particular periods of time. Another graphical set graphs flux vs. time. The later data plots show the significant drop in flux as the spacecraft passed through the heliopause region from the heliosphere dominated by the solar wind into interstellar space. Data from plots of flux of protons with energy from 0.6 to 1.13 MeV/nucleon vs. time for Voyager 2, for example, show a relatively constant flux of about 2.0 (1/cm^2 sr s MeV/nucleon) in a 4-month time interval up to the heliopause crossing. After heliopause crossing, the flux dropped rapidly to less than 0.01 (1/cm^2 sr s MeV/nucleon) in a short period of time. Helium ions of the same energy range had a flux of about 0.2 (1/cm^2 sr s MeV/nucleon) in the 4 months leading up to heliopause crossing dropping to less than 0.001 after crossing.

Cosmic Ray Subsystem (CRS)

The purpose of the cosmic ray subsystem was to detect and measure very energetic particles from outer space and in the radiation fields of the planets. Measurements include the energy spectrum of electrons and of cosmic ray nuclei from hydrogen through iron.

The CRS was mounted on top of the science boom and about midway along the length of the boom. The instrument included two high-energy telescopes, four low-energy telescopes, and one electron telescope. The view angles of the various telescopes were pointed in different directions to achieve wide spatial coverage. The high-energy telescope covered an energy range from 6 to 500 MeV/nucleon. The low-energy telescope covered the energy range from 0.15 to 30 MeV/nucleon. The electron telescope covered the energy range of electrons from 3 to 100 MeV/nucleon.

The high-energy telescope was a double-ended telescope with a field of view of 50° from both ends of the telescope. A stack of 11 detectors was located between the two apertures of the telescope. The detectors were cylindrical in shape with thickness ranging from 0.15 to 6 mm. The outer two detectors at each end had areas of 8.0 cm^2. The seven inner detectors had areas of 9.5 cm^2 each. Incoming particles in the energy range of 4–70 MeV/nucleon were stopped by the stack of detectors, and measurement was made of the energy loss, total energy, and range of penetration. For particles that passed through the stack of detectors, the incremental change in energy with incremental length of passage was measured. The atomic number of the ion could be determined from measurement of the energy lost in each of the first two detectors and the energy lost in the remaining detectors.

The low-energy telescope was a single-ended telescope. The input aperture was covered with aluminum plate 3 μm thick that acted as a collimator. The field of view of the telescope was 50°. The telescope had a stack of four detectors. Two detectors, each 35 μm thick with an area of 2.8 cm^2, followed the collimator plate. Two additional detectors 450 μm thick with an area of 4.5 cm^2 followed. Measurements were conducted in the same manner as in the high-energy telescope.

The electron telescope was single ended and incorporated a stack of eight detectors. Tungsten absorber plates of various thicknesses were inserted between detectors starting between detectors D2 and D3 and continuing to between D7 and D8. Measurements were made of the energy lost in the first two detectors and the number of detectors that were penetrated. Penetration of detector D7, as noted by response in detector D8, represented the highest energy.

The CRS instrument was still operating on both Voyager 1 and Voyager 2 at the end of the year 2023. A series of measurements made from November 2022 through December 2022 indicate that for cosmic ray particles with energy greater than 70 MeV/nucleon, the average count rate was about 2.25/s in Voyager 1 and about 2.4 counts/s in Voyager 2. Both spacecraft are now in interstellar space.

Magnetic Fields (MAG)

The purpose of the magnetic field experiment was to measure the planetary magnetic field of each of the outer planets and measure the magnetospheres of the planets and their interactions with the solar wind. The magnetometers used were fluxgate type, long used in spacecraft and in aircraft. A description of fluxgate magnetometers was given previously in Chap. 4 of this book. Voyager used three-axis versions of the fluxgate magnetometer in order to determine vector properties of the field. The spacecraft incorporated two low-field magnetometers (LFMs) and two high-field magnetometers (HFMs).

The very sensitive low-field magnetometers were mounted on a 13-m-long boom to distance them from incidental magnetic fields of the spacecraft. One low-field magnetometer was mounted at the end of the boom, and the other was mounted 5.6 m from the end. The use of two magnetometers provided redundancy, and the simultaneous data from the two allowed canceling incidental magnetic fields of the spacecraft in data processing.

The unit of magnetic field measurements in the International System of Units is the Tesla (T). For reference, the magnetic field at the surface of the Earth is about 40,000 nT where nT is 10^{-9} T. The measurement range of the LFM was ±50,000 nT. That range was divided up into eight smaller ranges. The instrument automatically changed ranges to operate in the most sensitive range that encompassed the data. The minimum measurement range was ±8.8 nT, and the maximum range was ±50,000 nT. Electronics associated with the magnetometers included 12-bit analog-to-digital converters that resulted in quantization for the ±8.8 nT range of ±0.0022 nT. The quantization was ±12.2 nT at the maximum range of ±50,000 nT. The data sampling interval was 0.6 s.

The two high-field magnetometers were mounted to the boom support struts near the body of the spacecraft. The two HFM magnetometers were mounted about 1 m apart. The measurement range of the HFM was $\pm 2 \times 10^6$ nT. The measurement range was divided up into two ranges: $\pm 5 \times 10^4$ nT and $\pm 2 \times 10^6$ nT. The quantization for the two ranges was ±12.3 nT and ±488 nT, respectively. The data was sampled at 0.6-s intervals.

The LFM was the only magnetometer used during interplanetary and interstellar cruises. Both the LFM and the HFM were used in the vicinity of planets. The closest approach to Jupiter during flyby of Voyager 1 was 280,000 km. The closest approach of Voyager 2 to Jupiter was 645,000 km. Magnetic field measurements near Jupiter by Voyager 1 were reported by Ness (1979). The data indicated field strength of 5 nT at the inbound bow shock, which occurred 85.7 Jupiter radius from the planet. The field strength was about 8 nT at the first crossing of the magnetopause at about 61 Jupiter radius from the planet. The field strength was about 3000 nT at the closest approach to the planet at a distance of 4.89 Jupiter radius.

The closest approach to Saturn by Voyager 1 was 1,323,910 km above the cloud tops. The magnetic field strength at that distance was 1093 nT. The magnetic field was tilted 0.7° from the spin axis of Saturn.

The closest approach to Uranus by Voyager 2 was 81,440 km above the cloud tops. Magnetic field strength measured at that distance was 413 nT. The angle between the axis of the magnetic field and the rotation axis of Uranus was found to be 60°.

The magnetic field of Neptune was measured by Voyager 2 to be about 10,000 nT at a distance of 4850 km above the cloud tops. The field was tilted 47° from the spin axis of Neptune.

Plasma Wave Subsystem (PWS)

The primary purpose of the plasma wave subsystem was to measure and develop spectrums of the plasma wave and low-frequency radio wave emissions in the magnetospheres of the outer planets. It also provided information on electron density in interplanetary and interstellar spaces. The instrument operated in the frequency range of 10 Hz to 56 kHz. It used two collapsible rod-type antennas 10 m long deployed in a V configuration. The PWS shared the two antennas with the planetary radio astronomy experiment. The flight data subsystem (FDS) controlled the instrument and processed its outputs.

An electronics unit accepted signals from each of the two antennas and fed the signals from each antenna to a pair of preamplifiers. One of the preamplifiers had a 40 dB attenuator in series with its input. The attenuator was provided to cope with the very high signal level anticipated near Jupiter. Either preamplifier could be selected by switches. The outputs from the selected preamplifier for each of the antennas were applied to a differential amplifier. The output of the differential amplifier was proportional to the voltage difference between the two antennas. The output of the differential amplifier was fed to a 16-channel spectrum analyzer and to a broadband channel.

The spectrum analyzer consisted of 16 individual filters that covered the frequency range of 10 Hz to 56 kHz. Filters numbered 1 through 8 were designated low-frequency filters, and filters numbered 9 through 16 were designated high-frequency filters. The outputs of individual low-frequency filters were selected by a switch that applied them to an output amplifier. The output of the switch for the high-frequency filters was applied to separate output amplifiers. The spectrum analyzer was scanned every 4 s. The outputs of the amplifiers were applied to the FDS.

The broadband channel contained amplifiers and a filter to accept an input frequency range of 50 Hz to 12 kHz. The output was sampled at a 28,800 Hz rate, and the samples were digitized by a 4-bit analog-to-digital converter. The data rate of the digital output was too high for direct transmission to the Earth, so they were applied to the digital tape recorder for later playback to Earth.

Detectable emissions from Jupiter were observed 12 million km from the planet as Voyager approached. Strongest signals were observed in the 56 kHz and 31 kHz channels of the spectrum analyzer. Later, at about 6.1 million km from the planet, the bow shock was encountered and the 10 lower frequency channels (10 Hz to 1.5 kHz) received strong signals simultaneously. As the spacecraft approached the point of closest approach, strong emissions were detected from 10 Jupiter radii inbound to 10 Jupiter radii outbound in the five spectrum analyzer channels from 100 Hz to 1 kHz. Sporadic signals were received in the upper frequency channels.

Planetary Radio Astronomy (PRA)

The planetary radio astronomy experiment was designed to measure radio wave emission from the outer planets over the frequency range of 1.2 kHz to 40.5 MHz. The emission measurements were used to investigate wave-particle-plasma interactions in the vicinity of the planets.

The PRA used the same 10-m-long pair of antennas as the plasma wave subsystem experiment. The PRA used two receivers to cover the frequency band of 1.2 kHz to 40.5 MHz. One receiver covered the band 1.2 kHz to 1.3 MHz, and the other receiver covered 1.2–40.5 MHz. The receivers were dual-conversion superheterodyne type. A digitally controlled frequency synthesizer set the local oscillator frequency and hence the operating frequency.

The synthesizer stepped through 70 different frequencies for the low-frequency receiver. The spacing of the frequency steps was 19.2 kHz, and the effective bandwidth of each channel was 1 kHz. The synthesizer stepped through 128 different frequencies for the high-frequency receiver. The spacing of the frequency steps was 307.2 kHz, and the effective bandwidth of each channel was 200 kHz. The two receivers together provided a total of 198 frequency channels. Two channels were allocated for housekeeping data for a total of 200 channels. The dwell time on each channel was 30 ms, and the total scan time to cover both receivers plus housekeeping was 6 s. The two receivers together acted as a sensitive 198-channel spectrum analyzer over the frequency band 12 kHz to 45.5 MHz.

Detectors at the output of the second converter of the receivers provided DC voltages proportional to the signal strength of particular frequency channels. The output was applied to an 8-bit analog-to-digital converter, and the resulting digital data was sent to the flight data system. Warwick et al. (1977) gives examples of data from the instrument. Data measured in the vicinity of Jupiter as Voyager 1 passed longitudes from 170° to 220° indicated strong signals in the 14 frequency channels between 20 and 308 kHz.

References

Behannon, K. W., et al, *Magnetic Field Experiment for Voyagers*, Space Science Reviews, 21, 1977

Folger, Tim, *Record-Breaking Voyager Spacecraft Begins to Power Down,* Scientific American, Volume 327, Issue 1, July 2022

Mauk, Barry H., *The Voyager Program at APL*, Johns Hopkins APL Technical Digest, Vol. 11, Number 1 and 2, 1990

Morrison, David, *Voyages to Saturn,* NASA Document NASA SP-451, 1982

NASA Report SP-420, *Voyager to Jupiter and Saturn*, 1977

NASA, Mission Status Bulletin VOYAGER, (No. 1 through No. 99), assembled by Planetary Society, https://www.planetary.org/outreach/voyager-mission

NASA News release-80-160, *Voyager Backgrounder*, Report No. N81-10091, 1980

NASA, *Triaxial Fluxgate* Magnetometer, NSSDCA/COSPAR-ID: 1977-084A-05

NASA, *Voyager 1*, NSSDCA/COSPAR-ID: 1977-084A

NASA, *Voyager 2*, NSSDCA/COSPAR-ID: 1977-076A

NASA/JPL, *Voyager-Fact Sheet-NASA*, https://voyager.jpl.nasa.gov/frequently-asked-questions/fact-sheet

Ness, Norman F., et al., *Magnetic Field Studies at Jupiter by Voyager 1*, Science, Vol. 204, No. 4396, 1979

Scarf, Frederick L., et al, *A Plasma Wave Investigation for the Voyager Mission*, Space Science Reviews, 21, 1977

Smith, B. A., et al, *Voyager 2 at Neptune: Imaging Science Results*, Science Vol. 246, Dec., 1989

Stone, E. C. and Lane, A. L., *Voyager 1 Encounter with the Jovian System*, Science, Vol. 204, No. 4396, 1979

Stone, E. C., *The Voyager Mission: Encounters with Saturn*, Journal of Geophysical Research, Vol. 88, No. A11, 1983

Tomayko, James E., *Computers in Spaceflight: The NASA Experience* NASA-CR-182505, 1988

Uri, John, NASA History, 30 Years Ago*: Voyager Explores Neptune*, Aug. 2019

Uri, John, NASA History, 35 Years Ago*: Voyager 2 Explores Uranus,* Jan. 2021

Uri, John, NASA History, 40 Years Ago*: Voyager 2 Explores Jupiter*, July 2019

Uri, John, NASA History, 40 Years Ago: Voyager 2 *Explores Saturn*, Aug. 2021

Warwick, J. W., et al, *Planetary Radio Astronomy Experiment for Voyager Missions*, Space Science Reviews 21, 1977

Chapter 6
Cassini-Huygens Spacecraft

The very capable Cassini spacecraft orbited Saturn for 13 years and returned over 450,000 images and a wealth of scientific information about the planet, its rings, and its intriguing moons. The accompanying Huygens probe descended to the surface of the large moon, Titan, and unlocked secrets of that large, haze-shrouded moon. An artist's rendering of Cassini in orbit around Saturn is shown in Fig. 6.1.

The Cassini-Huygens mission was an international effort involving the United States National Aeronautics and Space Administration (NASA), the European Space Agency (ESA), and the Italian Space Agency Agenzia Spaziale Italiana (ASI). Development of the Cassini orbiter was managed by the Jet Propulsion Laboratory (JPL) for NASA. ESA developed the Huygens probe, and ASI developed the high-gain antenna for Cassini, radar equipment, and some scientific instruments.

Background of Cassini Program

Cassini had its origin in 1982 with studies conducted by a joint working group of the Space Science Board of the National Academy of Science in the United States and the Space Science Committee of the European Science Foundation. The recommendation from the study was a joint effort to develop a spacecraft to orbit Saturn and carry a landing probe to the large moon, Titan. Further studies were carried out by NASA and ESA during the time period 1983–1988.

Funding for the Cassini Saturn mission, and for a comet-asteroid mission using the same spacecraft design, was approved by Congress in 1989. Later cutback in funding resulted in cancellation of the comet-asteroid portion of the mission, but a scaled-down Cassini mission to Saturn went forward. The European Space Agency

T. Lund, *Spacecraft Exploring Outer Planets*, Springer Praxis Books,
https://doi.org/10.1007/978-3-031-96544-9_6

Fig. 6.1 Artist's conception of Cassini in orbit around Saturn. (NASA graphic)

had approved the Huygens probe in 1988. Agreement to include Agenzia Spaziale Italiana in the Cassini-Huygens team was formalized in 1995.

The Cassini spacecraft program was managed by the Jet Propulsion Laboratory (JPL) of the California Institute of Technology, and the spacecraft was built by JPL. The Program Manager for Cassini at JPL was Earl Maize. The project scientist was Linda Spilker. Design began in 1989, and a successful Critical Design Review was held in December 1992. Assembly of the spacecraft began in 1995. The Cassini spacecraft was delivered to the Kennedy Spaceflight Center in April 1997 where it was integrated with the Huygens probe supplied by ESA.

Cassini-Huygens was launched on 15 October 1997 by a Titan IVB/Centaur launch vehicle. After gravitational assists from two flybys of Venus, a flyby of Earth, and a flyby of Jupiter, the spacecraft traveled to Saturn and entered orbit on 1 July 2004. It continued orbiting and downlinking observations to Earth until September 2017. The Huygens probe was released from Cassini on the third orbit, and it landed on Titan on 14 January 2005. Huygens data was received by Cassini and relayed to Earth.

Mechanical Configuration

A photograph of the Cassini-Huygens spacecraft being mated to the launch vehicle adapter at the Kennedy Spaceflight Center is shown in Fig. 6.2. The heat shield for Huygens is on the left side of the photo.

Fig. 6.2 Cassini-Huygens being mated with launch adapter. (NASA photo)

A NASA drawing of the spacecraft with labels identifying major components is given in Fig. 6.3.

The coordinate system sited the *Z*-axis along the longitudinal axis of the spacecraft with +*Z* in the direction of the rocket engines. The +*Y* axis was in the direction of the magnetometer boom. The +*X* axis was in the direction of the boresight of the stellar reference units. The telescopes for the stellar reference units are drawn as white cylinders at the left side of the drawing of the spacecraft above.

Cassini-Huygens was a large spacecraft. The Cassini orbiter was 6.8 m high and 4 m wide. A boom 11 m long held magnetometers away from the spacecraft, and three rodlike antennas 10 m long extended from the side of the spacecraft. A parabolic antenna with a reflector 4 m in diameter was mounted to the top of the spacecraft. Unfueled, the orbiter weighed 2125 kg. Propellants weighed 3132 kg. The Huygens probe was mounted to the side of the spacecraft. The heat shield for the probe can be seen on the left side of the spacecraft in Fig. 6.2. The Huygens probe was a disk-shaped space vehicle protected by a blunt conical heat shield. The probe

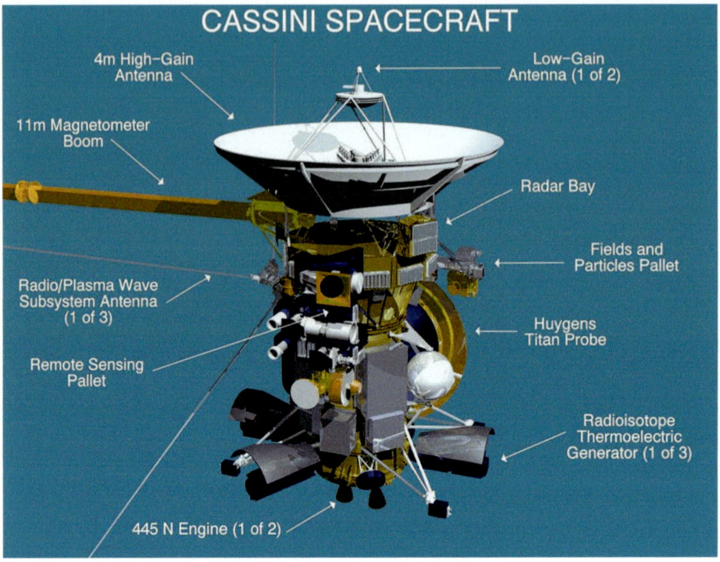

Fig. 6.3 Labeled drawing of Cassini-Huygens spacecraft. (NASA graphic)

was 2.7 m in diameter and weighed 320 kg. The total weight of Cassini-Huygens at launch, including propellants and launch vehicle adapter, was 5712 kg.

The Cassini spacecraft was composed of three main modules: upper equipment module, propulsion module, and lower equipment module. The three modules were assembled and tested separately and then stacked together. The propulsion module, which was the most massive and strongest of the modules, was in the center of the stack. The upper equipment module was mounted above it and the lower equipment module below it. The high-gain antenna was mounted to the top of the upper equipment module, and the launch adapter was connected to the bottom of the lower equipment module.

The upper equipment module included a 12-sided structure, referred to as the electronic bus. The electronic bus contained electronics in each of the 12 bays formed by the 12-sided structure. The width of the electronic bus across the flats was about 2.2 m, and it was about 0.5 m high. An upper shell structure was mounted to the bottom of the electronic bus. The lower end of the shell mated with the top of the propulsion module.

The propulsion module, which was designed and built by Lockheed Martin, contained two independent propulsion systems: the monopropellant system and the bipropellant system. The monopropellant system used a series of small thrusters for attitude control of the spacecraft and for small velocity corrections. The bipropellant system contained two redundant 445-N thrust rocket engines. The nozzles for the rocket engines are labeled in Fig. 6.3. That rocket system was used to impart large velocity changes to the spacecraft.

The monopropellant propulsion system used a series of small thrusters fueled by hydrazine that reacted with catalyst in the thrusters. The resultant hot gases from decomposition of hydrazine provided about 1 N of thrust (0.22 pounds) from each thruster. The thrusters, which were MR-103H type built by Aerojet, were about 34 mm in diameter and 15 cm long. The catalyst used was type S-405 in granular form and contained about 33% indium.

There were two redundant sets of eight thrusters, for a total of 16. Hydrazine was contained in a spherical tank mounted to the side of the propulsion module. The tank held 132 kg of hydrazine. Hydrazine was forced out of the tank to valves that fed the thrusters by pressurized helium in the hydrazine tank. The tank was recharged with helium from a separate small helium tank once during the mission.

The bipropellant system was used to impart large velocity changes to the spacecraft. There were two redundant bipropellant systems, each with a rocket engine providing 445-N thrust. The fuel was monomethyl hydrazine, and the oxidizer was nitrogen tetroxide. The fuel and oxidizer were contained in separate cylindrical tanks with domed ends. The tanks were mounted along the longitudinal axis (*Z*-axis) of the propulsion module. The fuel tank held 1131 kg of monomethyl hydrazine, and the oxidizer tank held 1869 kg of nitrogen tetroxide. A cylindrical tank holding helium, which was used to force fuel and oxidizer from their respective tanks, was mounted to the side of the propulsion module. The helium tank held 8.6 kg of helium.

A pellet, referred to as the remote sensing pellet, was attached to the spacecraft on the +*X* facing side as shown in Fig. 6.3. That pellet carried remote sensing science instruments, including two cameras. A second pellet, referred to as the fields and particles pellet, was mounted on the opposite side of the spacecraft. That pellet carried scientific instruments that made in situ measurements of fields and particles.

Spacecraft Systems

A functional block diagram of systems within the spacecraft is given in Fig. 6.4.

Electrical Power

Electrical power for the Cassini spacecraft was produced by three radioisotope thermoelectric generators (RTGs). The spacecraft was intended to operate at Saturn, which lies 1.43 billion km from the sun, 9.6 times farther from the sun than Earth. Solar arrays were not practical to deliver the large amount of power required by Cassini at that distance. The three RTGs produced a total of 882 W of electrical power at the beginning of the mission. The mission was planned to have 20-year duration, and the power output after 20 years was predicted to be about 600 W. The available power of 600 W provided ample margin above the 469-W electrical load of the spacecraft.

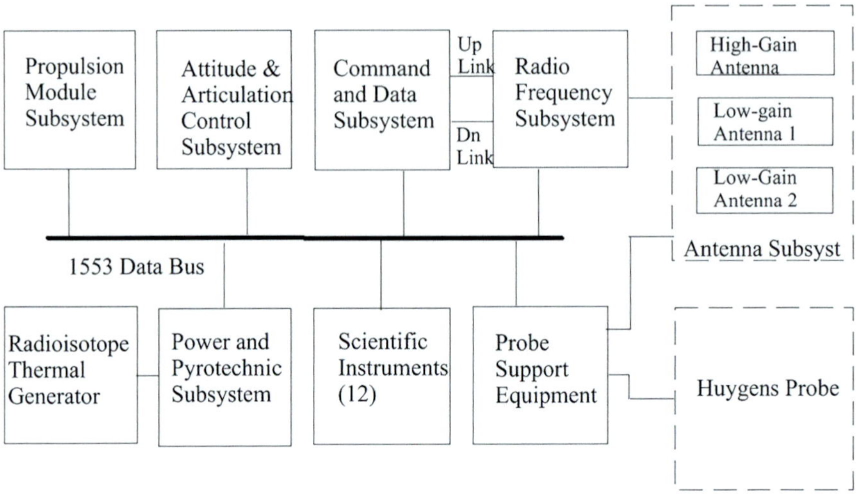

Fig. 6.4 Functional block diagram of Cassini spacecraft

The RTGs were cylindrical in shape, 42.2 cm in diameter, and 114 cm long. Each RTG weighed 56 kg. The three RTGs were mounted horizontally around the bottom of the spacecraft as illustrated in the drawing of Fig. 6.3. RTGs generate electrical power by applying heat from radioactive decay of a material such as plutonium-238 to a series of thermoelectric couples. Plutonium-238 decays into uranium-234 by emitting alpha particles. The half-life of the decay process is 87.7 years. The kinetic energy of the alpha particles is converted to heat when the particles strike surrounding material. Each RTG contained a total of 10.9 kg of plutonium oxide.

The Cassini RTGs used pellets of plutonium-238 oxide pressed into cylindrical shape 2.76 cm in diameter and 2.76 cm long. Each pellet was clad with a welded shell of iridium alloy. There were 72 such cladded pellets within a cylindrical region at the center of the RTG. A graphite cylindrical shell enclosed the pellets. A total of 572 silicon-germanium thermoelectric couples were mounted around the graphite shell. The devices were mounted such that one end was in contact with the hot surface and the other end was in contact with the much cooler outer case of the RTG.

Each of the RTGs had initial thermal output of about 4400 W, which resulted in initial electrical power output from the field of thermocouples of 294 W. The thermal to electrical power efficiency was 6.68%. The electrical outputs of the three RTGs were connected in parallel giving an initial total power output of about 882 W. The power output of the RTGs decreased with time. Telemetry data from the spacecraft indicated power available to be about 630 W in September 2017 after 20 years of flight. The average electrical load in the spacecraft was 469 W.

The voltage at the parallel output of the three RTGs was somewhat over 30 V. A regulated 30-V power bus was formed by using a shunt regulator set at 30 V. Excess power from the regulator was dissipated in resistors attached to radiator plates that radiated excess heat into space. Some excess electrical power was also used to warm the propellant tanks. The shunt regulator was included in the power and pyrotechnics subsystem (PPS) in the spacecraft. The PPS contained power supplies for various voltage forms along with a series of solid-state power switches that applied power to individual systems in the spacecraft.

Attitude Control

Cassini operated as a three-axis stabilized spacecraft under the control of the attitude and articulation control subsystem (AACS). AACS determined the attitude of the spacecraft, and it controlled the attitude as required for mission events. Operation of the AACS was controlled by the attitude flight computer. The computer was a radiation-tolerant microprocessor developed by Sandia National Laboratory for the U.S. Air Force. It was a 16-bit device that used IBM 1750A architecture and had 512 kB of RAM storage and 8 kB of programmable ROM.

The attitude flight computer received command data from the ground via the command and data subsystem. It could store the commands for future execution or immediately issue commands to the attitude control functions or main engine functions. The AACS controlled orientation of the spacecraft to point the high-gain antenna towards Earth for communication sessions, point scientific instruments to directions of interest when gathering scientific data, and align the spacecraft for thrusting during velocity changes.

Spacecraft attitude was controlled by the AACS by either firing selected thrusters or changing the speed of reaction wheels. Reaction wheels were used when accurate, and stable pointing of science instruments and cameras was required. Thrusters were usually used for attitude control during cruise.

The primary reference for attitude of the spacecraft was two redundant stellar reference units (SRUs). The SRUs were very accurate star trackers. A star catalog with 3500 stars was carried in each SRU. Four to five of the brightest stars were selected in the 15°-by-15° field of view of the star tracker during tracking. Sun sensors were used as required to orient the spacecraft initially until the star tracker acquired a star field that matched that in its catalog. The SRUs provided inertial attitude data at update rates from 1 to 5 s. The AACS combined that data with gyro data from the inertial reference unit to develop control signals for attitude control of the spacecraft.

An inertial reference unit (IRU) kept track of attitude between updates from the stellar reference unit. The IRU provided attitude data when the SRUs were not operating as when the telescopes were pointed at the planet or near the sun. Two redundant IRUs, known as IRUA and IRUB, were provided. Each IRU contained four hemispheric resonator gyros. Three of the gyros were primary, and the fourth was used as a parity check on the other gyros. The three primary gyros were mounted orthogonal to one another.

The hemispheric resonator gyro (HRG) consisted of a finely machined hemispheric shell of quartz with a stem as in a wine glass. An analogy often used in describing the operation of an HRG is the singing of a wine glass when a wet finger is run around the rim. In operation, the hemispheric shell was excited at its natural frequency electromagnetically, and the frequency was accurately measured by electronics in the IRU. Rotation of the IRU unit changed the frequency of the vibration in proportion to the rotation about the axis of symmetry of a given HRG. The changes of frequency of the three HRGs were accurately measured, and the data was resolved into rotation of the IRU.

The AACS also managed operation of the gimballed rocket engines of the redundant bipropellant propulsion system. That propulsion system, with a thrust of 445 N, was capable of performing large velocity changes. The rocket engines were gimballed ±12.5° in the pitch and yaw planes. The AACS-controlled actuators adjusted the angle of the rocket motor for each axis as required to thrust through the center of mass of the spacecraft.

The spacecraft was usually oriented to point the high-gain antenna towards the sun to shield the spacecraft from the sun during cruise when the spacecraft was at distances up to a few AU from the sun. Sun sensors with a field of view of 32° by 32° in the direction of the −*Z*-axis of the spacecraft were used to develop control signals for that pointing.

Reaction Wheels

There were three primary reaction wheels and a fourth wheel that was a spare. The primary reaction wheels were mounted so that their spin axes were orthogonal to one another. They were mounted at equal distances from the *Z*-axis of the spacecraft. Rotation of the spacecraft at about a particular axis of the reaction wheel set was induced by rotating the reaction wheel for that axis in the opposite direction. The reaction to changing speed of the wheel was an equal and opposite reaction on the spacecraft.

After several net speed adjustments in the same direction, the speed of the wheel could approach maximum and no further correction would be possible. To prevent this, at some point, the thrusters were engaged to hold that particular attitude, and the reaction wheel was run down to a stop. The reaction wheel was then able to control attitude again.

Thrusters

The thrusters were arranged in groups of four, and there were four groups. Two thrusters in each group were primary, and the other two were part of a redundant system. One redundant system was referred to as A-branch and the other as B-branch. The groups of thrusters were mounted to short supporting structures to distance them from the four corners of the spacecraft. The four groups of thrusters can be seen at the four corners at the bottom of the spacecraft in the photograph of Fig. 6.2.

Referring to the drawing in Fig. 6.3, the +*X*-axis was along the boresight of the stellar reference telescopes, and the +*Y*-axis was in the direction of the magnetometer boom. The *Z*-axis was the longitudinal axis of the spacecraft with +*Z* in the direction of the rocket motors. Pitch of the spacecraft was around the *X*-axis, yaw was around the *Y*-axis, and roll was around the *Z*-axis. The square pattern of groups of thrusters sited lines between thrusters parallel to the *X*-axis and the *Y*-axis.

One thruster in each redundant set of the four groups was pointed to thrust along the +*Z*-axis. Those four thrusters were fired together by the AACS to change the velocity of the spacecraft along the *Z*-axis. Pitch and yaw of the spacecraft were managed by the AACS by firing select pairs of thrusters. Looking at the spacecraft along the *X*-axis, the two thrusters on the right side were fired to cause pitch around the *X*-axis in one direction, and the two thrusters on the left side were fired to pitch in the other direction. Likewise, yaw around the *Y*-axis in one direction was achieved by firing the forward two thrusters, and yaw in the other direction was achieved by firing the aft two thrusters.

One thruster in each redundant set of each group was pointed outward parallel to the *Y*-axis. Roll of the spacecraft in one direction was achieved by firing one diagonal pair of thrusters, and roll in the other direction was achieved by firing the opposite diagonal pair of thrusters.

Telecommunications System

The telecommunications system received and demodulated command data uplinked from the very large antennas of the Deep Space Network (DSN) on Earth. The spacecraft transmitted scientific and engineering data on the downlink to the DSN. Communication was maintained with Cassini while it was in orbit around Saturn at distances up to 1.7 billion km from Earth. In addition, the Cassini orbiter received the uplink S-band signal from the Huygens probe that descended to Saturn's moon, Titan. It relayed Huygens data to Earth by the telecommunications system.

Communication with the DSN was carried out in the X-band frequency range. The telecommunications system included an antenna subsystem and a radio frequency subsystem.

Antenna Subsystem

The antenna subsystem consisted of a high-gain antenna (HGA) and two low-gain antennas (LGAs). The HGA was a Cassegrain type with a parabolic reflector 4 m in diameter. The antenna contained several feed horns that used the subreflector to transfer signals to and from the large reflector. Commands from Earth were received at X-band at a frequency of 7175 MHz by an X-band feed horn. Telemetry data was transmitted to Earth at a frequency of 8425 MHz using the same X-band feed horn. The antenna gain was 44.7 dB, and the beamwidth was 0.55° for the uplink signal. The corresponding values for the downlink signal were 46.6 dB and 0.64°.

The low-gain antennas operated at X-band at the same frequencies for up- and downlinks as the high-gain antenna. The low-gain antennas were used when the high-gain antenna could not be pointed towards Earth and when the range to the spacecraft was only a few AU.

One low-gain antenna, LGA1, was mounted on the outward-facing side of the subreflector for the high-gain antenna. Its boresight was along the $-Z$-axis. The gain of LGA1 was 8.4 dB, and the beamwidth was about 32° at the uplink frequency. The gain was 8.9 dB, and the beamwidth was about 24° at the downlink frequency. The other low-gain antenna, LGA2, was mounted on a short boom on the lower equipment module and below the Huygens probe. It was boresighted along the $-X$-axis of the spacecraft. The gain of LGA2 was 8.4 dB, and the beamwidth was about 40° at the uplink frequency. The gain was 9.0 dB, and the beamwidth was about 40° at the downlink frequency.

Radio Frequency Subsystem

The radio frequency subsystem (RFS) received uplink signals from Earth, detected commands on the uplink signal, and generated a carrier signal for the downlink that was coherent with the uplink carrier. The RFS amplified the downlink signal with a traveling wave tube amplifier and provided switches to switch among antennas. There were two complete and redundant receive/transmit channels. An ultra-stable oscillator provided reference frequency to both receive/transmit channels in case generation of the downlink frequency based on the uplink carrier was not successful.

Each receive/transmit channel contained a deep space transponder (DST) built by Motorola. The DST incorporated an X-band receiver and an X-band exciter. The exciter generated the downlink signal at low power. The receiver was connected via a diplexer through switches to the selected antenna. The diplexer allowed the radio frequency subsystem to transmit and receive at the same time from the same antenna. The receiver demodulated commands on the uplink signal and impressed the commands in biphase modulation form onto a 16 kHz subcarrier. The subcarrier was applied to a command detector unit (CDU) that was part of the RFS. The CDU demodulated the uplink data on the 16 kHz subcarrier, and the resulting digital data stream was sent to the command and data subsystem along with a bit-synchronization clock signal. The CDU also sent a lock signal to the CDS to indicate that the CDU was locked to a valid uplink signal.

The receiver demodulated an uplinked range modulation signal and arranged to modulate the exciter for the downlink carrier by the ranging signal. Equipment on Earth used the two-way range signal to determine the range to the spacecraft. The transponder also generated a one-way ranging function that consisted of two tones that was used to modulate the exciter signal when two-way ranging was not available.

The transponder contained a phase-lock loop that locked to the carrier of the uplink signal and generated a coherent signal offset in frequency by the ratio of 880/749. That signal was used to generate the downlink carrier in the exciter. The exciter could be modulated by signals generated by telemetry data, by the received ranging signal, or by the two tones of the one-way raging signal, or it could not be modulated whereupon just the downlink carrier was transmitted.

A telemetry control unit (TCU), which was part of the RPS, controlled operation of the RPS and controlled modulation of the downlink signal. Data to be downlinked was received from the command and data subsystem in digital format at rates from 5 to 248,850 bits per second (bps). The TCU managed modulation of the exciter by the downlink data. The TCU also selected which of the redundant elements were active and which were in standby, for example, DST-A/CDU-A On or Off and DST-B/CDU-B On or Off.

The output of the exciter was applied to the traveling wave tube power amplifier, which raised the output signal level to 20 W at X-band. The amplified signal was fed to the diplexer and from there to switches that selected which antenna to use for transmitting the downlink signal.

Command and Data Subsystem

The command and data subsystem (CDS) was the brains of the Cassini/Huygens spacecraft. Central to the brains were two redundant flight computers using IBM 1750A architecture. The CDS managed most aspects of operations of the spacecraft. It received uplink command data from the radio frequency subsystem, organized the data, and commanded response from various subsystems on the spacecraft. It also processed engineering data from the spacecraft and data from scientific instruments and organized that data for telemetry to Earth on the downlink.

The CDS included two redundant command and data electronics assemblies and two redundant solid-state recorders. Each assembly contained a flight computer, hardware for decoding uplink commands from the RFS, a Reed-Solomon encoder for downlink data to be furnished to the RFS, a 1553B data bus controller, and remote terminal for the data bus. The solid-state recorder was a dynamic random access memory type that could store 2.3 Gbits of data. The recorder stored a large amount of data from the scientific instruments that was downloaded to Earth during periodic communication sessions. It also stored flight software for the CDS.

The flight computer, which used 1750A architecture with 16-bit words, operated with 512 kbits of RAM and 8 kbits of ROM. The computer was programmed in Ada with 48,000 lines of assembly and source code.

Typically for data taking, the CDS would issue commands to cause the spacecraft attitude to point a scientific instrument in the desired direction and then issue a trigger command to cause the selected instrument to start taking data. Data from the instrument was gathered by the CDS, processed, and formatted for storage in the solid-state recorder. The stored data was later formatted for telemetry on the downlink to Earth.

Flight of Cassini-Huygens

The Cassini-Huygens spacecraft was launched from pad 40 at the Cape Canaveral Air Force Station by a Titan IVB/Centaur launch system on 15 October 1997. The two booster rockets of Titan were ignited first, and they burned for 2 min 23 s after liftoff. The spacecraft was at an altitude of 68,300 m at burnout. Stage 1 was ignited 2 min 11 s after liftoff at an altitude of 58,500 m and burned for about 3 min. Stage 1 was jettisoned at 5 min 23 s after liftoff at an altitude of 167,300 m, and stage 2 was ignited. The Centaur upper stage, with the spacecraft attached, was separated from the launcher at 9 min 13 s after liftoff at an altitude of about 206,700 m.

The first burn of Centaur began immediately after separation. The burn lasted for about 2 min and placed Cassini into an elliptical parking orbit around Earth, 170 m by 445 m in extent. After coasting in the parking orbit for 17 min, Centaur was fired for 6 min to increase the speed to escape velocity from Earth and place Cassini on a trajectory towards Venus. Centaur was then separated from the spacecraft.

A trajectory correction maneuver (TCM) was performed 25 days after launch to refine the trajectory. The attitude of the spacecraft was adjusted first so that the rocket thrust was along the direction for the required trajectory correction. The TCM changed velocity by 2.8 m/s.

Gravity assists were required from Venus (twice), Earth, and Jupiter to establish the Cassini/Huygens spacecraft on a trajectory to intercept Saturn. Several TCMs were made to adjust spacecraft trajectory to set up trajectories for gravitational assists from Venus, Earth, and Jupiter.

The spacecraft cruised towards Venus for the first gravity assist. Two TCMs were performed to set up the approach to Venus. The first gravity assist from Venus occurred on 26 April 1098 with the closest approach distance of 284 km. That gravity assist increased the speed of Cassini/Huygens by about 7000 m/s. The spacecraft traveled around the sun and approached Venus for a second gravitational assist. The second assist from Venus occurred on 24 June 1999 with the closest approach distance of 600 km.

The spacecraft then traveled towards Earth. It received gravity assist from Earth on 18 August 1999 with the closest approach of 1171 km. The gravity assist from Earth increased the speed of the spacecraft by about 5500 m/s. The encounter with Earth set up a trajectory to intercept Jupiter. The spacecraft received its final gravity assist from Jupiter, passing within 9,723,890 km of Jupiter on 30 December 1999. The gravity assist from Jupiter increased Cassini/Huygens's speed by 2011 m/s.

After a travel time from launch of about 6 years 7 months, Cassini/Huygens approached the Saturn system. A photograph of Saturn taken on 7 May 2004 from a distance of 28 million km by the narrow-angle camera of Cassini is shown in Fig. 6.5.

Cassini passed near an outer moon, Phoebe, 12.952 million km from Saturn, on 11 June 2004. The closest approach distance from Cassini to Phoebe was 2068 km. This was the only close flyby of Phoebe by Cassini because the Cassini mission

Fig. 6.5 Photograph taken by Cassini on approach to Saturn. (NASA pix)

positioned the orbital paths of Cassini closer to Saturn. Several photographs were taken of Phoebe while approaching and departing the moon.

Cassini/Huygens approached Saturn from below and crossed the ring plane in the 30,000-km-wide gap between the G-ring and F-ring on 1 July 2004. The attitude of the spacecraft was adjusted to place the high-gain antenna in the direction of travel to shield the spacecraft from impacts of stray ring particles.

A few minutes after the ring crossing, Cassini was oriented to place the thrust axis of the main engine along the velocity vector, and the engine was fired for orbit insertion. The main engine burned for 96 min and reduced the velocity by about 626 m/s. Just prior to orbit insertion, the speed of Cassini relative to the center of the planet was about 24 km/s. The speed continued to increase due to pull of gravity, but the reduction in velocity by the engine burn was enough to allow the strong gravity of Saturn to capture Cassini/Huygens and pull it into an elongated orbit. Orbit insertion occurred on 1 July 2004.

The inclination of the orbit was 17° to the equator of Saturn. Cassini rose to about 16,000 km above the ring plane before the orbit pulled it down to ring crossing in a downward direction about 3 h 20 min after the initial upward crossing. The initial orbit had an apogee of about 9 million km from the center of Saturn. Orbit trim procedures reduced the apogee to about 3.5 million km by the time of the third orbit. Perigee was about 1 million km.

The third orbit of Cassini was tailored to intercept the large moon, Titan, and deliver the Huygens probe. The Huygens probe was separated from Cassini on 25 December 2004 while in the third orbit. Springs were released, and Huygens moved along spiral rails to give it a slow spin as it left Cassini. Huygens continued on the orbit path and made a successful landing on Titan on 15 January 2005.

Scores of images were taken by Huygens during approach, landing, and while on the surface of Titan. Other scientific instruments gathered information about the clouds, atmosphere, and surface of Titan. The trajectory of Cassini was altered after release of Huygens to pass 60,000 km from Titan. Cassini's trajectory was arranged to allow it to receive data transmitted by Huygens during its descent to landing and while on the surface. Huygens data was relayed to Earth.

Cassini orbited well outside of the main rings of Saturn for most of the mission. The period of the orbits was about 3 weeks during the early years of the mission. The trajectory was changed at times to investigate a particular moon or to obtain a good vantage point to observe the rings. Small orbital trajectory changes were made by the thrusters. Large trajectory changes were accomplished by establishing particular flyby parameters past the massive moon, Titan.

Like Earth, Saturn has seasons because the spin axis is displaced from its orbital path. In the case of Saturn, the tilt is 25.7°. The orbital period of Saturn is 29.4 Earth years. Cassini entered the orbit of Saturn after winter solstice in the Northern hemisphere with the sun illuminating the rings from the southern direction. (The word "solstice" identifies the time when a planet's position in its orbit is such that the sun appears farthest north or farthest south of the equator.) Northern winter solstice on Saturn occurred in May 2002, and summer solstice occurred in May 2017.

The Primary Mission, which was planned to last for 4 years, was officially completed on 30 June 2008. Cassini made 75 orbits of Saturn during the 4-year primary mission. The rings were imaged extensively and investigated by scientific instruments. Several targeted flybys were made at the large moon, Titan. Titan was of particular interest because it has an atmosphere, clouds, and a solid surface with lakes of liquid methane. The icy moon, Enceladus, was also of high interest because water ice geysers were seen erupting from several regions near the south pole of the moon. A close flyby of Enceladus was made in March 2008, passing within 50 km of the moon and flying through one of the geysers. Emissions from Enceladus were found to be replenishing the tenuous, far-out E-ring of Saturn.

The spacecraft was performing well at the end of the Primary Mission, so the mission was extended for slightly over two additional years. That time period included the time of equinox of Saturn's orbit. At equinox, the sun was directly over the equator and the rings were illuminated edge-on. Equinox occurred on 12 August 2009. That portion of the mission was given the name Equinox Mission. The Equinox Mission ended on 11 October 2010. The spacecraft made 60 orbits of Saturn, 26 targeted flybys of Titan, seven flybys of Enceladus, and one flyby each of the icy moons Dione, Rhea, and Helene during the Equinox Mission.

Cassini was still performing well after the Equinox Mission, and the mission was again extended, this time for about 7 years. That portion of the mission was called the Solstice Mission because it would include the time of summer solstice in the northern hemisphere. The Solstice Mission lasted from 11 October 2010 to 15 September 2017 when, by design, Cassini impacted Saturn. Cassini made 155 orbits of Saturn, 54 flybys of Titan, and 11 flybys of Enceladus during the Solstice Mission.

Towards the end of the Solstice Mission, Cassini was maneuvered to pass by Titan to obtain a gravitational boost and change the orbit to a ring-grazing orbit where perigee of the orbit was 7800 km from the center of the F-ring of Saturn. The F-ring is the outer boundary of the main ring system of Saturn. The ring-grazing segment of the Solstice Mission began on 30 November 2016 and lasted until 22 April 2017. A total of 20 ring-grazing orbits were made. A targeted flyby of Titan was made at the end of the ring-grazing portion of the mission to set up a major orbit change for the grand finale to Cassini's mission.

The final segment of the Solstice Mission was called the Grand Finale. A major orbit change was accomplished by flying close to Titan to set up the Grand Finale orbits. The perigees of the orbits were trimmed to pass through the gap between the innermost ring (D-ring) and the cloud tops of Saturn. The first passage through the gap occurred on 26 April 2017, and 22 such orbits were made. A close flyby of Titan was made on the last orbit on 11 September to set up an impact trajectory. Cassini impacted Saturn on 15 September 2017, ending the remarkable Cassini mission.

During the productive 13-year mission, Cassini orbited Saturn 293 times. On the third orbit, it released the Huygens probe which went on to make a successful soft landing on the large moon, Titan. Cassini made 162 targeted flybys of Saturn's moons. Particular attention was given to the large moon, Titan, with 127 targeted flybys. Radar images were made of 43% of the surface of Titan during the mission, and science instruments probed the atmosphere and clouds. A total of 23 targeted flybys were made of the icy moon Enceladus. One of the flybys passed through a geyser plume of icy water on the southern portion of the moon. Six new moons were discovered. The rings of Saturn were investigated in detail. A total of 453,048 images were returned to Earth during the mission.

Scientific Experiments on Cassini

Cassini carried a total of 12 scientific instruments on its exploration of Saturn and its moons. The instruments all operated as intended and returned a wealth of new information about Saturn, its rings, and its many intriguing moons. With few exceptions, the instruments returned full and excellent data for the 13 years of orbit around Saturn. A labeled NASA drawing of the Cassini orbiter showing locations of the scientific instruments is given in Fig. 6.6. A list of experiments and a summary of purpose of each are given in Table 6.1.

Imaging Science Subsystem (ISS)

The imaging science subsystem (ISS) contained two imaging cameras. One, referred to as the narrow-angle camera, had a field of view of 0.35 deg^2, and the other, referred to as the wide-angle camera, had a field of view of 3.5 deg^2. The cameras were built by the Jet Propulsion Laboratory. The team leader for the imaging science subsystem was Dr. Carolyn Porco of the University of Arizona.

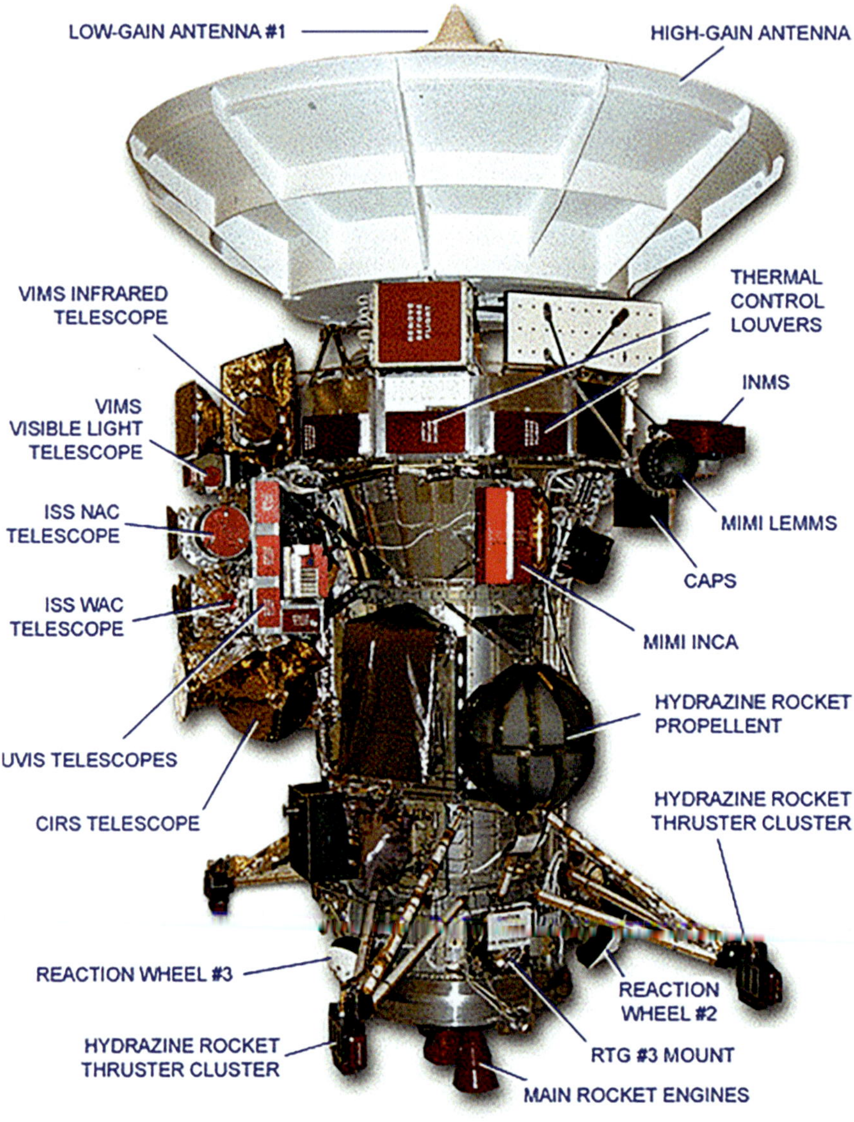

Fig. 6.6 Labeled drawing of Cassini orbiter showing the location of scientific instruments. (NASA graphic)

The two cameras were mounted to the remote sensing palette on the spacecraft. The location of the palette was shown in the drawing of the spacecraft in Fig. 6.3. The centers of the fields of view of the two cameras were parallel to the $-Y$-axis in the spacecraft coordinate system. They were fixed-mounted, so the spacecraft needed to be oriented to point the cameras at a particular direction in space.

Table 6.1 Experiments carried by Cassini

Experiment	Principal investigator	Purpose
Imaging science subsystem (ISS)	Carolyn Porco, University of Arizona	Generate multispectral images of cloud patterns and ring structures of Saturn and images of its moons using a narrow field of view camera and a wide field of view camera
Dual-technique magnetometer (MAG)	David Southwood, Imperial College	Measure the magnetic field of Saturn and investigate the interaction of the magnetic field with the solar wind, rings, and the moons
Magnetospheric imaging instrument (MIMI)	Stamatios Krimigis, JPL/APL	Generate global images and in situ measurements of Saturn's magnetosphere and investigate the interaction of the magnetosphere with the solar wind
Ultraviolet imaging spectrograph (UVIS)	Larry Espoosito, University of Colorado	Measure ultraviolet energy emitted from the atmosphere of Saturn and from its moons. Use UV emission from surfaces of the moons and from the rings to determine the composition
Cassini plasma spectrometer (CAPS)	David Young, SwRI	Investigate the plasma (electron and ion) environment surrounding Saturn and its magnetosphere
Cassini radar (RADAR)	Charles Elachi, JPL	Map the surface of Titan with synthetic aperture radar, radar altimeter, and radiometer sensors
Composite infrared spectrometer (CIRS)	Virgil Kunde, NASA Goddard SFC	Measure infrared emissions from the clouds, atmosphere, and rings of Saturn and the atmosphere and surface of its moons in the wavelength range of 7–1000 μm
Visible and infrared mapping spectrometer (VIMS)	Robert Brown, JPL	Map the mineral and chemical composition of the atmospheres of Saturn and Titan and of the rings of Saturn and surfaces of the moons
Ion and neutral mass spectrometer (INMS)	Jack Waite, SwRI	Determine the chemical and elemental composition of neutral particles and low-energy ions in Saturn's magnetosphere and ring environment and of Titan's atmosphere and ionosphere
Cosmic dust analyzer (CDA)	Eberhard Gruen, Max Planck Institute	Measure the composition, mass, and speed of dust grains and ice particles in the Saturn system
Radio and plasma wave science (RPWS)	Donald Gurnett, University of Iowa	Measure the electric and magnetic fields in the plasma in interplanetary space and in the Saturn system. Also measure the electron density and temperature
Radio science subsystem (RSS)	Arvydas Kliore, JPL	Use Cassini's telecommunications transmitter and high-gain antenna in conjunction with equipment on Earth to perform gravity and occulting experiments in the Saturn system

Both cameras used a 1024 by 1024 charge-coupled device (CCD) detectors, and both cameras were equipped with two filter wheels. The optics of the narrow-angle camera used a Ritchey-Chrétien reflector design with primary and secondary mirrors. The focal length was 2002 mm, and the *f*-number was 10.5. The diameter of the entrance aperture was 190 mm. The field of view of the camera was 0.35°. The spectral range of the camera was 200–1050 nm. The overall size of the narrow-angle camera was 95 × 40 × 33 cm. It weighed 30.5 kg.

The wide-angle camera used refractor-type optics. The focal length was 200 mm, and the *f*-number was 3.5. The entrance aperture was 57 mm in diameter. The field of view was 3.5°. The spectral range of the camera was 380–1050 nm. The overall size of the wide-angle camera was 55 × 35 cm × 33 cm. It weighed 26.4 kg.

The image in the narrow-angle camera was passed through segments in two filter wheels before being applied to the CCD detector array. Each filter wheel had 12 positions. One position in each wheel was clear. The two wheels were aligned so that when the clear filter was in place on one wheel, the other 11 segments could be selected on the other wheel. The two wheels together had 24 positions with central wavelengths ranging from 258 nm (ultraviolet) to 1092 nm (infrared). One of the wheels also contained four polarization filters. Three of the polarizers were effective in the wavelength range 350–750 nm and had three different polarizations of 0°, 60°, and 120°. The fourth polarizer operated at infrared wavelengths at an angle of 0°.

The wide-angle camera also contained two filter wheels in the same arrangement, but there were only nine filter positions on each wheel. One of the positions on each wheel was clear. The two wheels together had 18 positions with central wavelengths ranging from 420 nm (violet) to 1028 nm (infrared). One of the wheels contained two polarization filters, one at 0° and the other at 90° at infrared wavelengths.

A mechanical shutter was placed between the filter wheels and the CCD detector in both cameras. The shutter allowed 63 different commanded exposure times from 5 ms to 20 s.

The two cameras used the same type of CCD detectors and processing electronics. The CCD was a square array of 1024 by 1024 pixels. Readout of the CCD could be commanded to read each individual pixel or to sum a group of 2 × 2 pixels or a group of 4 × 4 pixels. Summing improved the signal-to-noise ratio at the expense of resolution.

Acquisition of images was achieved by first commanding a "prepare cycle" and then a "readout cycle." During the prepare cycle, the shutter blades were reset to the closed position, the filter wheels were set to selected settings, and the CCD was prepared for exposure. Preparing the CCD for exposure involved flooding the CCD area with light at a saturating level, followed by readout of the CCD. That process, which took about 1 s, erased any trace of the previous image. The required time for each image in the readout cycle was dependent on the amount of data from the CCD (whether or not pixel summing is used) and the pickup rate from the command data system.

The CCD was read out one line at a time. The output of each pixel, or sum of pixels, in the line was amplified and applied to a 12-bit analog-to-digital converter. The digital data was applied to an image memory, which was then read out and converted to science packets. The science packets were picked up by the command data system and transferred to the solid-state recorder. Image data was read from the solid-state recorder at a lower data rate and downlinked to Earth.

Images of Saturn, Rings, and Moon

A collection of images collected by the Cassini orbiter is given in the *Jet Propulsion Laboratory Photojournal*. An image of Saturn made at a time near its summer solstice is given in Fig. 6.7. The picture is a mosaic of images taken by the wide-angle camera at a distance of 3 million km from Saturn and from 30° above the ring plane. The picture combined images using the red, green, and blue spectral filters to render the colors shown. The NASA/JPL image was cropped by the author.

An image of the rings of Saturn taken by the narrow-angle camera from a distance of 2.05 million km is given in Fig. 6.8. Images using the red, green, and blue spectral filters were combined to generate the color image. Questions still exist about the cause of the various colors in the rings.

A unique image of Saturn and its rings made while Cassini was in the shadow of Saturn is given in Fig. 6.9. Saturn and the rings were backlit by the sun. The image is a mosaic of images from the narrow-angle and wide-angle cameras. Images taken through the red, green, and blue spectral filters were combined to make the color mosaic.

Saturn has 83 known moons, 20 of which have not yet been named. Cassini took images of many of the moons during its time in orbit around Saturn. The moons range in size from less than 2 km in diameter to 5152 km in diameter. The icy moon Enceladus was of particular interest to scientists because it was observed to be spewing jets of water ice from a source under the ice. Enceladus is 500 km in diameter.

Images of Enceladus were taken by the narrow-angle camera through four spectral filters spanning 338–930 nm (ultraviolet to infrared light) and at seven different

Fig. 6.7 Image of Saturn taken by Cassini near summer solstice of Saturn. (NASA image)

Fig. 6.8 Image of the rings of Saturn taken by the Cassini orbiter. (NASA image)

Fig. 6.9 Image of Saturn and rings backlighted by the sun taken by Cassini orbiter. (NASA image)

pointing angles. The four spectral filters were used at each of the seven angular positions. The picture shown in Fig. 6.10 was made by combining those 28 images. The distance from Cassini to Enceladus increased from 30,000 to 48,000 km in the time required to take the 28 images.

Fig. 6.10 Mosaic of images of icy moon Enceladus. (NASA image)

Geysers of water ice were observed spewing from the southernmost region of Enceladus. An image taken by the narrow-angle camera of the geysers is shown in Fig. 6.11.

There are a large number of very small moons orbiting Saturn. A typical small moon, Epimetheus, is irregular and about 113 km across.

Dual-Technique Magnetometer (MAG)

The dual-technique magnetometer was used to measure the magnetic field of Saturn and investigate interactions of the magnetic field with the solar wind, rings, and moons. The dual magnetometer contained a vector/scalar helium magnetometer (V/SHM) and a fluxgate-type magnetometer. The helium magnetometer was able to accurately measure very low levels of magnetic field. The sensor for the V/SHM was positioned at the end of an 11-m-long boom to space it away from the body of

Fig. 6.11 Geysers at the southern region of Enceladus. (NASA image)

the spacecraft. The sensor for the fluxgate magnetometer was mounted midway out on the boom.

The vector/scalar helium magnetometer and its electronics were supplied by the Jet Propulsion Laboratory. The fluxgate magnetometer and its electronics were furnished by the Imperial College of London. The principal investigators for the dual-technique magnetometer were Dr. David Southwood and Dr. Michel Dougherty of the Imperial College of London. Fluxgate magnetometers are commonly used on aircraft and spacecraft. The instrument on Cassini had four different dynamic ranges. The lowest range covered ±40 nanotesla (nT) with a resolution of 4.9 nT. The highest range covered ±44,000 nT with a resolution of 5400 nT.

The V/SHM contained two helium magnetometers. One was a vector helium type that measured magnitude and direction of the magnetic field. The other was a scalar helium type that made very accurate measurements of the magnitude of the magnetic field. The vector helium magnetometer had two measurement ranges: ±32 nT with 0.0039 nT resolution and ±256 nT with 0.0312 nT resolution. The scalar hydrogen magnetometer was required to cover the range 256–16,384 nT with a resolution of 0.036 nT. Test of the instrument during an Earth flyby showed accurate measurements up to the maximum field strength encountered of 21,194 nT. Principles of operation of helium magnetometers were given in Chap. 4 of this book.

The magnetic field of Saturn was found to be closely aligned with the spin axis. During the Grand Finale phase of the mission, Cassini's orbit was arranged to pass between the cloud tops and the inside of the closest ring. A total of 22 passes were made. The magnetic field strength was measured to be 21,000 nT a few thousand km above the cloud tops at the equator.

Magnetospheric Imaging Instrument (MIMI)

The purpose of the magnetospheric imaging instrument was to generate global images and in situ measurements of Saturn's magnetosphere and investigate the interaction of the magnetosphere with the solar wind. It explored the magnetosphere by detecting and locating energetic plasmas in the magnetosphere. The instrument contained three separate sensors: ion and neutral camera (INCA), low-energy magnetospheric measurement system (LEMMS), and charge-energy-mass spectrometer (CHEMS).

The CHEMS sensor was developed by the Space Physics Group at the University of Maryland. The INCA was developed by the Applied Physics Laboratory of the Johns Hopkins University. The LEMMS was developed by the Max Planck Institute for Solar Research. The principal investigators for MIMI were Stamatios Krimigis and Mitchell Donald of Johns Hopkins University.

Ion and Neutral Camera (INCA)

The ion and neutral camera was used to perform remote sensing and imaging of the magnetosphere of Saturn. Energetic ions in the magnetosphere can capture electrons, and the resulting neutral particles escape from the magnetosphere and travel out into space. Some of them were captured by the INCA to form images of the emitting region.

The INCA element measured the time of flight and direction of arrival of neutral particles by using penetrating foils and two microchannel plate detectors. It accommodated energetic neutral particles over the range of 7 keV/nucleon to 3 MeV/nucleon. It separately measured the time of flight and direction of arrival of ion species over the same energy range. The field of view was 120° by 90°, and the angular resolution was about 8°. The resolution of velocity measurements was 50 km/s. An image of the magnetosphere of Saturn, labeled by NASA, is shown in Fig. 6.12. Data for the image was taken by the ion and neutral camera from a distance of about 6 million km on the approach to Saturn.

Low-Energy Magnetospheric Measurement System

The low-energy magnetospheric measurement system was a two-ended telescope that measured the energy of ions and electrons. The forward direction was used to measure low-energy ions and electrons. Ions in the energy range of 0.03–18 MeV and electrons in the range of 0.15–0.884 MeV were measured. The backward direction of the telescope was used to measure higher-energy ions and electrons. It measured ions in the energy range of 1.6–160 MeV and electrons in the energy range of 0.1–5 MeV. The fields of view of the telescopes were 15° in the forward direction and 36° in the backward direction.

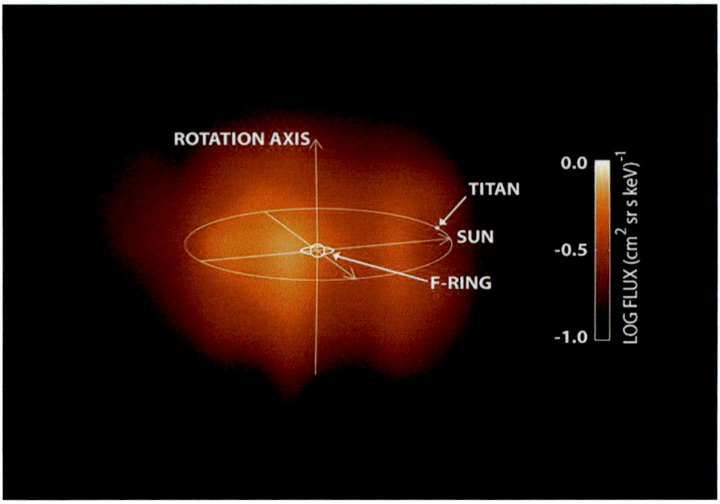

Fig. 6.12 Labeled image of Saturn's magnetosphere taken by Cassini ion and neutral camera. (NASA graphic)

Energy of ions or electrons was determined by noting the energy loss through successive semiconductor silicon detectors. There were six silicon detectors in the low-energy side of the instrument and five silicon detectors in the high-energy side.

The double-ended telescope was mounted on a turntable located on the side of the spacecraft with axis of rotation of the turntable parallel to the −*Y*-axis of the spacecraft. The telescope scanned in the *X*–*Z* plane. The field of view was obscured by the high-gain antenna in the upward direction and by a shield over one of the radioactive thermal generators on the downward direction. The 360° scan angle was divided up into 16 sectors, each 22.5° in extent. Each sector could be further divided into eight subsectors. Rotation of the turntable was performed by a stepper motor. The normal time for a complete revolution was 86 s. The time to pass through each of the 16 subsectors was 5.3 s.

Charge-Energy-Mass Spectrometer (CHEMS)

The charge-energy-mass spectrometer characterized the properties of the magnetosphere of Saturn by measuring the characteristics of energetic ions. The instrument employed electrostatic deflection combined with time of flight to determine energy, mass, and charge of ions. The apparatus for electrostatic deflection consisted of two parallel spherical segments, one with a radius of 10.55 mm and the other with a radius of 11.0 mm. The gap between the segments was 4.5 mm.

A positive voltage was applied to one segment, and a negative voltage was applied to the other segment. The voltage across the segments could be changed in 32 different steps that ranged from 0.238 to 18.4 V. Ions with a wide range of kinetic energies entered the deflection system. Ions with specific velocity and charge were deflected in an arc that followed the curvature of the segments and were detected. Ions with different velocities and charge struck the plates. The deflection arc was 125° from entrance to exit of the deflection system.

Velocity of ions at the output of the deflection system was determined by measuring the time of flight over a 10 cm distance by secondary electron emission from a start foil (start pulse) and detection by a microchannel plate detector (stop pulse). The energy of the particle was determined by a silicon solid-state detector. From knowledge of energy and velocity of the ion, the mass of the ion could be computed. Energy was measured over a range of 26–2466 keV. The field of view of the instrument was 4.0° by 159°.

Ultraviolet Imaging Spectrograph (UVIS)

The ultraviolet imaging spectrograph measured ultraviolet energy emitted and reflected from the atmospheres of Saturn and some of its moons. It also measured UV emission from surfaces of the moons and from the rings to determine composition. The instrument contained a high-speed photometer, a hydrogen deuterium absorption cell, and two ultraviolet spectrometers.

The hydrogen deuterium absorption cell was developed by the Max Planck Institute for Solar Research in Germany. Other elements of the UVIS were developed by the Laboratory for Atmospheric and Space Physics at the University of Colorado. The principal investigator for UVIS was Dr. Larry Esposito of the University of Colorado.

High-Speed Photometer

The high-speed photometer was used to view occultation of stars by the rings of Saturn to investigate the structure and density of material in the rings. It operated over a wavelength range of 1150–1900 Å. The photometer used a parabolic mirror-type telescope with a focal length of 200 mm. The field of view was 0.34° by 0.34°. A photomultiplier-type detector was used. The instrument was built by the Laboratory for Atmospheric and Space Physics at the University of Colorado.

Hydrogen-Deuterium Absorption Cell

The hydrogen-deuterium absorption cell was used to measure the ratios of hydrogen to deuterium in the Saturn environment. It operated in the 1150–2400 Å wavelength range. A telescope with a focal length of 150 mm and field of view of 3.3° was used to focus ultraviolet light onto three absorption cells in series. The first cell was filled with hydrogen, the second with oxygen, and the third with deuterium. A lens made of magnesium fluoride was located at the entrance end of the hydrogen cell. Transparent magnesium fluoride windows were located between the hydrogen and oxygen cells and between the oxygen and deuterium cells. The oxygen cell filtered out unwanted signals.

Tungsten filaments were located within the hydrogen and deuterium cells. When the filaments were heated by electrical current, H_2 and D_2 molecules were dissociated into individual atoms. Those atoms resonated and absorbed Lyman-alpha radiation from hydrogen and deuterium in the incoming ultraviolet beam of light. The resonant wavelength for hydrogen atoms is 121.57 nm, and the resonant wavelength of deuterium atoms is 121.53 nm. The current in the filaments was cycled to obtain a measurement of intensity of the hydrogen and deuterium Lyman-alpha lines.

The Max Planck Institute in Germany was responsible for the design and construction of the instrument, and the Laboratory for Atmospheric and Space Physics at the University of Colorado was responsible for electronics for the instrument.

Measurements in the atmosphere of Saturn indicated that the helium-to-molecular hydrogen ratio (He/H_2) was 0.11–0.16.

Ultraviolet Spectrometers

The ultraviolet spectrometers (UVISs) included two ultraviolet spectrometers. One operated in the far-ultraviolet wavelength range of 1115–1912 Å, and the other operated in the extreme ultraviolet range of 563–1182 Å. Both spectrometers used telescopes with a focal length of 100 mm and entrance aperture of 22 by 30 mm. The optics used an off-angle parabola reflector to focus the incoming ultraviolet signal onto a slit of variable width. The slit width for the far-ultraviolet spectrometer could be selected to be 75, 150, or 800 μm. The corresponding fields of view were 0.043°, 0.086°, and 0.46°, respectively, in the plane set by the slits. The field of view in the other plane was about 3.4° for both spectrometers. Slit widths of 100, 200, or 800 μm could be selected for the extreme ultraviolet spectrometer. The corresponding fields of view were 0.057°, 0.11°, and 0.46°. The 800 μm slit widths in both spectrometers were used during occlusion measurements.

The light at the output of the slits illuminated a refraction grating that in turn illuminated a multichannel plate detector. The detector contained 1024 pixels in the spectral direction and 64 pixels in the spatial direction. Data from the spectrometers was managed by the UVIS microprocessor electronics and control subassembly.

The ultraviolet imaging spectrograph discovered plumes of icy, salt-rich material being ejected from the moon Enceladus. It also determined that the rings of Saturn are made up of lumpy ice crystals and dust. The size of the particles varied from grains of sand to tens of meters. The UVIS imaged the northern and southern auroras. Two images of Saturn's auroras around the south pole are shown in Fig. 6.13.

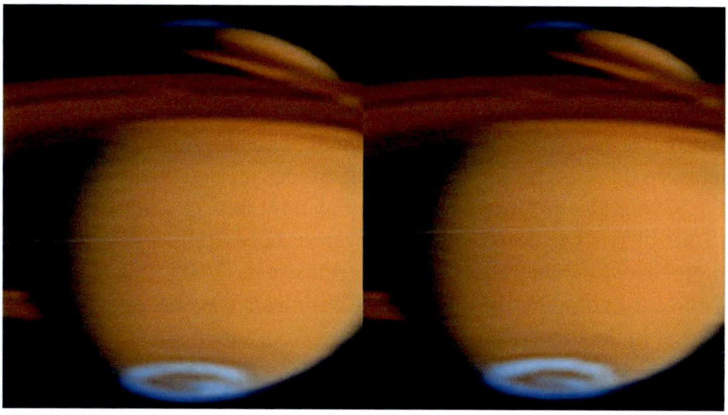

Fig. 6.13 Auroras at Saturn's south pole imaged by UVIS. (Image credit NASA/JPL/University of Colorado)

Cassini Plasma Spectrometer (CAPS)

The Cassini plasma spectrometer was an in situ collection of sensors that measured the flux of ions and electrons as a function of energy and mass per change. Measurements were made as the spacecraft passed through various regions. A notable finding by CAPS was that most of the ions in the Saturn system have origin in water ejected from the icy moon, Enceladus. It also discovered an atmosphere and ionosphere over the rings of Saturn.

The Cassini plasma spectrometer included three sensors: electron spectrometer, ion mass spectrometer, and ion beam spectrometer. CAPS was mounted on the fields and particles pellet. The three sensors were grouped together along with a data processing unit and mounted on an actuator. The actuator rotated the instrument so that the field of view of the sensors scanned parallel to the *X*–*Y* plane in the spacecraft coordinate system. The normal extent of rotation was 184°. The normal rotation rate was 1°/s.

The ion beam spectrometer and the ion mass spectrometer were developed by the Los Alamos National Laboratory in Los Alamos, New Mexico. The principal investigator for CAPS was Dr. David Young of the Southwest Research Institute in San Antonio, Texas.

Electron Spectrometer

The electron spectrometer measured the flux of electrons as a function of energy/charge and direction of entry into the sensor. Incident electrons passed through a baffle that established a field of view of 5° by 160° and then into an electrostatic analyzer. The analyzer included two concentric hemispherical plates. One plate was at ground potential, and the other had a variable positive voltage applied. Only electrons with a particular energy passed through the gap between plates at a particular voltage field. The voltage was varied to develop a spectrum of electron energies.

Electrons that successfully passed through the analyzer struck one of the two curved microchannel plates. The microchannel elements generated secondary electrons that were detected by a series of eight anodes arranged in an arc. Each anode covered a 20° angular sector, and together the eight anodes gave an overall field of view of 160° by 5°. The detector outputs were processed by a data processing unit that also processed the outputs of the other sensors of CAPS.

A significant finding by the electron spectrometer was that heavy organic negative ions exist in the ionosphere of Titan.

Ion Mass Spectrometer

The ion mass spectrometer measured the flux of positive ions as a function of energy/charge and arrival direction at the aperture of the sensor. Ions first passed through a baffle that limited the field of view to an annular region 12° by 160° in extent. The ions were then directed to the narrow space between two plates of a toroidal electrostatic analyzer. The plates had a variable DC voltage across them. Only ions with a particular energy level passed through the gap without hitting the sides for a given voltage level. The voltage was stepped to 64 different values to develop a spectrum of ion energy.

Positive ions at the output of the toroidal electrostatic analyzer were accelerated by a potential of −15,000 V and passed to a time-of-flight element consisting of a series of eight carbon foils arranged in an arc. Ions passed through those thin foils and generated secondary electrons which were detected and served as a start pulse. Ions passing through the start foil were further accelerated by an electric field and struck circular microchannel plates. Eight outputs of the microchannel plate corresponded to the eight angular channels set by the angular location of the start foils along an arc. The outputs of the microchannel plate provided a stop pulse allowing time of flight to be determined. The output of a given anode of the microchannel plate determined in which of the eight 20° elevation sectors the ion arrived. The voltage step of the analyzer, time of flight, and angle information were applied to the digital processing unit.

Ion Beam Spectrometer

The ion beam spectrometer measured the flux of positive ions as a function of energy/charge and direction of arrival. It contained a hemispherical electrostatic analyzer that consisted of two concentric metallic hemispheric plates with a small gap between them. A stepped voltage across the plates allowed developing a spectrum of energy/charge. Only ions with a given energy/charge were able to traverse the plates instead of hitting the sides for a given electric field within the analyzer.

Three entrance apertures, each with a field of view of 1.5° by 150°, directed ions to the analyzer. The three apertures were spaced 30° apart in elevation. Ions exiting the electrostatic analyzer struck one of the three channel electron multipliers, which were located in line with the three input apertures. The outputs of the channel electron multipliers were amplified and applied to the digital processing unit along with voltage step value and angle information.

Results of CAPS Experiment

A major finding from CAPS was that the moon Enceladus emits plumes of water-based ions that are the dominant source of plasma in the magnetosphere of Saturn. It discovered that the main rings of Saturn have an atmosphere that consists primarily of molecular oxygen and water ion products. The haze in the atmosphere of Titan was found to be caused by heavy negative ions with masses up to 13,800 amu/q and positive ions with masses up to 1000 amu/q.

Cassini Radar (RADAR)

The Cassini radar was used to penetrate the dense haze of the large moon, Titan, and map its surface with synthetic aperture radar and measure the topography of the moon by a radar altimeter. The radar had four operating modes: synthetic aperture radar imaging, radar altimetry, scatterometer, and radiometer. The scatterometer measured the scattering properties of the surface at the radar frequency. The radiometer measured the strength of natural radiation from the surface at the radar frequency to determine temperature. The scatterometer and radiometer modes were also used to probe the rings and the many moons of Saturn.

The radar, operating at a frequency of 13.78 GHz, shared the 4-m-diameter high-gain antenna with the telecommunication function. The antenna and portions of the radar were developed by the Italian Space Agency (ASI). The Jet Propulsion Laboratory also developed portions of the radar. The principal investigator for the radar was Charles Elachi of the Jet Propulsion Laboratory in Pasadena, California.

The radar was composed of three major subassemblies: radio frequency electronic subsystem (RFES), digital subsystem (DSS), and energy storage subsystem (ESS). In addition, it shared the high-gain parabolic antenna with the telecommunications function.

The radio frequency electronic subsystem contained the radar pulsed transmitter, a chirp generator that generated frequency modulation for the transmit pulses, and a receiver that received and preprocessed echo signals from the synthetic aperture radar, radar altimeter, and scatterometer. It also received the passive radiometric signal from objects in the antenna beam.

The digital subsystem controlled operation of the radar. It received commands from the Cassini command and data subsystem, generated signals to control transmitted pulse modulation characteristics, selected the antenna feed to be used, controlled timing functions of the radar, and packaged radar housekeeping and science data and packaged the data before sending it to the command and data subsystem.

The energy storage subsystem converted a portion of the 30-V power bus to a higher voltage and stored the energy in a capacitor bank. The capacitor bank stored sufficient energy to provide power to the high-power amplifier during a transmit pulse.

During its 13 years orbiting Saturn, Cassini passed by the large moon, Titan, 127 times. The radar probed the moon on 48 of those flybys. The radar was also used to gather information during flybys of eight of Saturn's icy moons.

The radar, which operated at a frequency of 13.78 GHz, shared the 4-m-diameter reflector of the high-gain antenna with the telecommunications function. The Cassini radar used a series of five feeds to illuminate the reflector. The feeds were positioned to provide five adjacent antenna beams in a line in the cross-track direction of spacecraft travel. The center feed was located at the focal point of the antenna, and the antenna beam was along the boresight of the reflector, which was along the *Z*-axis of the spacecraft. The beamwidth of the center beam was circular at 0.37°, and the gain was 50.7 dB. Two adjacent antenna beams were formed by their feed horns at beam angles displaced by 0.85° on either side of the center beam. Two additional adjacent beams were formed displaced by 2.2° on either side of the center. The widths of beams displaced from center were somewhat larger than that of the center beam. The center beam had the narrowest beamwidth and the highest antenna gain, so it was selected for radar altimeter and scatterometer modes of the radar. All five beams were usually used during imaging of the surface. They were selected and processed one at a time.

The narrow antenna beamwidth allowed radiometer measurements of Titan to be obtained at distances greater than 100,000 km. The beamwidth imprint was only about 12.5% of the diameter of Titan with Cassini at 100,000 km distance, giving the radiometer a beam-filling heat source. The radar altimeter and the scatterometer had adequate signal-to-noise ratio to function, although at low altimeter resolution, at ranges up to 22,500 km. Low-resolution imaging in the SAR mode could be gathered at ranges of 4000 km, and high-resolution imaging was obtained at ranges less than 1600 km.

A timeline for a typical flyby of Titan with closest approach distance of 1000 km is given in Table 6.2. The outbound path had the same events and time in reverse order. Information for the timeline was provided in graphical form by Elachi et al. The paper by Elachi also gives operating parameters of the radar.

Table 6.2 Timeline of measurements during typical flyby of Titan

Time to closest approach, min	Distance to surface, km	Measurements performed
300 to 78	100,000 to 25,000	Radiometer measurement
78 to 70	25,000 to 22,500	Calibrate
70 to 30	22,500 to 9000	Low-resolution altimetry, scatterometer, and radiometry
30 to 16	9000 to 4000	High-resolution altimetry and radiometry
16 to 6	4000 to 1600	Low-resolution SAR imaging and radiometry
6 to 0	1600 to 1000	High-resolution SAR imaging and radiometry

Synthetic Aperture Radar (SAR) Imaging Mode

The antenna beams were pointed to the side of the direction of travel in the SAR mode. The radar processed Doppler information in the received echo signals to achieve an effective (synthetic) aperture of the antenna much larger than the physical aperture in the along-track direction. The large synthetic aperture resulted in much better along-track resolution than would normally result from the beamwidth of the antenna.

The spacecraft was oriented so that the antenna was pointed 90° to the flight path with the five contiguous antenna beams in a vertical stack. The antenna was pointed at an appropriate angle in the elevation plane, so all five beams struck the surface at an acceptable angle from the local vertical considering both image quality and signal-to-noise ratio of the echo signal. The incidence angles of the five antenna beams were within 21–30° at altitudes below 1600 km and within 15–28° at altitudes above 1600 km.

"Chirp"-type pulse compression was used on the transmit pulse to achieve an effective narrow pulse width while retaining the same amount of energy as in the wide transmitted pulse. Pulse compression was achieved by imposing linear frequency modulation on the transmitted signal and processing the resulting modulation on the echo signal. An ultra-stable oscillator enclosed in an oven generated a reference signal at 30 MHz. The reference signal was linearly frequency modulated by a series of steps. The start and stop times of the frequency stepping controlled the pulse width of the radar. The frequency-modulated (chirped) pulse was up-converted to the radar operating frequency of 13.78 GHz.

Parameters of the frequency modulation were controlled to best match mission requirements at a particular time. The chirped pulse was amplified to a power level of 65 W peak power by a traveling wave tube amplifier. The width of the transmitted pulse was between 200 and 400 μs. The effective pulse width, which sets the resolution in the ranging, or cross-track direction, was considerably less, depending on chirp parameters.

The radar operated in a burst mode where several pulses were transmitted, and the burst was terminated before the first of the echo pulses was returned from the surface. The pulse repetition frequency (PRF) could be set between 1.6 and 6.0 kHz.

At a typical closest approach distance of 1000 km, the round-trip travel time of the pulse would be 6.66 ms. The receiver was activated after completion of the burst of transmit pulses.

Echo signals from the surface that appeared in the selected antenna beam were routed to the microwave receiver. The receiver was a double conversion type that amplified and down-converted the echo signals. The second conversion was made in quadrature to baseband. The quadrature conversion preserved phase information in the echo signal. The quadrature signals were sampled and digitized by an eight-bit analog-to-digital converters. The digital signals were sent to the command and data system (CDS), which formatted the data and sent the two quadrature channels to the solid-state recorder. Data rates from the radar were up to 360 kbps. The data from the recorder was transmitted to Earth at an acceptably low rate for adequate signal-to-noise ratio on the downlink channel.

Downlinked data from the spacecraft received on Earth included the two quadrature channels. SAR signal processing was performed on the ground at the Jet Propulsion Laboratory (JPL). Processing used unique software running on a powerful computer. The results were strips of images of the surface from the five contiguous antenna beams. The strips were up to 6600 km long and variable in width depending on distance to the surface. Images from the five beams were stitched together to make up the total width of the image of the surface. The narrowest portion of the five-beam strip was near the closest approach where the width was about 150 km, while the width near the ends of the strip was about 530 km. Resolution in the along-track direction (azimuth) was about 800 m at 4000 km distance to the surface, 330 m at 1600 m distance, and 300 m at 1000 m distance. Resolution in the cross-track direction was 500 m at a distance of 4000 km from the surface, 460 m at 1600 km distance, and 330 m at 1000 km distance.

A colorized rendering of a strip of the surface of Titan from SAR imaging showing some of the hydrocarbon lakes on Titan is shown in Fig. 6.14. A colorized SAR image of a large hydrocarbon sea on Titan named Ligeia Mare is shown in Fig. 6.15.

Radar Altimeter Mode

The radar altimeter mode used the center beam of the antenna pointed towards the nadir. High-resolution altitude measurements were made between 10,000 km and 4000 km altitude. A burst of 21 pulses was transmitted at a pulse repetition frequency (PRF) of 5 kHz. The pulse width was 140 μs, and the chirp bandwidth was 4.25 MHz. The resolution of altitude measurements was about 35 m.

Radiometer Measurements

Radiometers measure the natural electromagnetic radiation from a surface or body. Any object or surface at a temperature above absolute zero (0 K or −273 °C) radiates with intensity that is a function of temperature of the body and wavelength of

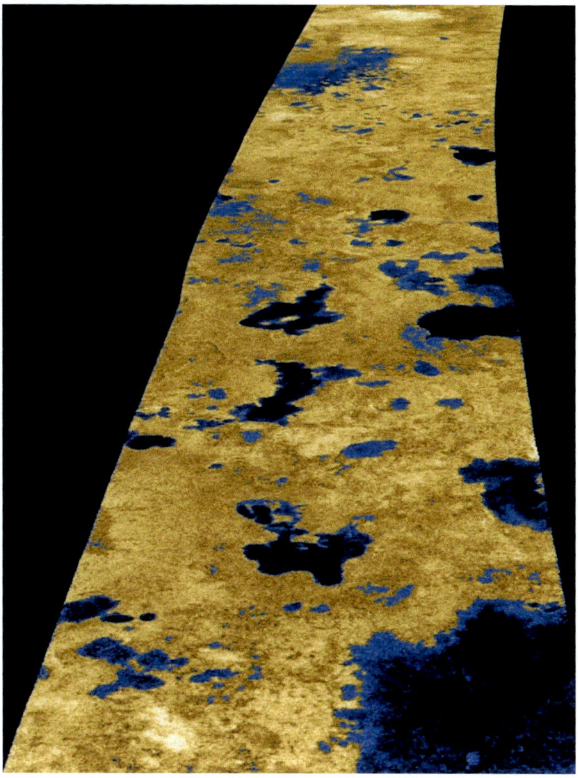

Fig. 6.14 SAR image of a strip of Titan showing hydrocarbon lakes. (Image was colorized by NASA)

the radiation. A black body, which is a perfect radiator and perfect absorber of energy, is the basis for radiometric work. Black body characteristics were investigated by the German physicist Max Planck, who derived what is now known as Planck's law in 1900. According to Planck's law, the spectral radiance per steradian emitted from 1 m^2 of surface involves the wavelength of radiation, temperature, velocity of light, Planck's constant, and Boltzmann's constant.

The Cassini radiometer used the center beam of the high-gain antenna. The radar was in a receive-only mode to receive natural thermal emissions from the surface. Temperature was determined from the strength of the radiation and antenna and receiver characteristics. At far distances from Titan, the radiometer mode was used alone. At closer distances, radiometer measurements were made after reception of bursts of radar echo pulses. The temperature of the disk of Titan was measured to be about 87 K (−186 °C). The temperatures were about the same in the northern and southern hemispheres.

The radiometer was also used to measure temperatures of several smaller moons of Saturn. The average temperature measured by 10 flybys of Enceladus was 36.0 K. The average temperature of four flybys of Dione was 50.5 K, and the average temperature of 10 flybys of Rhea was 346.9 K. The average temperature of

Fig. 6.15 SAR image of hydrocarbon sea named Ligeia Mare on Titan. (NASA image)

seven flybys of Iapetus was 64.4 K, and the average temperature of three flybys of Phoebe was 73.1 K.

Scatterometer Measurements

Scatterometer measurements were made of the surface of the large moon, Titan. The basic results were backscattering cross section per unit surface area as a function of incidence angle. The scatterometer mode used the center beam of the antenna and a relatively low chirp bandwidth of 106 kHz on the transmitter pulses. The pulse repetition frequency was 1200 Hz. A burst of eight pulses was transmitted at each cycle. The scatterometer was typically operated at ranges between 30,000 and 10,000 km. The spacecraft was maneuvered to give a raster scan of the scatterometer beam on the surface. Data from the scatterometer was processed by Stanford University.

Composite Infrared Spectrometer (CIRS)

The composite infrared spectrometer was mounted on the remote sensing pellet. It was the instrument farthest from the high-gain antenna on the pellet. The boresight of the telescope CIRS was parallel to the $-Y$-axis of Cassini.

The composite infrared spectrometer included two spectrometers; one operating in the mid-infrared wavelength range of 7–17 μm and the other in the far-infrared wavelength range of 17–1000 μm. The spectrometers measured the strength of infrared radiation as a function of wavelength to determine temperature as well as chemical composition of objects in the field of view of the instrument. Various molecules were identified by lines at particular wavelengths in the spectrums.

The CIRS used a Cassegrain-type telescope consisting of a 50.8 cm diameter parabolic primary mirror and a hyperbolic secondary mirror 7.6 cm in diameter. The parabolic mirror was made from beryllium, polished to optical quality. The output of the telescope was divided into two paths by two angled mirrors. One path fed the mid-infrared spectrometer, and the other path fed the far-infrared spectrometer. The field of view of the far-infrared channel was 3.9 mrad (0.22°) circular.

The far-infrared spectrometer was a polarization interference type that covered a wavelength range of 17–1000 μm. The reflected light from the telescope was directed at a collimator, which in turn sent the light to a polarizer. The reflected light from the polarizer, which was at an angle of 45°, was directed at a beam splitter. The beam splitter was polarized such that the reflected beam was polarized vertically and the pass-through beam was polarized horizontally. One of the paths of the beam splitter struck a moveable mirror, and the other was applied to a fixed mirror. The arrangement was such that the two light paths combined and the polarization of the combined beams was a function of path difference caused by motion of the movable mirror. A polarizer set at a 45° angle transmitted or reflected the beam, depending on polarization. The output of the polarizer was applied to a redundant pair of thermocouple detectors. The outputs of the two detectors were amplified and applied to an instrument data system.

The mid-infrared spectrometer, which was a Michelson interferometer type, operated over a spectral range of 7–17 μm. Light from the telescope was directed at a collimator which in turn sent the light to a beam splitter. One of the paths of the beam splitter struck a moveable mirror, and the other was applied to a fixed mirror. The arrangement was such that the two light paths combined generating interference patterns that were a function of position of the moveable mirror and wavelength. The combined signal was split into two optical paths. One path operated over a wavelength range of 7–9 μm, and the other operated over a range of 9–17 μm. Each of the two filtered paths contained a linear array of ten mercury-cadmium-telluride detectors. The detectors were cooled to between 76 and 87 K by a radiator that viewed deep space. Each of the detector channels had a field of view of 0.27 mrad. The angular spacing between detectors was 0.29 mrad. The ten adjacent fields of view, which fit within the 3.9 mrad field of view of the telescope, provided spatial information as well as spectral information of items of interest. The output of each detector was amplified before being applied to an instrument data system.

Results of the CIRS Observation of Saturn and Titan

Measurements by CIR determined vertical and latitude distribution of stratospheric hydrocarbons and determined zonal wind structure of polar vortices of Saturn. It also determined the abundance of methane. It measured the temperature of the main rings of Saturn.

Results from observation of Titan by the CIRS instrument included distribution of gases in Titan's atmosphere and vertical profile of water in the atmosphere. It characterized hydrogen cyanide ice above south pole.

Visible and Infrared Mapping Spectrometer (VIMS)

The visible and infrared mapping spectrometer provided high-resolution multispectral images in the visible and infrared wavelength ranges. It was used to study the atmosphere and clouds of Saturn, study the composition of Saturn's rings, and study the atmosphere and the surface of Titan and other moons of Saturn. The VIMS instrument was contained in two assemblies: the optical pellet assembly, which contained two spectrometers and support electronics, and the main electronics assembly. The optical pellet assembly was mounted on the front side of the remote sensing pellet, and the main electronics assembly was mounted on the backside.

The VIMS contained two grating-type imaging spectrometers. One spectrometer operated in a wavelength range of 0.35–1.07 micrometers (μm). This range included visible light (0.38–0.75 μm), and the spectrometer was referred to as the visible spectrometer. The other spectrometer, referred to as the infrared spectrometer, operated in near-infrared wavelengths from 0.85 to 5.1 μm. Separate telescopes were used to feed the spectrometers for the two channels. The VIMS was mounted on the remote sensing palette along with the ISS, CIRS, and UCIS science instruments. The boresights of the four instruments were aligned and parallel with the *Y*-axis of the spacecraft.

The visible spectrometer was developed by the Italian Space Agency (ASI), and the infrared spectrometer was developed by the Jet Propulsion Laboratory. The principal investigator for VIMS was Dr. Robert Brown of the University of Arizona.

Visible Spectrometer Channel

The visible spectrometer channel assembly was built by the Officine Galileo Company in Italy for the Italian Space Agency. The telescope for the visible channel used a primary concave mirror mounted on a scan unit. The scan unit was arranged to scan the mirror in 0.167 milliradian (mrad) steps in the *Z*-axis of the spacecraft. Each step formed a row of pixels along the *X*-axis. The pitch of the pixels of the CCD array, which was located at the focal plane of the optical system, was also 0.167 mrad. In the nominal mode, the signal from three pixels illuminated during the scan of the mirror was integrated, which resulted in a spatial resolution of 0.5 mrad. The total scan width was 31.9 mrad (1.83°). Sums of three pixels were made during the length of the lines giving resolution in the *X*-axis spatial direction of 0.5 mrad also. The total extent of the line was 1.83°.

The image from the mirror was focused onto a slit for the spectrometer. The image of the slit was spectrally dispersed by a convex spherical diffraction grating and imaged on a CCD detector array. Each column of the array had an image of the spectrally disbursed image of the slit. The CCD array was a frame transfer device with 512 by 512 pixels. Half of the array was used for frame transfer, and the resulting active area was 256 by 512 pixels. The spectral resolution per pixel was 1.46 nm. The outputs of groups of five detectors in the spectral direction were summed in the nominal mode resulting in spectral resolution of 7.3 nm. There were 96 groups of five pixels for a total spectral coverage of 700 nm for the visible channel.

The visible channel had a separate set of electronics in the optical pellet assembly to convert the analog outputs of the detectors to 12-bit digital form, control data sampling, and control stepping of the mirror. It preprocessed the data before sending it on to the signal processing electronics unit. The signal processing unit organized the data from the visible spectrometer channel and the infrared spectrometer channel and sent the data to the main electronics assembly via a 12-bit global data bus.

Infrared Spectrometer

The infrared spectrometer assembly was designed and built by the Jet Propulsion Laboratory (JPL). The infrared spectrometer channel used a Cassegrain-type telescope with a 23-cm-diameter primary mirror and a scanning secondary mirror. The secondary mirror was scanned in two orthogonal directions by a two-axis voice coil-type scanning mechanism. Scanning was parallel to the *X*-axis of the spacecraft, and then the scan retraced rapidly to the next *Y*-axis line position. The narrow optical beam was scanned over an angular range of 32 by 32 mrad (1.8° by 1.8°). Scanning was performed in 0.5 mrad steps.

The scanned beam was applied to a shutter located just before the entrance slit of a grating-type spectrometer. The shutter was closed during retrace at the end of each line of scan, and the dark current of the detectors was measured and subtracted from the image data during signal processing. The dispersed output of the diffraction grating was applied to a 256-element linear array of indium antimony (InSb) detectors. The detectors were cooled to 69 K or less by a passive cooler. The detector array was read out by two multiplexers. Each multiplexer handled 128 detectors. The outputs of the multiplexers were applied to the signal processing electronics where they were digitized by 12-bit analog-to-digital converter. The resulting digital signal was sent to the main electronics assembly via a 12-bit global data bus.

Main Electronics Assembly

The main electronics assembly was the link between VIMS and the Cassini spacecraft. The bus interface unit in the main electronics assembly communicated with Cassini systems by a Mil Std 1553 data bus. The main electronics assembly received command inputs from the flight software, and it controlled the operation of the two spectrometers of VIMS.

Images from VIMS

Thick haze around the large moon, Titan, prevented photographing its surface in visible light. Cassini was able to use certain wavelengths in the infrared coverage by VIMS to penetrate the haze and allow the surface to be viewed. A striking set of six images of Titan created from mosaics made during 13 years of observation of Titan at various distances, various angles, lighting conditions, and rotation of the moon is shown in Fig. 6.16.

The images show that Titan has a complex surface with different geologic features and different constituents. The brown-colored areas in the equatorial regions were thought to be dune fields.

Fig. 6.16 Different view of Titan from 13 years of observation by Cassini VIMS instrument. (NASA graphic)

Ion and Neutral Mass Spectrometer (INMS)

The ion and neutral mass spectrometer performed in situ measurements of neutral and charged particle in the atmosphere of Saturn and of its moons. Many passes were made through the upper atmosphere of the large moon, Titan. The instrument was designed to measure the chemical and isotopic properties of the particles. It was mounted on the fields and particles pellet, which was attached to the $-X$-axis facing the side of the spacecraft. The instrument weighed 9.3 kg. The input ports of INMS were in the direction of the $-X$-axis.

The INMS was developed by teams from the NASA Goddard Space Flight Center and the University of Michigan Space Physics Research Laboratory. The principal investigator for INMS was Dr. J. Hunter Waite of the University of Michigan.

INMS contained separate neutral and ion measurement channels referred to as the closed-source neutral (CSN) mode and the open-source ion (OSI) mode. The design of the CSN mode was optimized to measure neutral species although it could also measure ions. The term closed source refers to the fact that the input port was exposed to ram atmosphere at the high speed of the spacecraft and the exit was restricted to cause pressure in the channel. Incoming neutral gas particles were led into a spherical chamber where the temperature of the gas became the same as the walls of the chamber. The gas was led out of the chamber by a tube into an area where it was ionized by two electron guns. The resulting ions flowed through the metered exit port into a quadrupole switching lens. The quadrupole switch was used to select either the output of the CSN or the output of the OSI to be applied to a quadrupole mass analyzer.

The open source ion mode input port was followed by a four-segment set of electrodes in a cylindrical form that focused ions into a beam. The ion beam was passed through four open-source lenses set at ground potential before being applied to the quadrupole switching lens. The open-source ion mode could also be used to measure neutral particles. When processing neutral particles, voltage was applied to the four-segment input electrodes to trap ions and electrons. An electron gun was used to ionize the particles, and the resulting ions were focused by four open-source lenses and directed to the quadrupole switching lens.

The quadrupole mass analyzer contained four hyperbolic rods 10 cm long and separated by 0.58 cm. DC voltages and radio frequency excitation were applied to pairs of rods and that created an electrostatic field within the region enclosed by the rods. The combination of varying DC voltage and radio frequency excitation allowed selecting particular mass/charge ratio of ions to be passed by the analyzer. Ions at the output of the analyzer were applied to two secondary electron multipliers. The outputs of the electron multipliers were counted. The results were the number of counts as the analyzer was tuned over its mass/charge ratio measurement range.

An example of spectrums developed by the INMS is presented in the NASA Solar System Exploration website. That graphic gives a spectrum of chemical constituents in the plume of Enceladus during flyby through the plume in March 2008. A copy of that graphic is reproduced in Fig. 6.17.

The composition of the atmosphere in the southern hemisphere of Enceladus was measured by INMS during a flyby in July 2005. The relative amounts of constituents were H_2O at 91%, CO_2 at 3.2%, N_2 at 4%, and CH_4 (methane) at 1.6%.

The atmosphere of Titan was investigated during several flybys of the large moon. A paper by Cui et al. (2009) gives the globally averaged neutral constituents of Titan's atmosphere. Nitrogen was the most abundant constituent at 97.8%. Other important constituents were methane at 1.8% and hydrogen at 0.0037%. Acetylene, ethylene, and ethane were also present at much smaller amounts.

Cosmic Dust Analyzer (CDA)

The cosmic dust analyzer (CDA) measured the number of particles as a function of time in the Saturn system and determined the composition, mass, and speed of dust grains and ice particles. It made extensive measurements of the sizes and chemical composition of material in the rings of Saturn.

The cosmic dust analyzer was made up of a dust analyzer (DA) and a high rate detector (HRD). The DA was developed by the Max Planck Institute for Nuclear Physics in Germany with assistance from the University of Kent in the UK. The HRD was developed by the University of Chicago in the United States. The HRD was used to measure very high impact rates (up to 10,000/s), which would saturate

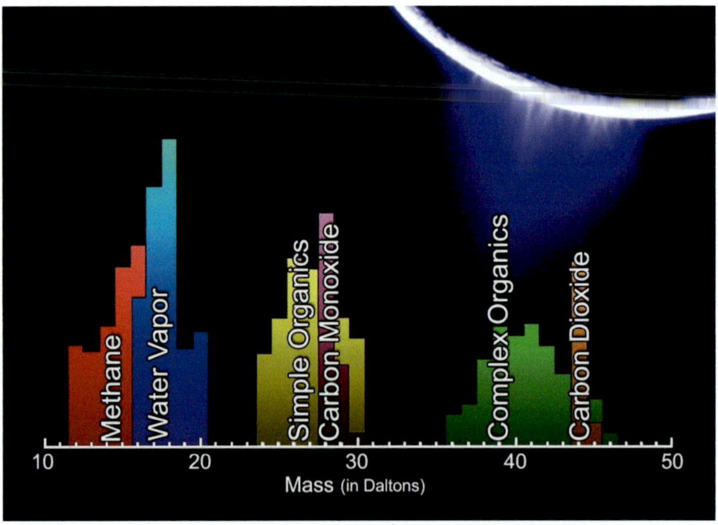

Fig. 6.17 Relative amounts of constituents in the plume of Enceladus vs. mass/unit charge (Daltons) measured by Cassini's INMS. (NASA image)

the DA portion of the instrument. The principal investigator for CDA was Dr. Ralf Stama of the Max Planck Institute.

A drawing, from the Cassini Final Mission Report (2019), of the Cassini dust analyzer showing the DA and HRD elements is given in Fig. 6.18. The high rate detector is contained in the rectangular package with two circular apertures attached to the frame of the dust analyzer. The boresights of the DA and HRD were in the same direction. The Cosmic dust analyzer was mounted to a turntable on the fields and particles pellet on the $-X$ side of the spacecraft. The axis of the turntable was 15° below the X–Y plane of the spacecraft and 30° from the $+Y$-axis towards the $-X$-axis of the spacecraft. The turntable was restricted to 270° rotation by the wrap of cables.

Dust Analyzer

The dust analyzer portion of the instrument was made up of a cylindrical can-shaped collector with grids across the opening and a hemispherical shell that functioned as an impact ionization detector (IID) at the far end of the cylinder. The input aperture was 432 mm in diameter, and the radius of the impact ionization detector was 229 mm. As per the scaling from drawings, the depth of the instrument from the entrance aperture to the center of the impact ionization detector hemisphere was about 540 mm.

The dust analyzer had an inlet screen at ground potential followed by two spaced screens mounted at an angle of 8.99° with electrodes attached. The two sensing screens were followed by another screen at ground potential. The charge of a particle was measured while it was transiting the space between sensing screens. The velocity of the particle was measured by noting the time of flight between screens.

Fig. 6.18 Drawing of Cassini dust analyzer. (NASA graphic)

The inclined mounting of the sensor grids caused an entrance angle-dependent shape of the charge pulse and allowed an approximate measure of the entrance angle.

High-velocity particles striking the spherical impact ionization detector resulted in impact fragments and impact plasma of ions and electrons. Positive ions were attracted to an ion detector located behind the hexagonal structure visible in the drawing. A screen in front of the ion detector was at a potential of −300 V to attract positive ions. The ion detector output was applied to a multiplier to raise the signal level for digitizing at 8-bit resolution.

A small region at the center of the impact ionization detector was open and covered by a screen at ground potential. A small chemical analyzer target was located three mm behind the screen. The target, which was made from rhodium, was at a potential of +1000 V. Positive ions resulting from the impact with the target were accelerated by the field between the +1000-V target and screen at ground potential. The accelerated positive ions were attracted to the ion detector. The flight time of the ion from the target to the ion detector was proportional to its mass.

High Rate Detector

The high rate detector (HRD) was used to measure very high impact rates (up to 10,000/s), which would saturate the dust analyzer portion of the instrument. Very high impact rates occurred when Cassini passed through the rings of Saturn. The HRD contained two separate detector channels. One detector had a circular aperture with an area of 50 cm^2, and the other channel had a circular aperture with an area of 10 cm^2. The sensors for both channels were of polyvinylidene fluoride material. A coating of Chemglaze Z306 was applied to the back of each sensor. The result was a permanently polarized material. When struck with a particle, the polarization was interrupted and a current pulse was generated. The amplitude of the current pulse was roughly proportional to the size and speed of the particle. The electronics for each of the two channels had four different current thresholds to identify particles of different sizes.

Results from the Cosmic Dust Analyzer Experiment

The CDA was capable of measuring the impact of nano-dust particles as small as 1×10^{-9} m in size. Such particles were measured in interplanetary space during the approach from about 100 million km to Saturn. Interstellar particles were also detected and identified by their high speed (72,000 km/h), direction, and composition.

Srama et al. (2011) give some results of the CDA experiment on Cassini. The *Cassini Final Mission Report (2019)* gives additional results. Some of those results are summarized in this section.

At distances between 1 and 2 AU from the sun, the CDA detected impacts of six interplanetary grains of dust. The electrical charges on the grains ranged from 1.3 to

5.4 femtocoulombs (fC). The corresponding particle masses were determined to range between 1.3×10^{-13} and 9.5×10^{-12} kg.

Emphasis was given to measurements of Saturn's E-ring by the CDA. The ring was found to extend out to at least 1.2 million km from Saturn, much farther than previously thought. The number density of particles was about $0.1/m^3$ at a distance of 6 R_S, where R_S is the radius of Saturn ($R_S = 60{,}280$ km). The density decreased to about $2 \times 10^{-4}/m^3$ at a distance of 15 R_S. The densest portion of the E-ring was at 3.98 R_S, just beyond the orbit of the icy moon, Enceladus. The E-ring was found to be mainly made up of micron-sized water ice particles. Some of the particles were composed of water ice infused with organic compounds and silicate minerals. About 6% were salt-rich ice particles. The ring was thought to be replenished from ice particles from geysers located in the southern hemisphere of Enceladus.

Measurements of particles at the edges of the G-ring were measured during a group of ring-grazing orbits through the ring plane. The CDA found that grains in the ring were ice particles, larger than encountered in the E-ring. Particles from the D-ring were measured during 22 orbits through the ring plane passing between the D-ring and cloud tops of Saturn. The particles were mostly water ice. Measurable amounts of silicates were also found. The particle sizes were smaller than 50 nm.

Radio and Plasma Wave Science (RPWS)

The radio and plasma wave science instrument was used to investigate plasma waves generated by ionized material (solar wind) flowing out from the sun and captured by the magnetic field of Saturn. It also measured naturally emitted radio frequency signals from the Saturn system. The RPWS instrument was developed by the University of Iowa with collaboration with the Observatoire de Paris Meudon, the Universite Versailles, and the Swedish Institute of Space Physics. Principal investigators for RPWS were Dr. Donald Gurnett and Dr. William Kurth of the University of Iowa.

The RPWS consisted of three main sensors and a main electronics assembly for the instrument. The three sensors were the electric field sensor, magnetic search coil sensor assembly, and Langmuir probe sensor assembly.

The electric field sensor included three 10-m-long deployable antennas, each orthogonal to one another. The antennas could be used as monopoles or pairs of them could be used as dipoles. Signals from the antennas were processed by five different receivers. The receivers and their frequency coverage were high-frequency receiver (HFR) (3.5 kHz to 16 MHz), medium-frequency receiver (MFR) (24 Hz to 12 kHz), five-channel waveform receiver (WFR) (1 Hz to 2.5 kHz), low-frequency receiver (LFR) (1 to 26 Hz), and wideband receiver (WBR) (60 Hz to 75 kHz).

The magnetic search coil sensor assembly contained three orthogonal metallic alloy cores. Each core had windings to produce flux in the core and another winding to sense the flux in the presence of external magnetic fields. The outputs of the three sensor coils were processed by the MFR, WFR, LFR, and WBR receivers.

The Langmuir probe sensor assembly consisted of a sphere 5 cm in diameter at the end of a 1-m-long narrow rod. A varying voltage was applied to the rod, and the resulting current flow through the plasma from the sphere to the frame of the spacecraft was measured. The voltage applied to the rod was swept in 256 steps between −32 and +32 V in less than 1 s. The current was measured at each step. Analysis of the current waveform yielded density and temperature of electrons and positive ions.

The RPWS instrument was controlled by a data processing unit that communicated with the Cassini data system. Data gathered by the three main sensors was also assembled for transmission to Earth by the data processing unit.

Results of RPWS Measurements

The Langmuir probe was used to measure the variation of electron density in the ionosphere of Titan as a function of solar zenith angle. Saturn's ionosphere was attributed to extreme ultraviolet (EUV) radiation from the sun ionizing the upper atmosphere of Saturn. At low zenith angles, the electron density was about 3000 electrons/cm^3. The density decreased to about 750 electrons/cm^3 at a solar angle of 90°.

A wide range of radio frequency emissions were observed in the Saturn system by RPWS. The most intense radio emission, given the name Saturn Kilometric Radiation (SKR), was observed emanating from the auroral regions. The emissions were strongest at latitudes from 50° to 60° and at frequencies between about 200 and 600 kHz. Strong narrowband radiation at about 5 kHz was found at higher latitudes.

Radio Science Subsystem (RSS)

The radio science subsystem was an experiment that used Cassini's telecommunications transmitter and high-gain antenna in conjunction with equipment on Earth to perform gravity and occulting experiments in the Saturn system. Team leaders for the RSS experiment were Dr. Arvydas Kliore of the Jet Propulsion Laboratory in Pasadena, California, and Dr. Richard French of Wellesley College in Massachusetts.

Both one-way and two-way Doppler tracking were used to determine the spacecraft velocity. The spacecraft contained an ultra-stable oscillator to control the transmitted frequency during one-way Doppler experiments. Two-way Doppler operation was obtained by transmitting a signal from Earth that was received by Cassini and transmitted back to Earth with a fixed frequency offset.

Gravity experiments involved monitoring the Doppler shift of the signal received on Earth as Cassini made flybys of various moons and orbited Saturn. The velocity of Cassini was altered by the gravitational field of Saturn or a moon as the spacecraft approached and then passed by. The variation in velocity was a function of the gravitational field and hence the mass of the body. The change in velocity resulted in a change in Doppler frequency, which was accurately measured on Earth. Gravitational measurements were made of the moons Titan, Enceladus, Mimas, Tethys, Dione, Hyperion, and Phoebe.

Occulting experiments were performed to explore the atmospheres of Saturn and Titan and explore the composition of the rings of Saturn. The transmitted signal from Cassini passed through the atmosphere of Saturn or Titan before Cassini's path took it behind the body. The amount of refraction (bending) of the signal passing through the atmosphere was dependent on the density of the atmosphere. The density, in turn, was dependent on the pressure and temperature of the atmosphere. The amount of refraction was measured on Earth, and vertical profiles of pressure and temperature of the atmospheres were determined.

Occulting measurements of the rings of Saturn were made at transmitted frequencies in S-band, X-band, and Ka-band. Use of three frequencies allowed determination of amounts of particles in bins of sizes from 1 mm to 20 m.

Huygens Probe

The Huygens probe was a capable spacecraft in itself. It was carried by the Cassini orbiter and released from the orbiter at the appropriate time to descend and land on Saturn's large moon, Titan. The Cassini-Huygens spacecraft was maneuvered by the propulsion module to a trajectory that would intercept Titan during the third orbit of Saturn. The Huygens probe was released from the Cassini orbiter during the third orbit and continued the trajectory towards Titan. The trajectory of the Cassini orbiter was adjusted to pass well above Titan and to be in a position to record data uplinked by Huygens as it entered the atmosphere of Titan, descended to the surface, and continued to operate on the surface.

Mechanical Configuration of Huygens Probe

The Huygens probe consisted of an entry assembly and a descent module. The entry assembly included an aeroshell and a back cover to protect the probe from the heat of entry into Titan's atmosphere. The aeroshell was a 60° half-angle conical surface 2.75 m in diameter. The outer surface was covered with tiles of ablative material to provide protection against heat of entry into Titan's atmosphere. The tiles in turn were covered with a thermal blanket to protect the spacecraft from heat of the sun while the spacecraft was still among the inner planets. The aeroshell was the large conical structure on the left side of the photograph of Cassini-Huygens shown in Fig. 6.19.

The Huygens descent module was mounted in the region between the aeroshell and the back cover. The back cover had a breakout patch through which the first drogue parachute was fired.

Fig. 6.19 Picture of Cassini/Huygens spacecraft showing the aeroshell for Huygens probe on the left side of the image. (NASA image)

A photograph of a model of the Descent module is shown in Fig. 6.20. The photograph was taken at Salon Européen de la Recherche in Paris by David Monniaux.

The main body of the descent module was 1.3 m in diameter. As per the scaling of the photograph, the main body was about 63 cm high. The black post sticking up on the top of the module is one of the two telemetry antennas. The boxlike structure on the top held parachutes.

A 3D rotatable drawing of the descent module was generated by NASA. A view from the bottom and a view showing the top and side of the module are shown in Fig. 6.21.

The descent module was composed of a bottom dome with vanes to rotate the spacecraft during descent, an equipment shelf, a top cover with equipment mounted, and a tapered cylindrical side member. NASA-generated cutaway views of the equipment shelf are shown in Fig. 6.22. Many of the labels in Fig. 6.22 refer to acronyms for science instruments that are described later.

Fig. 6.20 Model of descent module for Huygens probe residing in Salon Européen de la Recherche, Paris. (Wikimedia posting by David Monniaux)

Systems in Huygens Probe

Electrical Power

Huygens was powered by five lithium sulfur dioxide ($LiSO_2$) batteries. This type of battery was developed in the 1960s as a high-energy primary battery for military and space use. A primary battery is one that is not recharged. The $LiSO_2$ chemistry battery has recently been repurposed as a rechargeable battery, which is available commercially. The primary $LiSO_2$ battery cell had an energy density of 330 Wh/kg, significantly higher than other available primary batteries. It could operate over a wide temperature range, and it had long shelf life. The terminal voltage was typically 2.8 V under load.

Each of the five batteries in Huygens was made up of two modules of 13 $LiSO_2$ cells connected in series for a total of 26 cells per battery. Each cell was rated at 7 Ah. The nominal capacity when the five batteries were fresh was 2059 Wh. A power-conditioning distribution unit accepted power input from the batteries and converted it to a regulated 28-V power bus. A total of 28 switchable current limiters were connected to the power bus and distributed power to individual spacecraft systems and scientific instruments. The total power required from the batteries with all systems and instruments operating was 351 W. The nominal energy required from the batteries for the mission was 972 Wh.

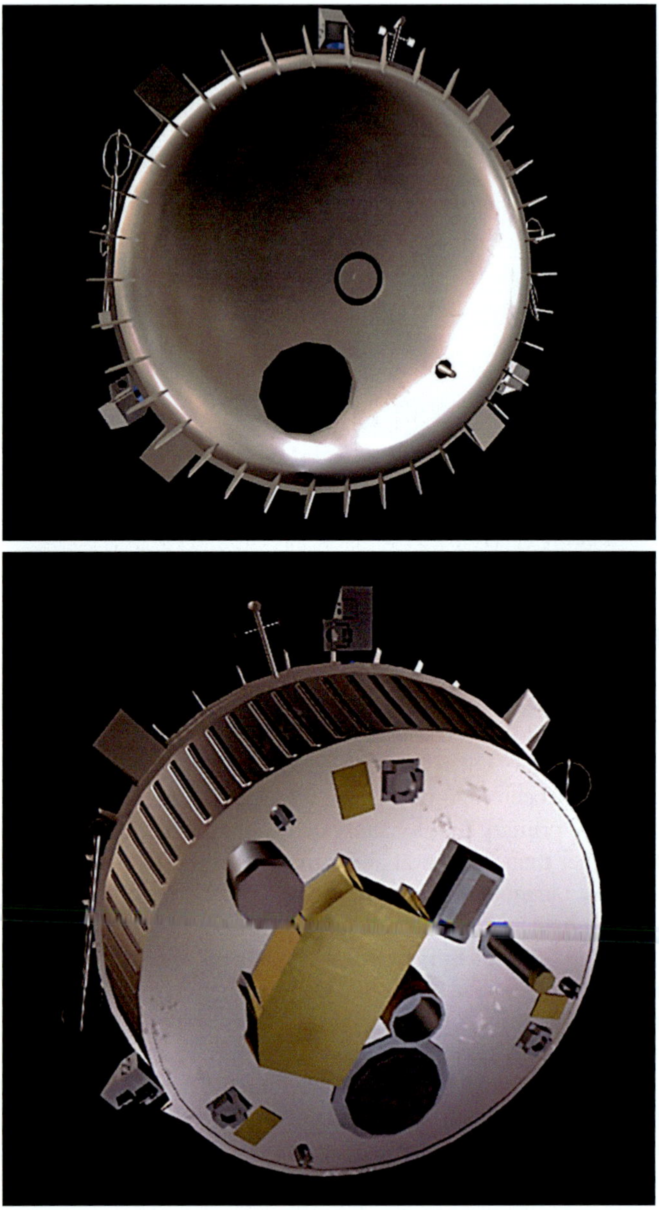

Fig. 6.21 NASA drawing of the descent module viewed from the bottom and the top

Control and Data Management Subsystem

Most of the other systems in the Huygens probe were grouped in the control and data management subsystem. Those systems included central accelerometer sensor

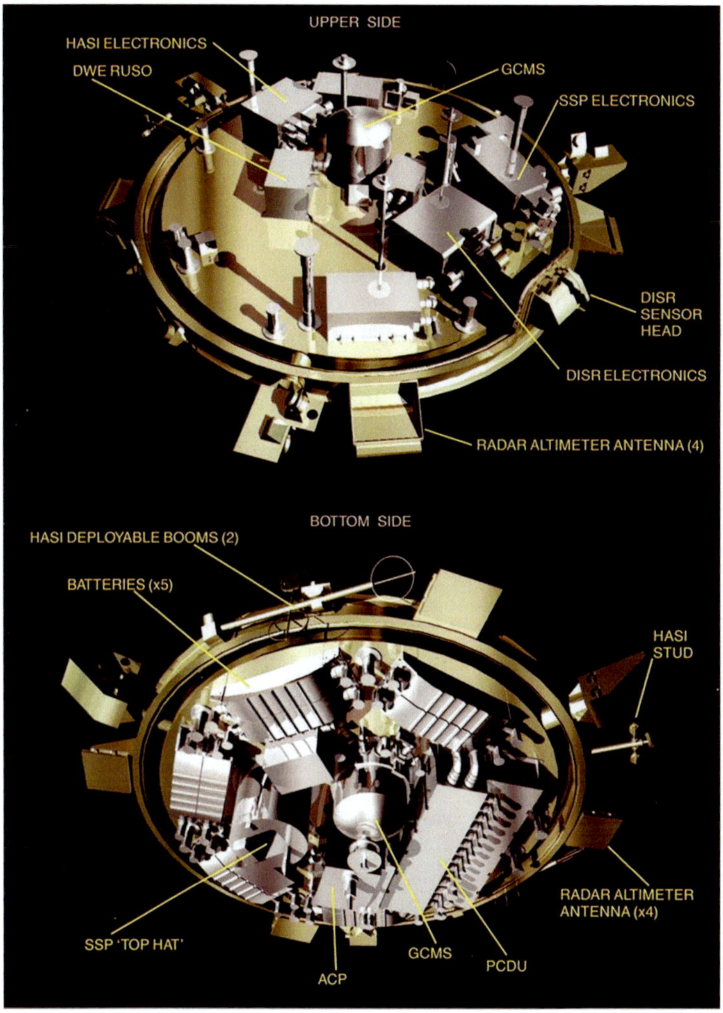

Fig. 6.22 Drawings of equipment shelf in Huygens descent module. (NASA graphic)

unit, radar altimeter, mission timer, command and data management unit, G-switches, and radial accelerometer sensor unit. Except for the accelerometer units, the other systems were dual redundant.

Central Accelerometer Unit

The central accelerometer unit measured acceleration in the axial direction of the Huygens probe. It had a range of 0–10 *G*s where 1 *G* is the acceleration due to gravity on Earth. It sensed deceleration effects of the atmosphere and established that as time zero (t_0) for the descent. The main parachute was deployed just after t_0. The accelerometer continued to provide acceleration data throughout the rest of the mission.

Radar Altimeters

The probe contained two redundant radar altimeters. They were turned on at an altitude of about 60 km. They were designed to provide altitude measurements from an altitude of 25 km to just above the surface. The radar altimeters were a linearly modulated FM type that was commonly used in spacecraft and aircraft of the day.

One radar altimeter operated at a frequency of 15.4 GHz, and the other operated at 15.8 GHz. The power output of the transmitter of each radar altimeter was 60 mW. Each altimeter had a transmit antenna and a receive antenna. Those antennas were waveguide planar arrays 125 mm by 162 mm in size. The beamwidth of the antennas was 7.9°, and the gain was 25 dB. The antennas were mounted at the edges of the probe with a downward view.

During Huygens' descent to Titan, the radar altimeters obtained intermittent detection of the surface at an altitude of about 45 km. Solid lock of the altimeter signal had occurred in both radar altimeters by the time the altitude had decreased to 28 km. Altitude data from the radar altimeter was used to turn on the cameras and other measurement instruments.

Mission Timer Unit

The mission timer unit contained three independent timers that were used to turn on power to probe systems at the end of the 22-day coast to Titan. The timers were set to the appropriate activate time, while Huygens was still attached to the Cassini orbiter. A voting circuit monitored the output of the three timers and output of two redundant G-switches in the central accelerometer unit. The G-switches activated when the spacecraft encountered the atmosphere. A vote that required two out of the three timers to have timed out or both of the redundant G-switches to have activated satisfied conditions for generating a mark signifying entry into the atmosphere. The mark started the descent sequence.

Command and Data Management Unit (CDMU)

Two redundant command and data management units were the brains of the Huygens probe. They each contained a MAS281 microprocessor built by Dynex Semiconductor in England. The microprocessor was a MIL-STD-1750A type, 16-bit central processing unit. It was built using silicon-on-sapphire technology to make it radiation resistant. The microprocessor ran at 10 MHz and used a

16,000-word PROM and a 64,000-word RAM. The microprocessors in both CDMUs ran the probe onboard software (POSW) simultaneously.

The software managed the probe mission, managed telemetry of engineering data and scientific instrument data, and accepted and managed data uplinked from Earth while the probe was still attached to Cassini.

The experiments operated mostly autonomously. They were sent mission data by the probe onboard software every 2 s. That data included time, spin data, and altitude. Data from experiments and from housekeeping functions (engineering data) were contained in packets 126 bytes long. The telemetry data was assembled into frames consisting of header information followed by seven packets of data and Reed-Solomon code words at the end. The frame size was 1024 bytes. Assignment of experiment data and housekeeping data to frames was controlled by a set of tables called mission timeline tables.

Probe Data Relay Subsystem

The probe data relay subsystem included two redundant S-band transmitters and two antennas mounted on top of the descent module. The frequency of each transmitter was set by a temperature-controlled crystal oscillator. One transmitter operated at a frequency of 2040 MHz, and the other operated at 2098 MHz. In addition, the frequency reference from one of the transmitters could be obtained from an ultra-stable oscillator. The ultra-stable oscillator allowed determining Huygens' velocity by Doppler processing of the received signal.

The power output of each transmitter was 10 W. The transmitters were phase modulated by a subcarrier signal. The subcarrier was modulated by digital data frames from the CDMU associated with that transmit channel. The data rate was 8192 bps. Data from experiments and housekeeping data was applied to both transmitters. Image data was alternated such that every other image was transmitted by a different transmitter.

The output of the 2040 MHz transmitter was applied to a helix antenna with left-hand circular polarization, and the 2098 MHz transmitter output was applied to a helix antenna with right-hand circular polarization. The high-gain antenna on the Cassini orbiter had separate antenna feeds for left-hand and right-hand polarizations at S-band. The antenna gain on boresight of Huygens' antennas was 5 dB at 2040 MHz and 3 dB at 2098 MHz. The antennas had a broad pattern and provided gain of at least 1.8 dB at angles up to 60° from boresight.

The probe support avionics in the Cassini orbiter had two receivers, one tuned to the 2040 MHz transmit signal of Huygens and the other tuned to 2098 MHz. The data from the two channels was extracted and stored in the solid-state memory. Gathering and storage of data from Huygens began at the time of turn on of Huygens' transmitters and continued through descent to the surface of Titan and for 72 min after landing. The signal was lost when Cassini passed over the horizon of Titan.

Unfortunately, one of the two receivers in the Probe support avionics in Cassini was not turned on due to an omission in the software. Since only one receiver channel was active, every other picture was not received. However, about 350 pictures were received. The receiver that was not turned on was also the receiver that was associated with the Huygens transmitter being supplied a frequency reference by the ultra-stable oscillator. As a result, accurate Doppler processing was not available.

Descent of Huygens Probe to Titan

The Huygens probe was deployed from the Cassini orbiter on 25 December 2004, while Cassini was in its third orbit around Saturn. The separation process imparted a speed of 35 cm/s to Huygens relative to Cassini and a spin rate of 7.5 rpm to the probe. The spin stabilized the probe during the 20-day cruise to Titan.

The trajectory of the third orbit was designed to impact Titan. Huygens continued the trajectory after separation. The trajectory of Cassini was altered after release of Huygens to pass 60,000 km from Titan and place it in a position to receive data transmitted by Huygens during its landing and while on the surface. Cassini was about 71,500 km from Titan when Huygens encountered the atmosphere of Titan.

Prior to separation of the Huygens probe, a timer was set to turn on essential systems of Huygens a few hours before the predicted time of entering Titan's atmosphere. There were three redundant coast timers, each powered by a separate battery. No other systems of the spacecraft were powered during 20-day cruise to Titan. The total power drawn by the three timers was 270 mW.

Lebreton et al. (2005) gave a timeline of the descent and landing of the Huygens probe on Titan. The main battery power was turned on to the spacecraft at 04:41:18 UTC, which was 4 h 29 min before reaching Titan's atmosphere. Probe support electronics was turned on 2 h 20 min before reaching the atmosphere.

Huygens reached a point 1,270 km above Titan that was referred to as the "interface altitude" at 09:05:53 UTC on 14 January 2005. The interface altitude was above the atmosphere of Titan. It had been NASA's responsibility to deliver the probe to that location, and thereafter it was the responsibility of ESA to navigate Huygens to a landing on Titan. Spacecraft events during descent and landing were controlled by the computer within Huygens.

Huygens entered the atmosphere of Titan 4 min 28 s after passing the interface altitude. The resulting deceleration caused accelerometer outputs to reach a preset threshold signifying entrance to the atmosphere. Entry time, t_0, began the descent sequence. Immediately after t_0, a 2.6-m-diameter pilot parachute was deployed that pulled off the back cover and extracted the main parachute. The 8.3-m-diameter main parachute deployed at $t_0 + 2$ s. The altitude of Huygens was 155 km, and vertical speed was 400 m/s at the time. The heat shield was jettisoned at $t_0 + 32$ s. The transmitter was turned on at $t_0 + 45$ s. Between $t_0 + 50$ s and $t_0 + 90$ s, caps were removed from four sensor ports, and the Huygens atmospheric structure instrument boom was deployed.

After 15 min of descent, the main parachute slowed Huygens to a velocity of about 35 m/s, and the altitude was about 110 km. The main parachute was then jettisoned and a 3-m-diameter stabilizing parachute was deployed. The velocity increased after switching parachutes, but by the time the probe had descended to 25-m altitude, the atmospheric drag had reduced vertical speed to about 5 m/s. The radar altimeter was turned on at t_0 + 32 min at an altitude of 60 km.

The probe had a series of canted vanes on the periphery that caused it to spin slowly. The spin allowed the imaging system in the probe to scan the terrain as the probe descended. A swivel on the parachute attachment lines allowed the probe to spin. It turned out that appendages on the spacecraft caused it to spin at an acceptable rate in the opposite direction.

The landing was targeted for the southern hemisphere on the sunlit side of Titan. The Titan coordinates at the landing site were latitude of 10.3° south and longitude of 167.7° east. The entry angle was 25° from the local vertical. The Huygens probe impacted the surface at a velocity of about 4.5 m/s at t_0 + 2 h 27 min 50 s. The probe continued to operate normally after landing. It continued transmitting data to the Cassini orbiter until the orbiter passed over the horizon about 1 h 12 min after landing.

An artist's rendering of the Huygens descent module on the surface of Titan is shown in Fig. 6.23. About 30 min after transfer of data from Huygens ceased, the Cassini orbiter was oriented to point the high-gain antenna towards Earth, and Huygens data, which had been stored in the solid-state memory, began to be transmitted to Earth.

Fig. 6.23 Artist's rendering of Huygens descent module on the surface of Titan. (NASA graphic)

Huygens Experiments

The Huygens probe carried six scientific instruments to measure the properties of Titan's atmosphere and surface. A list of instruments and their intent are given in Table 6.3. Also given in the table is the principal investigator for each instrument and the person's associated organization. It is interesting to see the international flair of the experiments.

Descent Imager/Spectral Radiometer (DISR)

The descent imager/spectral radiometer was a collection of optical instruments that took images and spectral measurements as Huygens descended to the surface of Titan. It consisted of a sensor head box and an electronic assembly box. The instrument was assembled by Lockheed Martin Astronautics, although many of the elements of the DISR were built by European companies. The Lunar and Planetary Laboratory at the University of Arizona was responsible for the overall DISR instrument.

The sensor head box was mounted near the edge of the experiment platform of the descent module so that sensors had a view upwards and downwards. Covers over the viewing ports were jettisoned after Huygens entered the atmosphere of Titan. The sensor head contained 11 different measurement functions. Sensors included three frame imagers, upward- and downward-looking visible spectrometer, upward- and downward-looking near-infrared spectrometer, and upward- and

Table 6.3 Scientific instruments on Huygens probe

Instrument	Purpose	Principal investigator
Descent imager/spectral radiometer (DISR)	Probe the atmosphere and image the cloud deck and surface by several sensors that covered a wide spectral range	M. Tomasko, University of Arizona, USA
Aerosol collector and pyrolyzer (ACP)	Collect aerosols for chemical analysis and transfer collection to ovens. Heat-decomposed organic material by vaporizing (pyrolysis)	G. Israei, at CNRS in France
Huygens atmospheric structure instrument (HASI)	Measure electrical and physical properties of the atmosphere and surface. A microphone picked up sounds of Titan	M. Fulchignoni, University of Paris, France
Gas chromatograph and mass spectrometer (GCMS)	Identify and measure chemicals in Titan's atmosphere. Analyze vapors from ACP instrument	N. Niemann, NASA Goddard Space Flight Center, USA
Surface science package (SSP)	Measure physical and electoral properties of the surface of Titan	J. Sarnecki, Open University, UK
Doppler wind experiment (DWE)	Measure wind speed during descent through Titan's atmosphere by measuring Doppler shift of received signal from the probe	M. Bird, University of Bonn, Germany

downward-looking violet spectrometer; four-channel solar aureole camera; and a sun sensor. Light from the lens systems of the sensors was coupled to detectors through fiber-optic bundles.

Frame Imagers

The three frame imagers operated in the visible spectral range of 660–1000 nm. The imagers were referred to as the high-resolution imager, medium-resolution imager, and side-looking imager. The three imagers had separate lens systems. The high-resolution imager used a lens system centered at 14° from the nadir with an angular range of 6.5–21.5° in the zenith plane and 9.6° wide in azimuth. The medium-resolution imager used a lens system centered at 31° from the nadir with a zenith plane coverage of 15.75–46.25° and 21.1° wide in azimuth. The side-looking imager lens system was centered at 71° from the nadir with an angular range in the zenith plane of 45.2–96° and 25.6° wide in azimuth.

A charge-coupled device (CCD) detector was used to detect the three visible wavelength images. The CCD was a 512 by 520 pixel frame transfer type, including a 256 by 520 pixel storage area. The high-resolution imager occupied a field of 160 by 254 pixels on the CCD. The medium-resolution imager used 176 by 254 pixels, and the side-looking imager used 128 by 254 pixels. The CCD also set aside fields of 8 by 240 pixels for the upward-looking visible spectrometer, 20 by 200 pixels for the downward-looking visible spectrometer, and four areas of 6 by 50 pixels for the four solar aureole cameras. Light from the three lens systems was brought to the face of the CCD by means of fiber-optic bundles. The CCD was read out and each pixel digitized by a 12-bit analog-to-digital converter. The process took about 2.2 s. The data format was reduced to 8 bits/pixel by a square root algorithm. The CCD subsystem was developed and built by the Max Planck Institute for Aeronomy in Germany.

As the Huygens probe rotated, images were taken with the three imagers each 30° in azimuth. A total of 36 separate images were taken in one rotation of the probe. Those images could be combined to produce a 36-image panoramic mosaic. An image mosaic of Titan constructed from images taken at an altitude of 10 km is shown in Fig. 6.24. The image mosaic is a Mercator projection formed by combining images taken throughout a 360° rotation of Huygens.

A view of the surface of Titan made after landing is given in Fig. 6.25. The picture was made up by combining images from the side-looking imager and the medium-resolution imager. What appear to be stones on the surface are thought to be made of water ice. The brighter flat "stone" near the center of the image was determined to be about 15 cm wide.

Fig. 6.24 Mosaic of images of Titan taken by Huygens at an altitude of 10 km. (NASA image)

Visible Spectrometer

Spectral measurements at visible wavelengths in upward-looking and downward-looking directions allowed determining the profile of solar absorption, vertical profile of methane, and measurement of properties of aerosols. The device used fiber-optic bundles to bring upward-looking and downward-looking light to adjacent positions on an entrance slit. Light at the exit of the entrance slit was conveyed to the CCD by fiber-optic bundles. The spectra were spread over 200 pixels of the CCD in the spectral direction. The spectral data was gathered when the CCD was read out as discussed for the imagers. The data was processed and made available for downlink to Earth where it was analyzed.

Infrared Spectrometer

The infrared spectrometer had an upward-viewing function and a downward-viewing function. The two measurements allowed determination of the thermal energy balance in the atmosphere as the spacecraft descended. The upper-looking aperture had a diffuser that allowed viewing half of the upper hemisphere. Infrared light was transferred from the entrance apertures to entrance slits for the spectrometer. The spectrometers for the upward- and downward-looking channels used transmission gratings. The output of each grating was applied to a linear array of InGaAs detectors. The outputs of the detectors were read out and downlinked to Earth for processing. Detectors and electronics for the instrument were provided by the Paris Observatory.

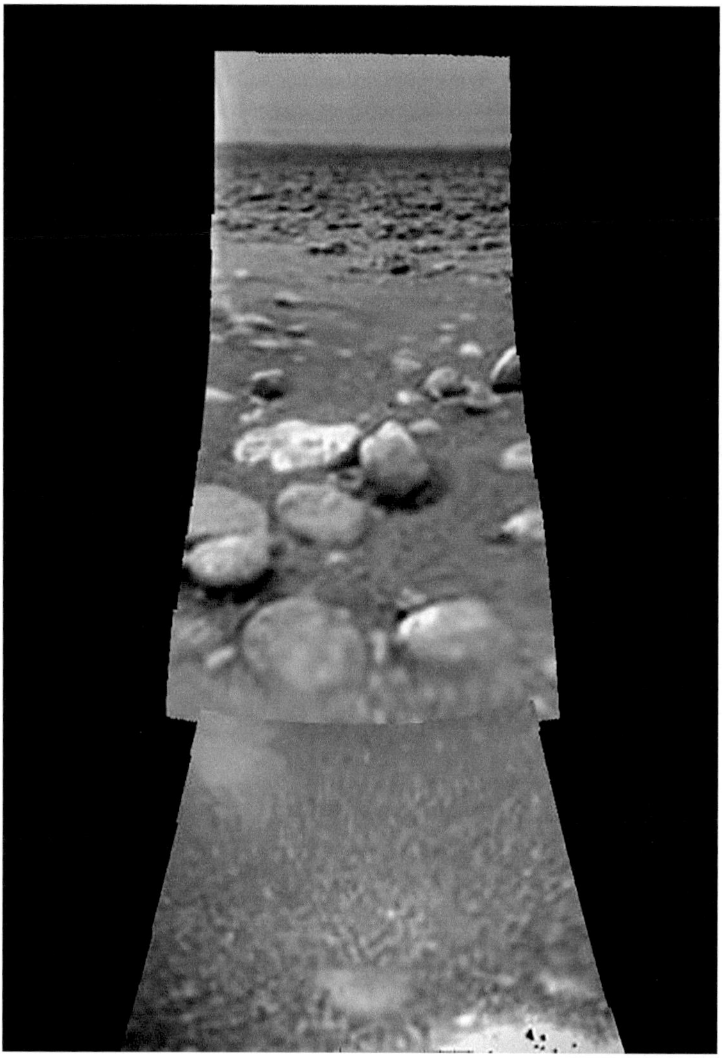

Fig. 6.25 View of the surface of Titan taken by Huygens after landing. (NASA image)

Solar Aureole Camera

Four cameras measured brightness in a strip of sky 6° wide in azimuth and 25–75° in extent in the zenith plane. Measurements were made at both polarizations at wavelengths of 500 and 935 nm. A fiber-optic bundle carried light from the four viewing apertures to the CCD. The angular extent of 6° by 50° for each camera was mapped into four separate 6 by 50 pixel areas on the CCD. The CCD was read out and data transmitted to Cassini for downlink to Earth as discussed previously.

Violet Photometer

The violet spectrometer extended the range of the visible spectrometer beyond 480–350 nm. Upward-looking and downward–looking channels were used to measure the absorption of violet energy by the atmosphere. Angular coverage was similar to that of the visible spectrometer. Light in each channel was detected, filtered, and digitized by a 12-bit analog-to-digital converter. The data was formatted for transmission to Cassini and then downlinked to Earth.

Sun Sensor

Knowledge of the azimuth angle of the sun was necessary for the upward-looking sensors of the descent imager/spectral radiometer. The lens system of the sun sensor had a field of view of 64° centered 47° from the nadir. The sensor operated at a wavelength of 939 nm. Light from the lens was applied to a reticle with three slits. The center slit was parallel to the line to the nadir, and the slits on either side were canted. The angle to the sun was established when signal was maximum in the center slit. The time delay between detection of signal in the canted slits relative to the center slit as the spacecraft rotated was used to determine the zenith angle of the sun. The amplitude of the solar signal in the center slit was measured to provide solar intensity as a function of altitude as modified by the atmosphere.

Aerosol Collector and Pyrolyzer (ACP)

The objective of the aerosol collector and pyrolyzer was to determine the chemical properties of aerosols in the atmosphere of Titan. To accomplish that, the instrument gathered a sample of the atmosphere, filtered it to isolate aerosols, subjected the filter to high heat (pyrolysis), and analyzed the volatiles. The volatiles were fed by a tube to the gas chromatograph mass spectrometer (GCMS) for chemical analysis. The inlet port to the ACP was a short tube that protruded through the bottom dome of the descent module. It is located at the lower right in the drawing of the bottom dome that was given in Fig. 6.21.

Two samples of the atmosphere were taken through the inlet port. The first sample was from an altitude of about 135 km down to 32 km, and the second sample was taken from about 22 km down to 17 km. The sample of the atmosphere was passed over a stainless steel filter by a fan to capture aerosols. After the sample period, the filter was transferred to an oven. Inlet and outlet of the oven were closed off, and the oven was heated to 250 °C. The volatiles were transferred from the oven to the GCMS by a stream of nitrogen-15 gas. The oven was then heated to 600 °C, and the volatiles were again transferred to the GCMS for analysis.

Analysis of the volatiles produced by the ACP identified molecular nitrogen, methane, ammonia, and hydrogen cyanide. The result was interesting since ammonia gas was not found in the atmosphere.

Huygens Atmospheric Structure Instrument (HASI)

The Huygens atmospheric structure instrument measured the physical and electrical properties of the atmosphere of Titan. It contained accelerometers, a temperature sensor, a pressure sensor, and an electric field sensor. The sensors were mounted on two deployable booms. The accelerometers provided a measure of density of the atmosphere as a function of altitude during the descent. The accelerometers also recorded the impact signature.

Temperature

Two redundant platinum resistance wire-type temperature sensors were employed. One was primary, and the other was a backup. Temperature was determined by measuring the resistance of the platinum wire in the sensor. The accuracy was about 0.2 °C. Temperature sensors were mounted on a short stub to space them away from the descent module. Fulchignoni et al. (2005) gave temperature and pressure profiles as Huygens descended to Titan. According to the curves in the study, the temperature was about −121 °C at 500 km, −89 °C at 250 km, −203 °C at 50 km, and − 180 °C on the surface. The author converted data given in Kelvin to degrees Centigrade.

Pressure

The vertical atmospheric pressure profile was measured during descent by a Kiel probe at the end of a boom. A Kiel probe is similar to a pitot tube but with a shield around it. A pitot tube is sensitive to attitude relative to the streamlines, whereas the Kiel probe can tolerate attitudes up to 45° from the streamlines. Pressure from the Kiel probe was transferred by a tube to a pressure sensor within the descent module. The pressure sensor was a "barocap" type that used dimensional changes in a silicon membrane to measure pressure. The change in the membrane moved an electrode of a capacitor relative to a stationary electrode. The pressure was about 2.8 mbar at 150 km, 10 mbar at 100 km, 80 mbar at 50 km, and 1.5 bar on the surface. The author converted pressure given in hectopascal in the reference to bars.

Electrical Properties

Electrical conductivity of the atmosphere due to free electrons and ions was measured during descent by two methods. The first method, referred to as the mutual impedance probe, used transmitter and receiver probes placed on two deployable booms. A fixed current was established between probes, and the voltage across the probes gave a measure of conductance. The probes, when disconnected from a voltage source, were also used to measure electric fields in the atmosphere.

A second approach to measuring conductivity, referred to as the relaxation probe, used relaxation probes consisting of a flat disks 7 cm in diameter. A potential of 5 V was applied to the probes for a short time and then disconnected. The disks collected electric charges from the atmosphere. The potential of the disks returned to equilibrium levels following an exponential law with a time constant, or relax time, related to conductivity of the atmosphere.

Conductivity is measured in siemens per meter (S/m) where one siemens is the reciprocal of one ohm. Data taken during the descent of Huygens reported by Morente et al. (2009) indicated over an order of magnitude higher conductivity measured by the relaxation probe method compared with the mutual impedance probe method. However, plots of conductance vs. altitude had about the same shape. Data from the mutual impedance probe method indicated conductivity of about 1×10^{-12} S/m at an altitude of 100 km, about 1×10^{-10} S/m at 70 km, and 1×10^{-12} at 40 km.

Radar Altimeter

The HASI processed the signals from the two redundant radar altimeters that were described previously. The radar altimeters were linearly modulated FM types that were commonly used in spacecraft and aircraft of the day. The difference in frequency between that transmitted and received at an instant of time was proportional to the distance above the terrain. This difference signal was processed to generate a digital representation of altitude.

Gas Chromatograph and Mass Spectrometer (GCMS)

The gas chromatograph and mass spectrometer was designed to analyze the atmosphere during the descent to Titan and on the surface of Titan after landing. The instrument identified chemical constituents and the quantity of each. It was also used to analyze products generated by the aerosol collector and pyrolyzer instrument.

The GCMS included (1) a gas chromatograph for batch sampling of the atmosphere gas at given times; (2) a mass spectrometer subsystem consisting of ion sources, mass analyzer, and an ion detector; (3) a direct atmospheric gas sampling system to introduce atmospheric gas into the ion source of the mass spectrometer; and (4) a transfer element to transfer products of the pyrolyzer to the mass spectrometer. The instrument had provisions to collect a sample and analyze it later or to sample and analyze continuously.

The inlet port for the GCMS protruded through the bottom dome of the descent module near the center of the dome. It is the circular port near the center of the drawing of the bottom of Huygens in Fig. 6.21. The gas chromatograph and the mass spectrometer were both fed from the inlet port.

Gas Chromatograph

The gas chromatograph identified gases in the atmosphere by virtue of their transit time through a long narrow column filled with special mesh material. Identification of the gases was based on different strengths of interaction of a particular gas with the mesh material. The stronger the interaction, the longer the time it took. A hydrogen gas carrier was used to move the sample of the atmosphere through the columns. A detector at the output of the column produced an electrical signal displaced in time by the transient time through the column. The amplitude of the electrical signal was proportional to the abundance of the particular gas.

The GCMS included three gas chromatograph columns to cover the range of expected gases. Column 1 was 2 m long and had a diameter of 0.75 mm inside. Column 2 was 10 m long and had a diameter of 0.18 mm inside. Column 3 was 14 m long and had a diameter of 0.18 mm inside. The columns were wound on a 178 mm diameter form, and each was contained in a separate oven for temperature control.

The time required for a complete sample was about 10 min. At altitudes above 60 km, Huygens was traveling too fast to perform repeated sampling. Instead, three samples spaced in time were taken and stored. Those samples were analyzed later. At lower altitudes, samples were repeatedly taken and analyzed.

Mass Spectrometer

The mass spectrometer was a quadrature mass spectrometer type. The analyzer portion included four narrow rods stacked close to each other in a square pattern. The inner-facing surfaces of the rods had a hyperbolic cross section to create a symmetrical electric field in the space between rods.

Atmospheric gas molecules that passed through the inlet port of the GCMS were conducted through a tube to the mass spectrometer. The gases were ionized in an ionization chamber by bombarding by electrons. The electron bombardment broke up the molecules into fragments of positive ions. Those fragments were accelerated and directed to the entrance of the mass analyzer by electric field lenses.

Diagonally opposite pairs of rods were connected together electrically, and a combined DC voltage and RF voltage was imposed between the pairs. A particular combination of DC and RF voltage caused an ion of a particular mass-to-charge ratio (m/z) to set up a resonant oscillation in space without touching the rods and travel through the center of the rod arrangement to strike the detector. Ions with a different m/z struck a rod or escaped between the rods. The DC and RF voltages were swept to selectively pass ions with different m/z ratios.

Ions that passed through the mass spectrometer were applied to an electron multiplier. The output of the electron multiplier was detected and applied to a pulse counter. The counts were accumulated for an integration time and the results sent to the telemetry system.

Results of GCMS Measurements

Some results of GCMS measurements are given in a work by Niemann et al. (2005). Measurements were made from an altitude of 140 km to the surface. Nitrogen (N_2) was the most abundant substance in the atmosphere. Other references put the abundance of nitrogen of over 97% of the total. Methane was the next prevalent gas with a mole fraction to nitrogen of 1.4×10^{-2} at altitudes around 130 km. The fraction remained at about that level until the altitude had decreased to about 30 km when it began rising. The fraction was about 2.2×10^{-2} at 20 km and about 4.9×10^{-2} from an altitude of 8 km to the surface. Argon (^{40}Ar) was detected at altitudes below about 18 km. The mole fraction was about 4.3×10^{-5}. Hydrogen (H_2) and carbon dioxide (CO_2) were also present at lesser amounts.

Surface Science Package (SSP)

The surface science package contained nine instruments designed to characterize the surface of Titan. Several of the instruments also made atmospheric measurements during the descent. Some of the instruments were designed to determine the properties of liquid substances should the probe land in a body of fluid. A list of the instruments in the SSP and brief descriptions are given in Table 6.4.

Table 6.4 Instruments and their purpose in the surface science package

Instrument	Purpose
External accelerometer	Determine the characteristics of the soil at the landing site by using an accelerometer mounted on a stub located below the module. The stub would penetrate soft terrain, and acceleration readings would help in determining the properties of the surface
Internal accelerometer	Measured acceleration during the descent and landing
Acoustic properties sounder	The sounder consisted of a sonic transmitter and a receiver to process the echo signal at short distances from the surface. In case of solid terrain, it would measure the acoustic cross section of the terrain. In case of a liquid landing, it could measure the depth of the liquid
Density sensor	In case of a liquid landing, a float would be used to determine the density of the fluid
Permittivity sensor	In case of a liquid landing, the permittivity of the fluid would be measured by noting the difference in electrical capacity of a series of stacked plates
Refractive index sensor	In case of a liquid landing, the refractive index would be measured by means of critical angle refractometer
Thermal properties sensor	The instrument contained two separate cylindrical cavities with an axial sensor wire in each. One cavity measured thermal properties of the atmosphere, and the other would measure thermal properties of a liquid in case of a liquid landing. The wire was heated by applying a known current for a known period of time. The resistance of the wire was then measured as heat was conducted away from the wire by the thermal conductivity of the atmosphere or liquid. The resistance of the wire was proportional to temperature
Tiltmeter	The tiltmeter was a two-axis type that measured spacecraft attitude during descent and during and after the landing event

References

Asmar, S. W. et al., *Cassini Radio Science User's Guide*, Jet Propulsion Laboratory, 2018

Cui, J., et al., *Analysis of Titan's neutral upper atmosphere from Cassini Ion Neutral Mass Spectrometer measurements*, Icarus, Vol. 200, Issue2, April 2009

Elachi, C. et al., *Radar: The Cassini Titan Radar Mapper*, Space Science Reviews 115, 2004

Elachi, C., *Cassini Radar Views the Surface of Titan*, Science, Vol. 348, Issue 5724, 2005

Fulchignoni, M., et al., *In Situ Measurements of the Physical Characteristics of Titan's Environment*, Nature Vol. 438, 2005

Henry, Curt A., *An Introduction to the Design of the Cassini Spacecraft*, Space Science Review, 104, 2002

Hsu, Hsiang-Wen, et al., *In situ collection of dust grains falling from Saturn's rings into its atmosphere*, Science, Vol. 362, Issue 6410, Oct 2018

Jet Propulsion Laboratory, *Cassini Final Mission Report 2019, Volume 1, Mission Overview, Science Objectives and Results*, NASA Jet Propulsion Laboratory, California Institute of Technology, 2020

Kliore, A. J., *Cassini Radio Science*, Space Science reviews, Vol 115, 2005

Kliore, A. J., et al., *Cassini Radio Science*, Space Science reviews, Vol. 115, 2004

Krimigis, S. M., et al., *Magnetosphere Imaging Instrument (MIMI) on the Cassini Mission to Saturn/Titan*, Space Science Reviews, Vol 114, 2004

Lebreton, Jean-Pierre, et al., *An Overview of the descent and Landing of the Huygens Probe on Titan*. Nature, Vol. 438, 2005

Lockheed Martin, *Final Technical Report GPHS-RTGs for the Cassini Mission*, Lockheed Martin Document No. RR18, 1998

Lopes, R. M. C., et al., *Titan as Revealed by the Cassini Radar*, Space Science Review (2019) 215.33

Morente, John A., et al., *An analysis of VLF electric field spectra measured in Titan's atmosphere by Huygens probe*, Journal of Geophysical Research: Planets, Vol. 114, Issue E6, 2009

NASA NSSDA, Cassini, COSPAR ID: 1997-061A

NASA, *Cassini Final Mission Report, Volume 1: Mission Overview and Science Objectives and Results,* 2019

NASA Press Kit, *Cassini Launch*, October 1997

NASA Press Kit, *Cassini-Huygens Saturn Arrival*, June 2004

NASA, Planetary Data System, Instrument Host Information: *Cassini Orbiter*

Niemann, H. B., et al., *The abundances of constituents of Titan's atmosphere from the GCMS instrument on the Huygens probe*, Nature, Vol. 438, Dec 2005.

Niemann, H. B., et al., *The Gas Chromatograph Mass Spectrometer for the Huygens Probe*, Space Science reviews 104, 2002

Srama, Ralf, et al., *The Cosmic Dust Analyzer Onboard Cassini: Ten Years of Discoveries, CEAS Space Journal* (2011) *2:3*

Striepe, Scott A., et al., *Huygens Titan Probe Trajectory Reconstruction Using Traditional Methods and the Program to Optimize Simulated Trajectories II*, AAS/AIAA Space Flight Mechanics Meeting, Sedona, AZ, 28 Jan–1 Feb 2007.

Tomasko, M. G, et al., *The Descent Imager/Spectral Radiometer (DISR) Experiment on the Huygens Entry Probe of Titan*, Space Science Reviews 104, 2002

Waite, J. H., et al., *The Cassini Ion and Neutral Mass Spectrometer (INMS) Investigation*, Space Science Reviews, Vol 114, 2004

Webster, Julie L., *Cassini Spacecraft Engineering Tutorial*, Jet Propulsion Laboratory, March 2006

West, Richard D., et al., *Cassini Radar Sequence Planning and Instrument Performance*, IEEE transactions on Geoscience and remote Sensing, Vol. 47, No. 6, June 2000

Wong, Edward G., and Breckenridge, William G., *An Attitude Control Design for the Cassini Spacecraft*, https://doi.org/10.2514/6.1995-3274

Ye, S. Y., et al., *Dust Observations by the Radio and Plasma Wave Science Instrument During Cassini's Grand Finale*, Geophysical Research Letters, Vol. 45, Issue 19, 2018

Chapter 7
Galileo Spacecraft

The Galileo spacecraft orbited Jupiter for nearly 8 years and made several close flybys of Jupiter's four largest moons. The spacecraft returned hundreds of images and a wealth of scientific information about Jupiter and its substantial moons. Galileo dispatched an atmospheric probe while on approach to Jupiter. The probe penetrated deep into the hot, hostile atmosphere of Jupiter and sent important findings back to Galileo, which relayed them to Earth.

The Galileo mission overcame a serious setback when the high-gain antenna failed to deploy and the low-gain antenna had to be used for the entire mission. Ingenious workarounds in data handling and transmission timing were performed by scientists to salvage most of the planned science investigations.

A photograph of Galileo being prepared for mating with the upper stage booster at Kennedy Space Center (KSC) is shown in Fig. 7.1. The large circular plate near the top of the spacecraft was a sun shield. The vertical structure above the plate was folded up ribs of a large wire-mesh high-gain antenna.

Background of Galileo Program

The Galileo mission had its formal origin in studies conducted by NASA's Outer Planet Working Group in 1969. Another program, the "Grand Tour," was also being studied. A fortunate positioning of the outer planets Jupiter, Saturn, Uranus, and Neptune occurred in the period late 1970s and early 1980s. The alignment allowed taking advantage of gravitational assist from each planet in turn to set up a trajectory to the next planet as the spacecraft traveled outward. The Grand Tour program was approved for funding in 1972, and funding for a Jupiter mission was deferred. The

T. Lund, *Spacecraft Exploring Outer Planets*, Springer Praxis Books,
https://doi.org/10.1007/978-3-031-96544-9_7

Fig. 7.1 Galileo spacecraft at Kennedy Space Center (NASA image)

grand tour mission was accomplished by the Voyager 2 spacecraft described in Chap. 5 of this book. The Voyagers were launched in 1977.

Several NASA-sponsored groups, including the Outer Planets Working Group, studied types of missions to Jupiter. A Jupiter orbiter carrying a deep-entry probe that would be launched into Jupiter's atmosphere was favored. The spacecraft was called the Jupiter Orbiter Probe. The Jet Propulsion Laboratory (JPL) of the California Institute of Technology was tasked by NASA in the fall of 1975 to perform a conceptual design of the Jupiter Orbiter Probe so that the weight of the spacecraft and cost of the program could be estimated. NASA planned to use the Space Shuttle to launch planetary spacecraft during that time period, and weight of the payload was an important consideration.

NASA's budget was modest at the time, and it was difficult to get funding from Congress for any substantial space mission. After many months of discussions between NASA personnel and Congress and extensive lobbying of Congress by NASA, congressional approval for the Jupiter Orbiter Probe program was given on 1 October 1977.

After congressional approval of the program, NASA assigned the Jupiter Orbiter Probe mission to the Jet Propulsion Laboratory. JPL would manage the program and build the spacecraft. The project manager at JPL was John Casani. The West German company Messerschmitt-Bölkow-Blohm (MBB) developed and built the propulsion module for the spacecraft. Hughes Aircraft Company was selected to develop the atmospheric probe in June 1978.

The name of the spacecraft was changed to Galileo to honor the great scientist, Galileo Galilei, who made important scientific findings in many fields. He had used a telescope of his own design to discover and study the moons of Jupiter in 1610.

The Galileo spacecraft was designed and assembled by JPL. After extensive performance and environmental testing, it was delivered to the Kennedy Space Center in December 1985. Launch was planned from the Space Shuttle Atlantis in May 1986.

A major tragedy of the Space Shuttle program occurred on 28 January 1986 when the Space Shuttle Challenger exploded 73 s after liftoff, killing all aboard. Subsequent Space Shuttle flights were put on hold while a detailed investigation was made of the accident and means to improve the safety of the Space Shuttle program. Space Shuttle flights did not resume until September 1988.

During the stand-down, a major change that affected Galileo was that NASA decided not to use the liquid hydrogen-fueled Centaur upper stage to launch spacecraft from the Shuttle for concerns of crew safety. An upper stage that had been used to launch satellites from the Space Shuttle, the inertial upper stage (IUS), was used instead. IUS did not have the capability to place Galileo in a direct trajectory to Jupiter. A modified trajectory using gravitational assist from Venus and two gravitational assists from Earth was adopted instead. The new trajectory, which took Galileo nearer to the sun, required changes to the thermal design. A large sunshade was added in the area below the high-gain antenna, and a smaller sunshade was placed at the upper end of the spacecraft.

Galileo was launched from the Space Shuttle Atlantis on 18 October 1989. After gravitational assists from a flyby of Venus and two flybys of Earth, the spacecraft traveled to Jupiter and entered equatorial orbit on 7 December 1995. The atmospheric probe was released from Galileo on 13 July 1995, and the probe plunged into Jupiter's atmosphere on 7 December 1995. Galileo continued orbiting and downlinking observations to Earth until 21 September 2003 when it was deliberately allowed to burn up in Jupiter's atmosphere. It had explored Jupiter and its major moons for nearly 8 years.

Mechanical Configuration

Stabilization of the Galileo spacecraft in inertial space was unique in that a large portion of the spacecraft was spun about its longitudinal axis for attitude stability. The rest of the spacecraft, referred to as the despun section, remained at a selected inertial angle in the azimuth plane. The despun section held cameras and other optical instruments.

The Galileo spacecraft was composed of three major parts: spun section, despun section, and atmospheric probe. A NASA/JPL-generated labeled sketch of the spacecraft is shown in Fig. 7.2. The drawing includes sunshades that were added after the trajectory to Jupiter was changed to include gravity assists from Venus and Earth. The atmospheric probe was mounted at the center bottom of the despun section so that it could be easily deployed. The drawing was cropped by the author to allow the text to be larger. A magnetometer sensor and a plasma-wave antenna were at the end of the boom to the right of the drawing.

Galileo was a large spacecraft, measuring 5.3 m from the bottom of the atmospheric probe to the top of the center low-gain antenna. A deployable magnetometer boom extended 11 m from the centerline of the spacecraft, and booms for the radioisotope power source extended 5 m on either side of the spacecraft. The Galileo orbiter and atmospheric probe together weighed 2223 kg at launch. That weight included 118 kg of scientific instruments and 959 kg of rocket propellant. The atmospheric probe weighed 331 kg.

The coordinate system sited the Z-axis along the longitudinal axis of the spacecraft with +Z in the opposite direction to the high-gain antenna. The Y-axis was in

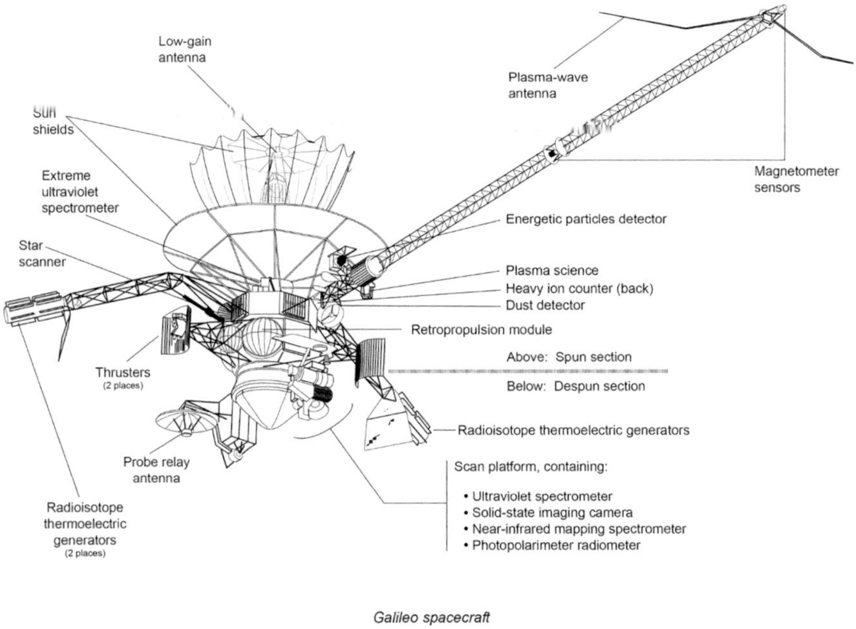

Fig. 7.2 Drawing of Galileo spacecraft (NASA graphic)

the direction of the magnetometer boom with +Y in the boom direction. The X-axis was perpendicular to the Z- and Y-axes with +X on the same side of the spacecraft as the relay antenna.

Spun Section of Spacecraft

The spun section was the largest section of the spacecraft. It included the high-gain antenna, sunshades, radioisotope power sources, science boom, magnetometer boom, propulsion module, electronic equipment bays, and six scientific instruments. The section was spun at a nominal rate of 3 rpm. The six scientific instruments investigated particles and magnetic fields. A remote sensing element, the extreme ultraviolet spectrometer, was also attached to the spun section.

The electronic equipment bays consisted of an eight-sided structure mounted below the high-gain antenna. Faces of three of the bays can be seen in the drawing of the spacecraft in Fig. 7.2. Structural longerons separated at each bay. Usually, one bay was dedicated to a particular subsystem. Electronics for science instruments often occupied 1/3 of a bay each. Lower power telecommunications electronics occupied one bay, and the TWT power amplifiers occupied a separate bay. One bay held a data memory subsystem, which was a four-track tape recorder.

The science boom, which extended 3 m from the centerline of the spacecraft, held four scientific instruments and the magnetometer boom. The magnetometer boom, when unfurled, extended 11 m from the centerline of the spacecraft. The magnetometer boom held two magnetometers, one at the end of the boom and the other at the midpoint. Science instruments mounted to the science boom were energetic particle detector, dust detector, heavy ion counter, and plasma science.

The spun and despun sections of the spacecraft were connected mechanically by a spin bearing assembly. That assembly included 48 slip rings for power and low-frequency electrical connection and rotary transformers to transfer high data rate signals. Piping for the main rocket engine was fed through the center of the spin bearing assembly. The exit of the rocket engine was close to the top of the atmospheric probe at the bottom center of the despun section. The rocket motor fired through the opening after the atmospheric probe was deployed.

Despun Section of Spacecraft

The despun section contained the atmospheric probe, a scan platform that held remote sensing instruments, and electronic bays to support various subsystems. Electronics were contained in five disbursed bays.

The scan platform mounted to the side of the despun section. The platform rotated in a plane parallel to the longitudinal axis of the spacecraft. The platform could be scanned through an angle of 210° to point the boresights of the remote sensor instruments in the plane of the longitudinal axis (elevation scan). The entire despun section could be rotated to any desired angle in azimuth. This allowed the boresights of the remote sensing instruments to be pointed at any azimuth angle and any elevation angle within the scan limits. The remote sensing instruments included a camera called solid-state imager, a near-infrared mapping spectrometer, an ultraviolet spectrometer, and a photopolarimeter-radiometer.

The atmospheric probe fit into the center bottom portion of the despun section. The probe included a heat shield, a descent module, and a back cover. The conical heat shield was 125 cm in diameter. It protected the descent module during the extreme high-speed entry into Jupiter's atmosphere. The overall height of the atmospheric probe was 87 cm.

The main engine, which had a thrust of 400 newtons (89.9 pounds), was located above the atmospheric probe in the despun section. Piping for fuel and oxidizer was run through the center of the spin bearing assembly. The engine could not be fired until after the atmospheric probe was deployed.

Spacecraft Systems

A functional block diagram of systems within the spacecraft is given in Fig. 7.3.

Electrical Power

Electrical power for the Galileo spacecraft was produced by two radioisotope thermoelectric generators (RTGs). The spacecraft was intended to operate at the planet Jupiter which lies 779 million km from the sun, 5.2 times farther from the sun than

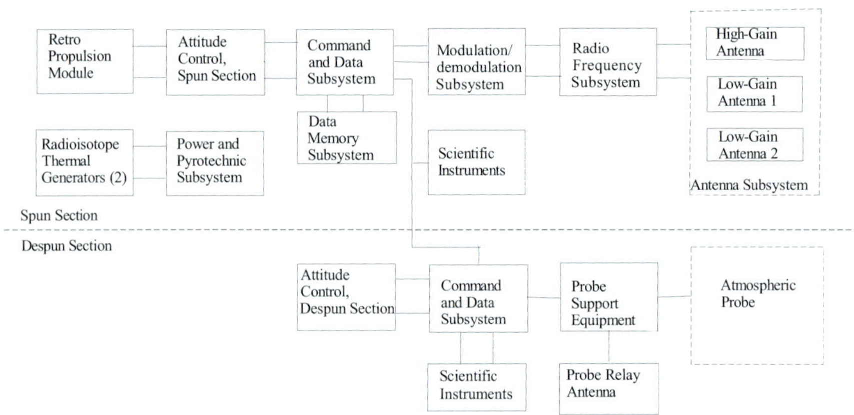

Fig. 7.3 Functional block diagram of Galileo spacecraft

Earth. At the time, solar arrays were not practical to deliver the amount of power required by Galileo at that distance. The RTGs were developed by General Electric for the Department of Energy for space use. They were named "general-purpose heat source-radioisotope thermoelectric generator" (GPHS-RTG). Essentially, the same GPHS-RTGs were used to power the Ulysses, Galileo, Cassini, and New Horizon spacecraft.

Originally, the Galileo RTGs were required to produce a total of 510 W of electrical power 4.7 years after the beginning of the mission. The mission was planned to begin with launch in May 1986. However, the tragic accident of Space Shuttle Challenger delayed the launch by about 4.3 years, and a new trajectory to Jupiter with gravitational assists from Venus and Earth required longer travel time than the direct trajectory originally planned. As a result, the power requirement from the RTGs was changed to 470 W at the end of the mission (about 8.1 years).

The RTGs were cylindrical in shape, 42.2 cm in diameter, and 114 cm long and weighed 55.9 kg. Each RTG was mounted on the end of a 5-m-long boom to space them away from sensitive areas of the spacecraft. The neutron emission rate was about 5.0×10^3 neutrons/s/g of plutonium-238. The neutron dose rate at a distance of 1 m was between 20 mrems/h and 50 mrems/h depending on the angle. The two booms supporting the RTGs were mounted on the opposite side of the spacecraft from the magnetometer boom and ±45° from the Y-axis of the spacecraft.

In addition to the RTGs, the Galileo orbiter contained 103 radioisotope heater units (RHUs) that used a small amount of plutonium-238 to produce about 1 W of thermal power. The units were 2.5 cm in diameter and 3.3 cm long. The RHUs were strategically placed to warm sensitive components during the long, cold mission. There were 17 RHUs on the atmospheric probe.

RTGs generate electrical power by applying heat from radioactive decay of a material such as plutonium-238 to a series of thermoelectric couples. Plutonium-238 decays into uranium-234 by emitting alpha particles. The half-life of the decay process is 87.7 years. The kinetic energy of the alpha particles is converted to heat when the particles strike the surrounding material. Each RTG contained a total of 8.1 kg of plutonium-238.

The GPHS-RTG used pellets of plutonium-238 oxide pressed into cylindrical shape 2.76 cm in diameter and 2.76 cm long. Each pellet was clad with a welded shell of iridium alloy. There were 72 such cladded pellets within a cylindrical region at the center of the RTG. A graphite cylindrical shell enclosed the pellets. A total of 572 silicon-germanium thermoelectric couples were mounted around the graphite shell. The devices were mounted such that one end was in contact with the hot surface and the other end was in contact with the much cooler outer case of the RTG.

Each of the GPHS-RTGs had initial thermal output of about 4400 W. Initial electrical power output from the field of thermocouples was 290 W. The resulting thermal to electrical power efficiency was about 6.6%. The electrical outputs of the two RTGs were connected in parallel giving an initial total power output of about 580 W. The power output of the RTGs decreased with time. Telemetry data from the spacecraft indicated power available to be about 550 W after about 1000 h and about 482 W after 71,000 h (8.1 years).

The voltage at the parallel output of the two RTGs was somewhat over 30 V. A power management and distribution system regulated and conditioned the RTG power for spacecraft systems. In addition, there were two pyrotechnic switching assemblies that drove 148 different squibs associated with 47 pyrotechnic devices on the orbiter.

Retro Propulsion Module

The retro propulsion module (RPM) was a truss-like structure with booms at either end that held clusters of thrusters and a center section that contained tanks for propellants and a main rocket engine. The retro propulsion module was provided to the Galileo program by the Federal Republic of Germany. It was developed and built by Messerschmitt-Bölkow-Blohm (MBB) in Germany. A photograph of the RPM being handled by workers at MBB is shown in Fig. 7.4. The perspective of the photograph does not show the second boom, which extends on the right side of the retro propulsion module in the photograph.

The retro propulsion module was a fully tested module shipped from Germany ready to install in the spacecraft. It became a load-carrying member of the lower portion of the spinning section of the spacecraft. The module was 0.84 m high and 4.32 m long across the thruster booms. The empty weight of the module was 179.6 kg. When loaded with 959 kg of propellants and helium pressurizer, the total weight was 1131 kg. The weight of useable propellant was 925 kg.

Propellants for the RPM were the hyperbolic combination of monomethylhydrazine (MMH) fuel and nitrogen tetroxide (NTO) oxidizer. The fuel and oxidizer

Fig. 7.4 Retro propulsion module being worked on at MBB in Germany (NASA image)

ignited upon contact. Fuel was contained in two tanks, and the oxidizer was also contained in two tanks. In addition, two smaller tanks held helium used to pressurize the propellant system. A series of valves allowed either fuel tank or oxidizer tank to supply propellants to the thrusters and to the main engine.

The two booms that held the clusters of thrusters extended 2 m from the Z-axis (centerline) of the spacecraft. There were six thrusters in a cluster at the end of each boom. Each thruster had a thrust of 10 newtons (2.25 pounds). The thrusters were oriented such that pairs of them could be activated to increase or decrease rotation speed. Other thrusters could be used for precession of the spin axis and to increase or decrease velocity along the longitudinal or lateral axes. Thruster firing was controlled by valves activated by electrical signals from the attitude and articulation control subsystem.

The main engine had a thrust of 400 newtons (89.9 pounds). The engine was located above the atmospheric probe in the despun section. Piping for fuel and oxidizer was run through the center of the spin bearing assembly. The engine could not be fired until after the atmospheric probe was deployed. Burn time of the main engine was controlled by data from accelerometers. A given velocity increment was commanded, and the engine fired until integrated accelerometer data indicated that the commanded velocity change had been achieved.

The main engine performed well during its three operational burns during the mission. A two-second "wake-up" burn was made on 24 July 1995, 11 days after the atmospheric probe was deployed. The first operational burn was an "orbiter deflection maneuver" to change the path of Galileo from that of the probe. The new trajectory positioned Galileo to fly past Jupiter and receive data transmitted from the probe as it entered the atmosphere. The orbiter deflection burn, which occurred on 27 July 1995, was 5 min in duration and changed spacecraft velocity by about 70 m/s.

The second burn was the critical orbit insertion burn that put Galileo in orbit around Jupiter. That burn, which was made on 7 December 1995, lasted for 49 min and decreased spacecraft velocity by 643 m/s.

The third and final burn was a "perijove raise maneuver" that raised perijove (lowest point of the orbit) to 715,000 km above Jupiter to lessen the effects of Jupiter's radiation on spacecraft systems. That burn lasted 29 min and increased speed by 378 m/s.

Command and Data Subsystem

The command and data subsystem (CDS) was the executive manager of the Galileo spacecraft. It was a software-based manager with six RCA 1802 microprocessors arranged in two redundant paths with three microprocessors each. During the long cruise to Jupiter, one of the paths was primary and the other was backup. Each path was connected by a high-speed data bus to functions internal and external to the CDS. Four of the microprocessors were located in the spun section of the spacecraft, and two were located in the despun section. After arriving at Jupiter, all six microprocessors were used as a single microprocessor-based computer to conduct science and telemetry functions.

Galileo used special 1802 microprocessors fabricated in silicon on sapphire, which made them resistant to radiation. They were built by Sandia National Laboratories in collaboration with RCA. The microprocessors were clocked at a frequency of 1.6 Megahertz. The total RAM memory of the microprocessors and bulk memory on the spun side was 400,000 words. RAM memory on the despun side was 32,000 words. The CDS decoded and processed commands and operational sequences sent up from Earth. The commands were sent to spacecraft systems for immediate use or stored for future use. The CDS executed stored command sequences to control spacecraft systems at designated times. It orchestrated science data collection from various instruments and collected, organized, and formatted science data and engineering data for transmission to Earth. It conducted system-level fault-protection services. It also moved data between subsystems on the spacecraft using its high-speed data buses.

Attitude and Articulation Control Subsystem

The attitude and articulation control subsystem (AACS) was responsible for the control of orientation of the spin axis, spin rate, thruster firing, main engine firing, and control of the scan platform upon which remote sensing instruments were mounted.

The spin rate was nominally 3 rpm except for times when high stability was needed, such as launch of the atmospheric probe and during firing of the main engine. In those events, the spun and despun sections were locked together and the spacecraft was spun at a 10-rpm rate. The atmospheric probe retained the 10-rpm spin rate after it was launched to aid its stability.

The primary source of attitude information for the spacecraft was obtained from a star scanner. The star scanner was a V-slit type that timed the passage of the image of a selected star from one arm of the V to the other as the spacecraft rotated. It determined the cone angle and the brightness of the star. Software associated with the scanner identified the star and provided spacecraft attitude in inertial coordinates.

Star scanner information was processed by a computer in the AACS, and corrective signals were sent to the retro propulsion module where pairs of thrusters were fired to maintain desired orientation of the spin axis in space. During communication sessions with Earth, the spin vector and associated boresights of the high-gain and low-gain antennas were pointed towards Earth. Sun sensors located on the spun section provided backup spin rate and sun direction information. Sun sensor outputs could be used for initial orientation of the spacecraft.

The heart of the AACS was two redundant attitude control electronic (ACE) units. Each ACE contained an Itek Advanced Technology Airborne Computer (ATAC) using a type 2901 radiation-hardened microprocessor built by Sandia National Laboratory. The microprocessor was a 16-bit machine with a cycle time of

250 ns. Cycle time of the memory slowed the operational rate to 143,000 cycles/s. Each microprocessor was supported by 31,000 words of 16-bit RAM and 1000 words of 16-bit ROM.

Data Memory Subsystem

Much of the data gathered by Galileo during its mission was stored in memory and then transmitted to Earth at a later time and at a lower data rate. A four-track tape recorder that could store 914 megabits of data was used to store large data fields. The tape recorder was a reel-to-reel type that held 564 m of ¼ inch (0.64 cm) Mylar tape. It could be controlled from the ground.

A malfunction of the high-gain antenna precluded direct transfer of large data fields such as images from the solid-state imager over the downlink. Instead, the data was recorded on tape and played back later in the orbit at a lower data rate. The data transmitted to Earth by the low-gain antenna was at a sufficiently low data rate that acceptable bit error rate was achieved. The mission would have been severely curtailed without the tape recorder.

Telecommunications System

The telecommunications system received and demodulated commands and other data uplinked from Earth. Uplinking was conducted through the large antennas of the Deep Space Network (DSN) on Earth. The telecommunications system also transmitted data from scientific experiments and engineering data to the DSN. A major disruption to the telecommunication system happened when the high-gain antenna on the spacecraft could not be deployed. The low-gain antenna had to be used for the entire mission. Communications were maintained with Galileo in orbit around Jupiter at distances up to 730 million miles from Earth despite failure of the high-gain antenna. Work-arounds, including implementing sophisticated error correction coding in the spacecraft and arraying groups of large antennas on Earth, allowed receiving good-quality images and science data at the Deep Space Network, although at low data rates.

The telecommunications system included an S–X-band antenna subsystem (SXA), radio frequency subsystem (RFS), modulation/demodulation subsystem (MDS), and an X- to S-band downconverter (XSDC).

Antenna Subsystem

The Galileo antenna subsystem included a high-gain antenna and two low-gain antennas. The high-gain antenna was a deployable parabolic antenna with a reflector 4.8 m in diameter. It consisted of a series of 18 ribs that folded up for launch like an umbrella and when deployed formed ribs that supported a gold-plated wire-mesh parabolic reflector. The high-gain antenna contained feeds for both X-band and S-band. The boresight of the antenna was along the −Z-axis of the spacecraft. The high-gain antenna was designed to have 50 dB gain at X-band and 38 dB gain at S-band. The frequency of the X-band downlink signal was to be 8415 MHz. Unfortunately, the high-gain antenna could only be partially opened during flight even after many tries. It was not useable during the mission.

There were two low-gain antennas, LGA-1 and LGA-2. LGA-1 was mounted above the feed section of the high-gain antenna with its boresight along the −Z-axis of the spacecraft. LGA-2 was mounted on a short boom, which when deployed hung from one of the booms holding a radioisotope thermoelectric generator. Its boresight was parallel to the +Z-axis of the spacecraft. LGA-2 was not used during the Galileo mission. The low-gain antennas could only be used in S-band. LGA-1 had a gain of 8.1 dB and a beamwidth of about 120°.

Radio Frequency Subsystem

The radio frequency subsystem contained redundant receivers and redundant transmitters for both X-band and S-band frequencies. A network of switches allowed selection of particular receivers or transmitters. The high-gain antenna could not be deployed during the mission, and the low-gain antennas had only S-band capability. As a result, the X-band equipment was not used, and only the S-band equipment will be described here.

Receiver

S-band uplink signals from the low-gain antenna were passed through a diplexer, which allowed receiving and transmitting through the same antenna, and on to switches to one of the two redundant S-band receivers. The receiver processed the uplink S-band signal and outputted a narrowband signal consisting of a 16 kHz subcarrier modulated with uplinked data. The receiver output was sent to the modulation/demodulation subsystem.

The receiver contained a phase-lock loop that developed a reference signal coherent with the carrier of the uplink signal. That reference signal could be used by the exciter to generate a transmit frequency that was coherent with the received carrier but displaced in frequency from it. The coherent downlink signal allowed two-way Doppler processing on the ground that was used for accurate tracking of the spacecraft. The receiver also detected and processed a ranging data on the uplink signal. The ranging signal was retransmitted on the downlink for two-way ranging.

Transmitter

The two transmitters consisted of redundant exciters that drove redundant S-band traveling wave tube power amplifiers. Each exciter developed the transmitted carrier that was biphase modulated with science and engineering data. The exciter units received a subcarrier from a telemetry modulation unit (TMU) that was biphase modulated with the telemetry data. The modulating signals for the downlink were generated by the command and data subsystem. The TMU applied convolution encoding to the incoming bit stream from the CDS.

Each exciter had a crystal oscillator that could be used to set the frequency of the downlink signal. Alternatively, the frequency could be set by a coherent reference signal developed by the receiver, or the output of an ultra-stable oscillator could be used. The ultra-stable oscillator was used when one-way Doppler data was to be processed on the ground. The coherent reference signal from the receiver allowed two-way Doppler processing on the ground. The exciter included a ranging signal generator for one-way ranging.

The output of each exciter could drive a selected traveling wave tube (TWT) power amplifier. The TWT developed output power of 20 W at S-band. The output of either TRW amplifier could be selected for application to a diplexer. Switches after the diplexer allowed selecting either of the two low-gain antennas for transmission. Low-gain antenna 1 (LGA-1) was used throughout the mission.

Work-Arounds to Return Data Without the High-Gain Antenna

The mission had been planned with the use of the high-gain antenna, which would allow 134,400 bits per second (bps) data transmission rate to Earth from Jupiter. In contrast, using the low-gain antenna, original spacecraft configuration, and original ground antenna configuration, a data transmission rate of only 8–16 bps could be achieved from Jupiter. This was much too low to transfer image data.

An image from the solid-state imager was made up of an array of 800 by 800 pixels. Each pixel was digitized to eight-bit words resulting in an image size of 5.12 megabits. At 10 bps, it would take 5.9 days to transfer one picture. The 134,400 bps data rate with the high-gain antenna could transfer an image in 38 s.

The problem with the antenna was discovered in April of 1991. Galileo would not arrive at Jupiter until December 1995, giving time to study and devise work-arounds to make the best use of available resources. Data transmission would have to use the low-gain antenna. Work-arounds included optimizing observation schedules for the scientific instruments and prioritizing data gathering by particular scientific instruments. Data processing on the spacecraft was improved, and data compression was implemented. Changes to data processing and implementation of data compression were uploaded to Galileo's computers from the ground. The Galileo telecommunications system had the ability to transmit in a suppressed carrier mode. This mode, which puts all the transmitted power in the information-carrying sidebands, was selected to work with the low-gain antenna.

Other changes to work with the low signal levels on the ground from the low-gain Galileo antenna included arraying DSN antennas on the ground and improving the DSN receivers. Antennas used during the Galileo mission were the 70-m diameter antenna at Goldstone in California; a 70-m-diameter and two 34-m-diameter antennas at Canberra, Australia; a 64-m-diameter antenna at Parks, Australia; and a 70-m-diameter antenna at Madrid, Spain. The outputs of the three antennas at Canberra were arrayed together to increase the effective size of the receiving aperture. The 70-m Goldstone antenna was also arrayed with the Canberra antennas at times when both Goldstone and Canberra had a view of Galileo.

The result of the changes implemented to accommodate Galileo's low-gain antenna increased the allowable downlink bit rate to up to 160 bps. Scientists estimated that about 70% of the originally planned data gathering was achieved.

Relay Radio Equipment on Galileo Orbiter

An antenna and a receiver to gather signals from the atmospheric probe were mounted on the despun section of the orbiter. The antenna was a parabolic type 1.1 m in diameter. The beamwidth was 25°, and peak gain was 21 dB. The antenna had two feeds: one to receive right-hand circular polarized signals and the other to receive left-hand circular polarized signals from the probe. The outputs of the two antenna feeds were applied to separate receivers. One receiver was tuned to a frequency of 1387.0 MHz and the other to 1387.1 MHz. The outputs of the two receivers were processed by the command and data subsystem of the orbiter and applied to the tape recorder for later relay to Earth.

Flight of Galileo Spacecraft

Galileo was carried into Earth orbit by Space Shuttle Atlantis (STS-34) on 18 October 1989. Atlantis was launched from the Kennedy Space Center in Florida at 12:53 pm EST on 18 October 1989. Galileo was launched from the cargo bay of Atlantis on a path towards Venus at 7:23 pm EST on 18 October 1989. The launch of Galileo was originally planned for May 1986, but the tragic loss of Space Shuttle Challenger and its crew in December 1986 resulted in a 32-month delay in flights of the Space Shuttle. A detailed investigation was made of the accident and means to improve safety of the Space Shuttle program during the downtime. Flights of the Space Shuttle resumed in September 1988.

A photograph taken by the crew of Atlantis of the Galileo spacecraft raised to launch position in the cargo bay is shown in Fig. 7.5. Galileo was attached to the inertial upper stage booster that was often used to launch satellites from the Space Shuttle. The limb of Earth is on the left side of the photograph. Launch from Atlantis involved releasing springs that thrusted the booster and spacecraft away from the cargo bay. An hour after deployment from Atlantis, the first and second rocket stages of the inertial upper stage fired in succession to place Galileo on a trajectory towards Venus for the first gravity assist. Galileo separated from the inertial upper stage after burnout of the last stage.

Galileo did not have sufficient fuel to power the spacecraft on a direct trajectory to Jupiter. Instead, gravity assists from Venus and Earth were used to set a lengthy trajectory to Jupiter. Gravity assist is achieved by flying a spacecraft past a massive body on a path that results in gravity of the body giving the spacecraft a boost of energy and a change of direction. A NASA graphic showing the trajectory of Galileo from launch to arrival at Jupiter is given in Fig. 7.6. The graphic also shows the orbit of Jupiter (in red).

Fig. 7.5 Galileo spacecraft and inertial upper stage booster raised to launch the position in cargo bay of Space Shuttle Atlantis (NASA image)

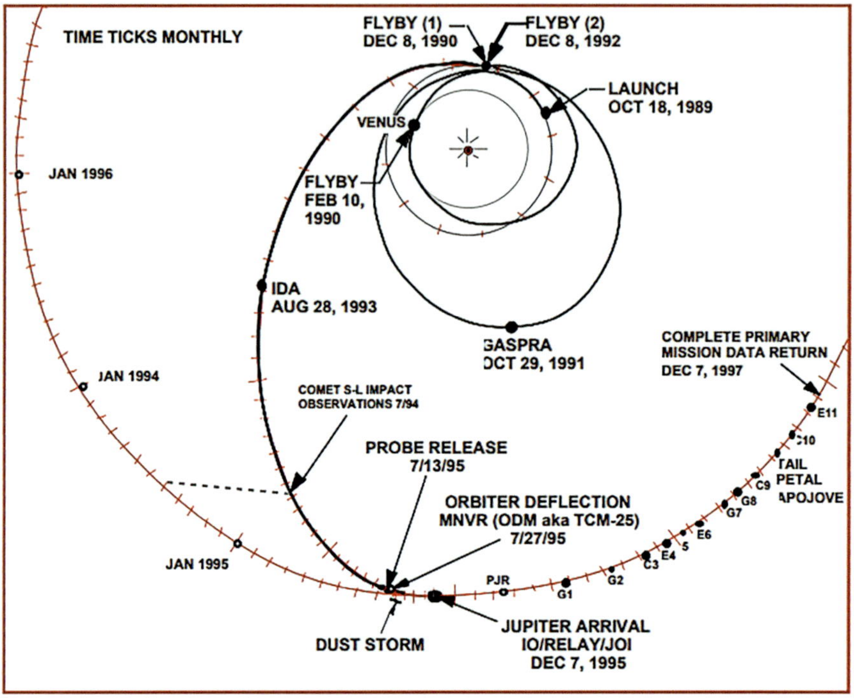

Fig. 7.6 Trajectory of Galileo on its journey to Jupiter and orbit of Jupiter (NASA graphic)

Galileo passed 16,000 km from Venus during a flyby on 10 February 1990. Science instruments were activated to gather information and images of the planet. The encounter with Venus increased the speed and bent the trajectory to intercept and fly by Earth. The first Earth flyby occurred on 8 December 1990 with Galileo passing about 960 km above the West Atlantic Ocean. Science information was gathered, and about 1000 photographs were taken of Earth. One photograph taken of Earth is shown in Fig. 7.7. Deployment of the high-gain antenna was attempted during the flyby, but the deployment failed because ribs supporting the wire-mesh reflector would not unfold. The low-gain antennas had to be used for the entire mission.

The flyby of Earth added another increment of velocity to Galileo and sent it on a path to orbit the sun and make a second flyby of Earth. Galileo's trajectory after the first flyby of Earth took it through the asteroid belt and within 1600 km of the asteroid Gaspra. An image of Gaspra was taken by Galileo at a distance of about 5300 km. Gaspra is about 18.2 × 10.5 × 8.9 km in size.

The second flyby of Earth took place on 8 December 1992 with the closest approach distance of 305 km. The encounter provided the final increment of velocity needed for a flight to Jupiter. The trajectory took Galileo again through the asteroid belt and within 2414 km of the 56-km-long asteroid Ida. Images were taken of Ida during passage.

Fig. 7.7 Earth imaged by Galileo during first flyby in 1990 (NASA image)

Galileo happened to be about 240 million km from Jupiter when fragments of the comet *Shoemaker-Levy 9* impacted Jupiter. Instruments on Galileo observed the impacts as they happened. The impacts occurred on a portion of Jupiter not visible to Earth although rotation of Jupiter brought the impact scars into view later.

The atmospheric probe was deployed on 13 July 1993 while Galileo was about 80 million km from Jupiter. The probe reached the upper atmosphere of Jupiter on 7 December 1995. The probe's velocity was 170,550 km/h at the first encounter with the atmosphere. The entry point was at latitude of 6.5° and longitude of 4.4° west in the Jupiter coordinate system. The probe was slowed by a heat shield and then parachutes. It descended about 180 km into the atmosphere before radio transmission from the probe stopped, likely due to extreme temperature or pressure of the atmosphere.

The Galileo orbiter continued on a path to Jupiter until 27 July 1995 when a burn of the main engine for 308 s was made to slow the velocity by 61 m/s. The burn changed the trajectory to pass by Jupiter rather than impacting it. It also set up favorable conditions for orbit insertion. A NASA graphic showing the approach to Jupiter by both the atmospheric probe and the Galileo orbiter is given in Fig. 7.8. The path of Galileo took it past the moon, Io, at a distance of 892 km at 17:45 UT

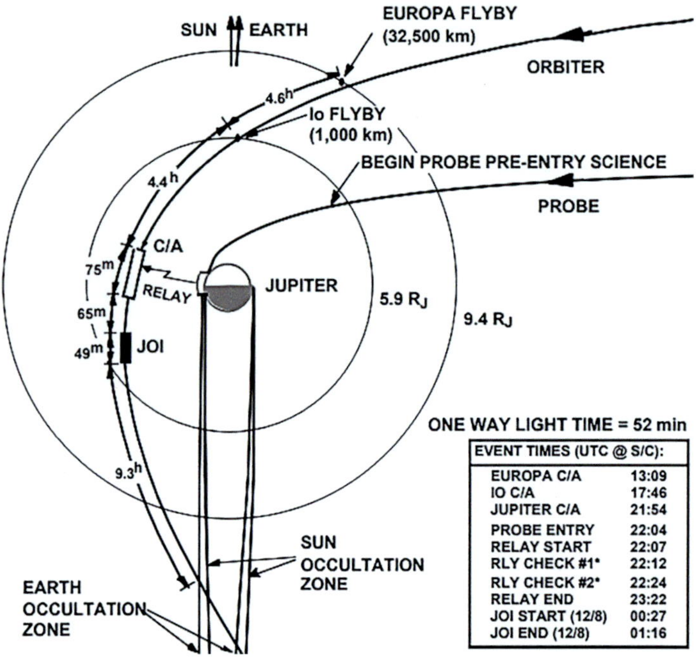

Fig. 7.8 Close-up of approach to Jupiter by Galileo orbiter and probe (NASA graphic)

on 7 December 1995. The encounter slowed Galileo by about 200 m/s relative to Jupiter. The closest approach to Jupiter occurred at 21:53 UT on 7 December 1995 at an altitude of 214,570 km.

The atmospheric probe entered the atmosphere about 8 min after the time of the orbiter's closest approach, so the orbiter was in good position to receive transmissions from the probe as it descended. The probe descended 160 km below the cloud tops into the hostile atmosphere of Jupiter before transmissions from it ceased.

The Galileo orbiter fired its main engine for orbit insertion on 7 December 1995. The burn lasted 49 min and decreased the orbiter's velocity by 630 m/s. The velocity reduction allowed Jupiter to capture Galileo and pull it into orbit. The initial orbit had a period of 200 days. A burn of the main engine was made at apojove (the greatest distance from Jupiter) of the first orbit to raise perijove of the orbit and of subsequent orbits to about nine Jupiter radii. The orbit was raised to lessen the radiation exposure to the spacecraft.

Primary Mission

A NASA/JPL graphic showing the 11 orbits of the primary mission is given in Fig. 7.9. The first orbit was used for initial adjustment of the trajectories. The last 10 orbits were each tailored for a close flyby and gravity assist from one of the moons, Ganymede, Calisto, or Europa.

Impressively, the trajectories were computed in advance and selected so that the flyby of a moon set up a trajectory to the next moon in the sequence with minimum use of spacecraft fuel. A summary of the orbits for the primary mission is given in Table 7.1. Orbit data was obtained from tables generated by NASA/JPL. Values given for the closest approach distance vary slightly between NASA documents. The orbit numbers in the table are preceded by the first letter of the particular moon to be investigated. For example, G1 refers to the first orbit, and the moon, Ganymede, was investigated. No data was returned during orbit 5 because communication was blocked due to solar conjunction.

The perijoves of the orbits are listed in the table. Perijoves of the orbits ranged from five million km to nearly 20 million km. Galileo's scientific instruments imaged each moon and determined the characteristics when near the closest approach. The encounter time with each moon was about 7 days. The primary mission included a setup orbit and ten orbits that targeted particular moons. The primary mission officially ended on 14 December 1997. It had lasted 24 months.

A montage of images of the four largest moons of Jupiter taken by Galileo during the primary mission is given in Fig. 7.10. The montage was assembled by the

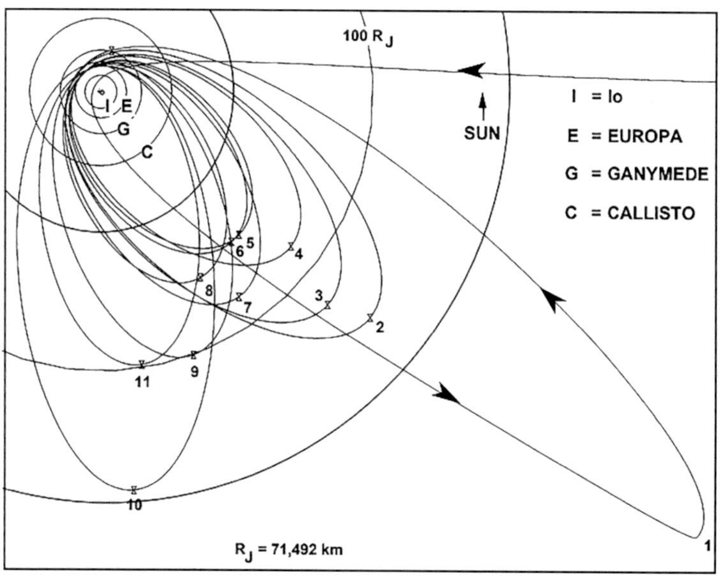

Fig. 7.9 Galileo orbits during prime mission (NASA graphic)

Table 7.1 Orbits of Galileo around Jupiter and the moons investigated during primary mission

Orbit	Perijove, Jupiter radii	Moon	Date of flyby	Altitude of flyby, km
J0	4	Initial orbit		
G1	11.03	Ganymede	27 June 96	844
G2	10.65	Ganymede	6 September 96	262
C3	9.21	Callisto	4 November 96	1100
E4	9.16	Europa	19 December 96	695
E5A	9.05	Europa	20 January 97	27,419
E6	9.12	Europa	20 February 97	588
G7	9.12	Ganymede	5 April 97	3065
G8	9.27	Ganymede	7 May 97	1584
C9	10.77	Callisto	25 June 97	416
C10	9.17	Callisto	1 September 97	524
E11	8.97	Europa	6 November 97	1119
G1	11.03	Ganymede	27 June 96	844

Fig. 7.10 Images of the four largest moons of Jupiter taken by Galileo (NASA graphic)

German Aerospace Center. From the left, the moons are Io, Europa, Ganymede, and Callisto. The moon sizes are presented to scale. Ganymede, with a diameter of 5260 km, is the largest moon in the solar system. It is larger than the planet Mercury.

Galileo Europa Mission (GEM)

The Galileo Europa Mission, which followed the Primary Mission, lasted from 15 December 1997 to 2 May 1999. It included seven orbits by Galileo from E12 to E19. The Galileo Europa Mission was divided into three phases: Europa Campaign, Perijove Campaign, and Io Campaign. The Europa Campaign consisted of eight orbits (E12 to E19) with Europa as the targeted moon. Closest approach to Europa ranged from 201 km to 3582 km. The Europa Campaign lasted from 15 December 1997 to 11 February 1999.

The Perijove Campaign consisted of four low-altitude flybys of the moon Callisto. The purpose of the campaign was to use gravity assist from Callisto to reduce perijove of Galileo's orbit from 9.4 Jupiter radii to 5.5 Jupiter radii. The trajectory was set up for a later encounter with the moon, Io, at 5.9 Jupiter radii. The Perijove Campaign, which lasted from 2 May 1999 to 10 October 1999, comprised Galileo orbits C20 to C23.

The Io Campaign consisted of three flybys of the moon, Io, and one flyby of Europa. Orbit numbers for the campaign were I24, I25, I27, and E26. The altitudes of the closest approach to Io ranged from 200 km to 611 km. Closest approach to Europa was 360 km. Science measurements were conducted at each encounter, with emphasis on magnetic field measurements. The Io Campaign lasted from 10 October 1999 to 20 May 2000.

Galileo-Cassini Phase

The Cassini spacecraft made a flyby of Jupiter en route to Saturn while Galileo was orbiting Jupiter. The closest approach of Cassini to Jupiter occurred on 30 December 1999. The arrival of Cassini allowed study of the Jupiter system by two spacecraft at the same time. The special time was called the Galileo-Cassini phase. That phase consisted of two orbits of Galileo that targeted the moon, Ganymede (G28 and G29). Closest approach to Ganymede was 808 km during orbit G28 and 2338 km during orbit G29. The Galileo-Cassini phase lasted from 17 May 2000 to 5 February 2001.

Io Campaign

The Io Campaign consisted of four orbits (C30, I31, I32, and I33). Orbit C30 targeted the moon Callisto with flyby at an altitude of 123 km. The last three orbits targeted the moon, Io, with closest approach distances of 200 km, 181 km, and 100 km, respectively. The Io Campaign lasted from 22 May 2001 to 20 January 2002.

Inner-Jupiter Phase

The Inner Jupiter Phase began with Galileo being placed into a trajectory to impact Jupiter. It was decided to terminate the mission because Galileo's fuel was nearly gone, and it was deemed important not to have an uncontrollable spacecraft make unintentional impact with any of the moons. The phase consisted of one partial orbit (A34) before impact. It was a very elongated orbit with apojove of about 360 Jupiter radii. The orbit included flyby of the moon, Amalthea, early in the orbit on 5 November 2002. Closest approach to Amalthea was 914 km. Impact with Jupiter occurred on 21 September 2003. Galileo entered the cloud tops 0.25° south of the equator at an angle of 20° from the local horizon at a speed of 173,520 km/h.

Conclusion from Flight of Galileo

The Galileo mission was a great success with most of the objectives achieved in spite of failure of the high-gain antenna. Work-arounds to succeed after hardware failure were testament to human ingenuity and perseverance.

Science Instruments on Galileo Orbiter

The Galileo orbiter carried six field and particle instruments on the spun section of the spacecraft and four remote sensing instruments on the despun section. The instruments all operated as intended and returned a wealth of new information about the Jupiter and its four largest moons. A list of field and particle instruments and summary of purpose of each are given in Table 7.2. A list of remote sensing instruments and summary of purpose of each are given in Table 7.3.

Field and Particle Instruments

Dust Detection System

The dust detection system (DDS) was designed to characterize dust particles in the Jupiter system. Characteristics of particles measured included number of particles, mass, speed, direction of motion, and electrical charge. Volcanic activity of the moon, Io, was found to be the source of most of the dust in the Jupiter system. The

Table 7.2 Field and particle instruments carried by Galileo orbiter

Experiment	Purpose	Principal investigator
Dust detection system (DDS)	Measure the amounts and characteristics of interplanetary dust and dust and particles in the Jupiter system	Eberhard Grūn, Max Planck Institute, Germany
Energetic particle detector (EPD)	Determine angular distribution, intensities, and compositions of energetic particles in Jupiter's magnetosphere	J. Williams, Johns Hopkins University, USA
Magnetometer (MAG)	Measure magnetic fields in Jupiter's magnetosphere and in vicinity of moons	Margaret Kivelson, University of California, Los Angeles, USA
Plasma subsystem (PLS)	Measure the characteristic of plasma in Jupiter's magnetosphere and near its moons	Lou Frank, University of Iowa, USA
Plasma wave subsystem (PWS)	Measure the intensity of plasma waves in Jupiter's magnetosphere	Donald Gurnett, University of Iowa, USA

Table 7.3 Remote sensing instruments carried by Galileo orbiter

Experiment	Purpose	Principal investigator
Solid-state imaging camera	Image important areas and objects during the Galileo mission	Michael Belton, National Optical Astronomy Observatory
Near-infrared mapping spectrometer (NIMS)	Determine the composition of Jupiter's atmosphere and of the surface of its moons	Robert Carlson, JPL
Photopolarimeter-radiometer (PMR)	Measure the intensity of thermal energy radiation in Jupiter system	J. Hansen, Goddard Institute for Space Science
Ultraviolet spectrometer (UVS)	Locate cloud layers and analyze cloud particles. Analyze Jupiter's atmosphere and search for atmosphere of moons	C. Hord and Ian Stewart, University of Colorado

dust detector was developed by the Max Planck Institute for Nuclear Physics in Heidelberg, Germany. The principal investigator was Eberhard Grūn.

The dust detector was described in detail in a paper by Eberhard Grūn et al. (1992). The instrument consisted of a cylindrical can-shaped dust collector with grids across the opening and a hemispherical shell referred to as an electron collector at the far end of the cylinder. The hemispheric shell also functioned as a stop for particles. The cylinder was 42.2 cm in diameter and 30.1 cm long. A photograph of the dust detector and its electronics unit is shown in Fig. 7.11. The picture was taken by Dusteg and posted on Wikimedia.

The field of view of the instrument was 140° circular, and the effective aperture was about 0.1 m^2. The dust detector instrument was mounted on the spinning section of the Galileo orbiter. The center of the field of view was directed 55° from the spin axis and pointed in the opposite direction from the high-gain antenna.

A cylindrical element holding an ion collector and a channeltron ion detector was mounted by struts at the center of the entrance to the dust detector. Scaling from drawings, the cylindrical element was about 8 cm in diameter. Three screens, referred to (in order of approaching particle) as entrance grid, charge grid, and shield grid, were located near the entrance as shown in the drawing. The entrance grid and shield grid were connected to the frame and were at zero voltage. The charge grid was connected to a charge-sensitive amplifier.

When a charged dust particle entered the instrument, it would pass through the grids and induce a charge on the charge grid. That charge was converted to a voltage and amplified by the charge amplifier. If the incoming particle were an electron, it would pass through the grids and strike the hemispheric electron collector. The output of the electron collector was applied to a charge-sensitive amplifier.

If the incoming particle was a heavier ion, it would pass through the grids and impact the hemispheric electron collector. The impact would result in impact fragments and impact plasma of ions and electrons. Grids in front of the ion detector had

Fig. 7.11 Photograph of dust detector by Dusteg)

a potential of −350 V applied to attract positive ions. A current pulse of positive ions developed on the grid was applied to a charge-sensitive amplifier.

Positive ions that passed through the grids were gathered by a channeltron photomultiplier. A channeltron photomultiplier is typically a hollow horn-shaped structure coated on the inside with electron-emissive material. When an ion struck the inside walls, it liberated electrons that in turn liberated additional electrons and resulted in an avalanche of electrons at the far end of the tapered structure. The current pulse at the output of the channeltron was applied to a charge-sensitive amplifier.

Velocity of an electron was determined by measuring time delay between the pulse from the charge grid and the pulse from the electron collector. Ion velocity was determined by measuring time delay between the pulse in the charge grid and the pulse from either the ion collector grid or the channeltron.

The output of each of the four charge-sensitive amplifiers was digitized and applied to a signal conditioner. The outputs of the four signal conditioners were applied to a data bus controlled by a microprocessor. The microprocessor was an 1802 type with 3000 bytes of ROM and 2000 bytes of RAM. The microprocessor managed the measurement cycle, collected measured data, and processed the data.

Output data from the dust detector was contained in an experiment data frame. The frame consisted of 128 bits of data and 80 bits of engineering and housekeeping data for a total of 208 bits. One experimental data frame was transmitted every 8.7 s for a data rate of about 24 bps.

Dust Measurements by the Dust Detector

Dust measurements were made within an intense dust storm in space while still 177 million km out from Jupiter in August 1995. The dust detector counted up to 20,000 dust particles per day during the storm. After passing through the storm and approaching closer to Jupiter, the count was about 100 per day. After flyby of the moon, Io, on the approach on 6 December 1995, the count dropped to a very low value because the field of view no longer included Io. This indicated that most of the dust was streaming from Io.

In total, there were 21,224 complete datasets recorded for dust impacts during the entire mission from 1989 to 2003. The datasets included charge, signal rise times, and direction. There were 18,340 complete datasets recorded while Galileo was exploring the Jupiter system from 1996 to 2003.

Energetic Particle Detector (EPD)

The energetic particle detector (EPD) measured the number and energies of high-energy ions and electrons in the magnetosphere of Jupiter. The EPD contained two measurement systems: composition measurement system (CMS) and low-energy magnetosphere measurement system (LEMMS). Both systems used bidirectional telescopes with shielded detectors. The energetic particle detector was mounted on the science boom of the spinning section of the spacecraft.

A drawing from a NASA/JPL document, cropped by author to only show the EPD assembly with the CMS and LEMMS telescopes and the electronics box, is given in Fig. 7.12. The CMS telescope was located on top of the compartment holding the LEMMS telescope. The ends of the bidirectional telescopes were referred to as the 0-degree and the 180-degree ends. The 0-degree end of both telescopes is facing outward in the drawing.

The CMS had a time-of-flight telescope on the 0-degree end and two smaller ion detector telescopes on the 180-degree end. The time-of-flight telescope had a field of view of 18°. One ion detector telescope had a field of view of 23°, and the other had a field of view of 54°. The LEMMS telescope had a field of view of 15° at the 0-degree end and 45° field of view at the 180-degree end.

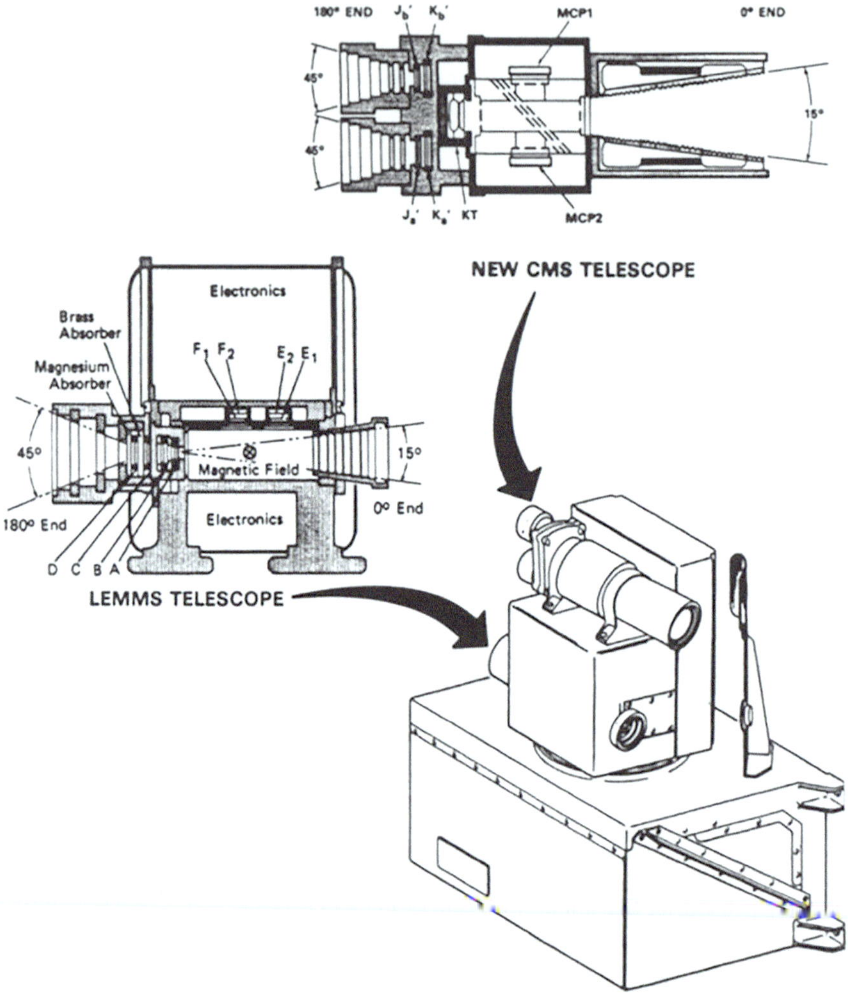

Fig. 7.12 Drawing of EPD assembly with CMS and LEMMS telescopes (NASA /JHU graphic). The upper arrow points to the CMS telescope and the lower arrow points to the LEMMS telescope. The original image was cropped by the author to eliminate other material

The main electronic box under the EPD assembly contained a stepper motor that rotated the structure supporting the telescopes to one of eight different positions in a plane along the top of the main electronics box. The axis of rotation was 90° to the spin axis of the spacecraft. The steps were 30° apart. Angular steps were typically made at the start of each spin revolution. The combination of spacecraft spin and eight angular steps resulted in complete spherical coverage of the environment by both ends of the telescopes. The nominal spin rate of the spinning section was 3.25 rpm.

At one angular step per revolution and eight steps, complete spherical angular coverage was obtained at each 147 s.

CMS Telescope

The CMS telescope was designed to determine compositions of ions in the Jupiter magnetosphere with energies between 10 keV/nucleon and 10 MeV/nucleon. The 0-degree end of the CMS telescope used time-of-flight and total energy to identify elements helium through iron. A magnet at the entrance deflected electrons and prevented them from entering the system. Time of flight was measured between time an ion struck and passed through an entrance foil and when it struck a detector some distance away. Ions striking the entrance foil generated a shower of electrons, which were detected by a microchannel plate to generate a start pulse. A stop pulse was generated when the ion struck the detector. The amplitude of the signal from the detector was a function of energy of the ion. The mass of the ion could be calculated from knowledge of total energy and velocity. The mass would identify a particular element.

The 180-degree end of the CMS telescope held two small telescopes to measure the energy of ions. Two were provided for redundancy. One had a field of view of 23°, and the other had a field of view of 54°. Those telescopes were arranged to measure substantially higher energies than the time-of-flight telescope. The telescopes measured energy loss in passing through a detector and then residual energy of the ion when it struck another detector. Each of the telescopes had a thin detector at the front and a thick detector a short distance behind it. The front detector was about 5.5 μm thick, and the following detector was 185 μm thick. Measurements of energy lost in the front detector and residual energy measured by the second detector were used to determine ion composition.

LEMMS Telescope

The low-energy magnetospheric measurement system (LEMMS) telescope was designed to measure lower-energy ions and electrons than the CMS telescope. It measured electron energies from 15 keV to 11 MeV and ions from 22 keV to 55 MeV. LEMMS contained back-to-back telescopes. The telescope in the 0-degree end had a field of view of 15°, and the telescope in the 180-degree end had a telescope with a field of view of 45°.

Particles entering the 0-degree telescope passed through baffle plates that established seven hexagonal channels that were directed towards a detector. That detector, which faced the input aperture, was referred to as detector A. It was 102 μm thick. Detector B was located behind detector A. Magnets beyond the deflection plates deflected electrons towards detectors E and F, which were located perpendicular to the boresight of the telescope.

Ions striking detector A were measured in eight different particle rate channels ranging in energy from 22 keV to 3.5 MeV. Additional ranges were developed by noting the coincidence between outputs of detector A and detector B. The output of detector A was digitized to provide a 46-channel energy spectrum covering the energy range of 22 keV to 290 keV.

The telescope in the 180-degree end was arranged for measurement of higher-energy particles. The telescope contained two detectors referred to as C and D. Both detectors were 100mm^2 in area and 500 µm thick. Particles struck detector D first, passed though it, and struck detector C. A magnesium shield 2 mm thick was placed in front of detector D, and a brass shield 3.2 mm thick was placed between detectors D and C. The detectors were each followed by two discriminator circuits with threshold settings for electrons of 2 Mev for detector D and 11 MeV for detector C. Ion (proton) thresholds were 14.5 MeV for detector D and 51 MeV for detector C.

Results of EPD Measurements

Parameters of the radiation belts of Jupiter as measured by the EPD experiment were given by H. R. Garrett and I. Jun (2020). A few data points from that research are given in this section. The study provides plots of counts of electrons with energies >11 MeV as a function of distance in Jupiter radius (R_J) from 8 to 25 for all 34 orbits of Galileo. At a distance of 8 R_J, the average count was about 10^5/s. At 12 R_J, the count was about 3×10^3/s, and at 20 R_J, it was about 30 counts/s.

A plot was also provided for differential flux of protons over the same distance range. The dimension of flux was protons/(cm^2-s-MeV).

The average flux from 34 orbits was about 10^6 at 8 R_J, 5×10^4 at 12 R_J, and 5×10^2 at 20 R_J.

Magnetometer (MAG)

Magnetometers on the Galileo spacecraft were used to measure the ambient magnetic field of Jupiter and fluctuations of the field. It was also used to map Jupiter's far-ranging magnetosphere and determine whether the moons of Jupiter have a magnetic field. Two sets of triaxial magnetometers were mounted on a deployable magnetometer boom. One set was mounted at the end of the boom, 11 m from the spin axis. The other magnetometer was mounted inboard on the boom, 6.9 m from the spin axis.

The magnetometers were fluxgate types. Fluxgate magnetometers have been successfully used for decades on aircraft and spacecraft to measure magnetic fields. Typically, flux gate magnetometers employ a highly permeable toroidal core that is driven in and out of saturation by an alternating current in a drive winding around the core. A sense winding around the core detects a change in the symmetry of core saturation as would be caused by an external magnetic field. The presence of an external magnetic field induces distortion of the waveform in the sense coil and generates harmonics. The amplitude of the second harmonic is proportional to the strength of the external magnetic field. Another winding around the coil, referred to as the feedback winding, is energized by current to null the influence of the external magnetic field. The amount of current necessary to null the field is proportional to the strength of that component of the magnetic field. The amount of feedback current is measured and digitized and becomes the output of the magnetometer.

Each magnetometer had a set of three orthogonal sensors to measure the three orthogonal axes of the magnetic field. One of the sensors was aligned with the spin axis, and the other two were orthogonal in the spin plane. The Galileo magnetometers had a "flipping" feature that rotated two of the sensors and allowed the sensor aligned with the spin axis to be exchanged with one of the sensors in the spin plane. The flipping feature could be used to determine residual offset in the sensor aligned with the spin axis.

Galileo used a deployable boom 8.2 m long mounted at the end of the science boom to space the magnetometers away from incidental magnetic fields of the spacecraft. The boom had two independent and deployable masts that were extended in series to form the 8.2-m-long boom. The two masts were coiled and contained in a launch canister before deployment. The launch canister, which was a cylinder 0.6 m long, was attached to the science boom. The inboard mast was deployed first from the canister, and then the outboard mast was deployed from the end of the inboard mast. The outboard magnetometer was attached to the end of the outboard boom. The inboard magnetometer was attached to the inboard mast near the junction of the two masts.

An electronic unit, to service the magnetometers, contained circuits to energize the drive and feedback windings and to process signals from the sense winding. The six analog outputs of the fluxgate sensors were digitized by a 12-bit analog-to-digital converter and further processed. The electronics unit was controlled by an 1802-type microprocessor with 4 kB of ROM and 4 kB of RAM. Output data from the electronics unit was in the form of a 160-bit word. The word contained nine 16-bit data words and a 16-bit sub-commutator word.

Each of the magnetometers had two different settings for dynamic range. The outboard magnetometer had dynamic ranges of ±512 nT and ±16,384 nT. The inboard magnetometer had dynamic ranges of ±32 nT and ±512 nT. The expression "nT" stands for nanoTesla or 10^{-9} Tesla.

Onboard calibration of the magnetometers was performed twice a year while in orbit around Jupiter. A precise local magnetic field was generated by a coil of wire located near the base of the boom with its axis aligned with that of the boom. A field strength of 4.5 nT was generated at the location of the outboard magnetometer. The field could be square-wave modulated at frequencies of 0.5 or 0.75 Hz to distinguish the calibration signal from ambient magnetic fields. A frequency of 1.5 Hz was also available with a 240-degree portion of a sine wave transmitted at each cycle. Calibration was performed for each of the flip positions of each magnetometer unit.

Measurements by the Magnetometers

Magnetometers in the Galileo spacecraft detected the magnetosphere of Jupiter while still nine million km out from the cloud tops. The magnetosphere extends one to three million km in the direction of the sun where it is compressed somewhat by the solar wind. The magnetosphere has a long tail that extends more than one billion km in the anti-sun direction.

Magnetometer measurements during flybys of Ganymede indicated that the large moon has its own magnetosphere within the huge magnetosphere of Jupiter. Analysis of magnetometer data indicated that the magnetic field of Ganymede is a dipole type with axis through the north pole tilted 10° from the spin axis of the moon. The magnetic field strength at the surface at the equator of the moon was determined to be about 750 nT.

Magnetometer measurements during 11 flybys of the icy moon Europa indicated disruption of the magnetic field of Jupiter in the vicinity of Europa. Scientists concluded that the most likely cause was currents induced in an electrically conductive medium, such as a salty ocean below the surface, and those currents induced a magnetic field.

Galileo found that the moon, Io, does not have a magnetic field despite its roiling molten core. The immense tides experienced during Io's orbit of Jupiter deform the core and cause extreme heating.

Plasma Subsystem (PLS)

The purpose of the plasma subsystem (PLS) was to measure the density, temperature, direction of flow, and composition of low-energy plasmas in the Jupiter system. The plasma torus associated with the moon, Io, was investigated in detail. The PLS instrument contained two electrostatic analyzers with different fields of view. Each determined energy per unit charge for positive ions and for electrons. The analyzers also determined the direction of flow. The electrostatic analyzers were followed by seven sensors and three mass spectrometers. Each sensor and each mass spectrometer responded to different angles of arrival of particles. The mass spectrometers measured mass per unit charge and were used to determine species of positive ions.

There were two electrostatic analyzer systems with different fields of view, referred to as "A" and "B." Each electrostatic analyzer contained three segments of a spherical shell nested together with gaps between them. The shells were 70-degree segments of a sphere. The inner and outer shells were at ground potential, and the center shell had a stepped positive voltage applied. The gap between the inner shell and center shell was used to analyze positive ions and the gap between the outer shell and the center shell was used to analyze electrons. An electron or a positive ion entering the analyzer with a particular energy-to-charge ratio would pass through the curved plates only if the voltage applied to the center plate was the proper value to deflect the ion or electron in a curved path that did not strike the walls. The voltage was stepped through 64 logarithmic values to develop a spectrum of energy-to-charge values from 0.9 to 52,000 eV.

The spherical shells of the two electrostatic analyzers were nested together to make a more compact design. The radius of the inner shell of the innermost electrostatic analyzer was 9.68 cm, and the radius of the outer surface of the outer shell of the outermost analyzer was 12.57 cm.

Particles at the output of the electrostatic analyzer struck one of seven sensors for positive ions and one of seven sensors for electrons. Each sensor system had a different field of view. The result was a fan-shaped viewing angle (with gaps) in the vertical plane of the spin axis. The plasma subsystem was mounted on the science boom on the spinning section of the spacecraft. As the spacecraft spun, about 80% of the sky was viewed by the fan beam of the plasma subsystem.

Electrons and positive ions exited the electrostatic analyzer at angles corresponding to their arrival angles at the apertures of the electrostatic analyzer. Seven sensors were placed at seven different angular positions. Each position had separate sensors for the electron channel and for the ion channel of the analyzer. The particles were applied to a spiral-type electron multiplier that had a gain of about 1×10^8. The charge at the output of the electronic multipliers was collected and applied to a pulse amplifier. A minimum threshold of the amplifiers was set at 1×10^6 electrons.

The instrument contained three mass spectrometers. Two were associated with electrostatic analyzer B, and one was associated with electrostatic analyzer A. Charged particles at the output of the electrostatic analyzer at angles corresponding to the locations of the mass spectrometers were processed. One of the mass spectrometers of channel B had its center of the field of view directed normal to the spin axis, and the other was directed upward about 25° from the positive spin axis. The mass spectrometer associated with channel A looked downward about 25° from the negative spin axis.

The mass spectrometers used magnetic deflection to determine mass per unit charge of ions. A circular electromagnet with a gap in the core was used to generate a controllable magnetic field. Ions passing through the gap were deflected by an amount proportional to the ratio of mass to charge of the ion. A varying magnetic field was set up in the gap by controlling the current through the coil of the electromagnetic in 64 steps.

In the absence of a magnetic field, ions passed through the gap undeflected and passed through a slit in front of a sensor. The sensor was a spiral-type electron multiplier followed by a detector. That channel was referred to as the integral channel because it responded to all ions. A second slit located 3.3 mm from the first was used as a capture point for deflected ions. That slit was followed by a spiral-type electron multiplier followed by a detector. That channel was referred to as the differential channel because it only responded to ions of a particular mass-to-charge ratio.

Results of PLS Measurements

The PLS surveyed plasma trapped by the magnetic fields in the magnetosphere of Jupiter at distances from 5 Jupiter radii (R_J) to 30 R_J. A sharp decrease was found in plasma density with distance. The plasma density was about 3000/cc at 6 R_J and only 0.05/cc at 30 R_J. A flyby of the huge moon, Ganymede, showed that it had a magnetic field and associated small magnetosphere. Galileo PLS detected and measured charged particles in that plasma.

Eruptions from the moon, Io, contribute much of the particle matter around Jupiter. The material becomes ionized by ultraviolet radiation and forms a dense plasma torus around Jupiter. The torus is centered on the orbit of Io, which is at a distance of 5.9 R_J from the center of Jupiter. The Galileo spacecraft passed about 1000 km from Io and on through the plasma torus on 7 December 1995. This was a few hours before Galileo was inserted into orbit around Jupiter. A paper by L. A. Frank et al. (1992) gives some results of the initial pass by Io. A summary of 13 sets of data were presented between distances of 7.62 R_J and 6.09 R_J. A few data points from Table 7.1 of that reference are given in Table 7.4. Oxygen, O^+, is the dominant ion by far at a closer range to Jupiter.

Plasma Wave Subsystem (PWS)

The plasma wave subsystem measured the intensity of plasma waves in Jupiter's magnetosphere. Plasma waves are generated by changes with time of electrostatic and magnetic fields within the plasma surrounding Jupiter.

The PWS used a dipole antenna mounted near the end of the magnetometer boom to detect electric fields. The span of the dipole was 6.6 m from tip to tip. The axis of the dipole was perpendicular to the magnetometer boom and perpendicular to the spin axis of the spacecraft.

The elements of the dipole, which were built from graphite epoxy, were hinged so they could be folded to fit within the launch shroud. A separate preamplifier for each element was located at the base of the antenna.

Magnetic fields were detected by two magnetic antennas mounted to the feed structure for the high-gain antenna. Both magnetic antennas consisted of high-permeability rods wound with wire search coils. One magnetic antenna was designed

Table 7.4 Densities of electrons and heavy ions in the Io plasma torus

	Density, number per cubic centimeter				
Distance, R_J	Electrons	S^+	S^{++}	O^+	O^{++}
7.62	583	0–70	0–34	25–226	99–230
6.09	3778	103–531	118–389	1783–2299	255–673

Here, R_J is the radius of Jupiter = 71.492 km, S^+ and S^{++} are positive sulfur ions, and O^+ and O^{++} are positive oxygen ions for low frequencies in a range of 5 Hz to 3500 Hz. The other magnetic antenna was designed for high frequencies in the range of 1 kHz to 50 kHz. The length of the high-permeability rod for the low-frequency antenna was 25.5 cm, and the length was 27.5 cm for the high-frequency antenna. The high-frequency magnetic antenna was oriented perpendicular to the electric dipole antenna. The low-frequency magnetic antenna was oriented parallel to the electric dipole antenna.

A main electronics unit, located near the base of the magnetometer boom, accepted signals from the electric and magnetic antennas and processed the signals. Signal processing was performed by two functional systems. One was a low-rate system that consisted of three spectrum analyzers, and the other was a high-rate system that provided waveform measurements in three frequency bands.

The spectrum analyzer channels were identified as high frequency, medium frequency, and low frequency. Characteristics of those spectrum analyzer channels are given in Table 7.5. The outputs of the spectrum analyzers were digitized by analog-to-digital converters and assembled into 18 8-bit words. An 8-bit engineering word and an 8-bit status word were added for a total output of 20 8-bit words sent to the spacecraft command and data subsystem.

The high-rate system measured signal waveforms in three frequency bands: 5 Hz to 1 kHz, 50 Hz to 10 kHz, and 50 Hz to 80 kHz. The sample rates for the three bands were 3150, 25,200, or 201,600 samples/s, respectively. The sampled signal was digitized by a 4-bit analog-to-digital converter and sent to the spacecraft tape recorder. The particular frequency band used was commanded by the spacecraft command and data subsystem.

Switches in the electronics unit allowed selecting either the electric dipole antenna output or the outputs of the magnetic antennas for analysis. The electric field signals were applied to the three spectrum analyzers and to the wideband waveform processor. The magnetic antennas did not support high frequencies, so only the medium-frequency and low-frequency spectrum analyzers and the waveform analyzer were used for magnetic field signals.

Table 7.5 Characteristics of PWS spectrum analyzers

Channel	Frequency range	Number of channels	Bandwidth of channel, Hz	Time for sweep, s
High	100.8 kHz to 5.65 MHz	42	1340	18.67
Medium	42.1 Hz to 160.8 kHz	4 bands with 28 channels in each band	Bandwidth 1: 14.3 Bandwidth 2: 6.7 Bandwidth 3: 120 Bandwidth 4: 1510	18.67
Low	5.6 Hz to 31 Hz	4 logarithmically spaced	Channel 1: 0.83 Channel 2: 1.86 Channel 3: 2.75 Channel 4: 4.79	2.67

Results of PWS Measurements

Frequency measurements by the spectrum analyzers of PWS were used to determine electron density at select points in the Jupiter system. Measurements were made in the vicinity of some of the moons. The spectrum data was transmitted to Earth where computation for electron density was done.

The upper hybrid frequency, f_{UH}, was measured, which is a mix of electron plasma frequency, f_P, and electron cyclotron frequency, f_C. The expression relating these frequencies is $f_{UH} = (f_P^2 + f_C^2)^{1/2}$. The electron cyclotron frequency, f_C, is equal to 28xB, where B is the magnetic field in nanoteslas and the constant 28 includes the mass and charge of an electron. The electron plasma frequency, f_P, is equal to 8960 $(Ne)^{1/2}$, where Ne is the electron density per cubic centimeters (cc) and the constant 8960 includes the mass and charge of an electron.

Electron densities were determined during the flight of Galileo through the Io torus in the vicinity of the moon, Io, on 7 December 1995. Measurements were made from a distance of 7.7–5.4 R_J inbound to Jupiter and from a distance of 5.3–4.3 R_J outbound. The electron density was 628/cc at a distance of 7.7 R_J and 3775 at a distance of 6 R_J. A sharp peak of 40,766/cc was measured at the time of closest approach to Io of 900 km.

Seven flybys of the moon, Callisto, were made by Galileo. Measurements made during orbit C10 resulted in a broad peak of electron density of about 4000/cc near the closest approach to Callisto. Flybys of the large moon, Ganymede, indicated evidence of a magnetosphere. PWS measurements were used to determine a broad peak of electron density of about 40/cc in the vicinity of the closest approach to Ganymede.

Remote Sensing Instruments Carried by Galileo Orbiter

There were four remote sensing instruments mounted on a scan platform on the despun section of the Galileo orbiter. The instruments were solid-state imaging camera, near-infrared mapping spectrometer, ultraviolet spectrometer, and photopolarimeter-radiometer. The scan platform, including a scan actuator, was mounted on a short boom on the despun section of the spacecraft. The scan actuator contained two torque motors for redundancy and a 16-bit optical encoder that reported scan position. A honeycomb plate was mounted to the flange of the rotor of the scan actuator, and the four remote sensing instruments were mounted to the plate. The boresights of the four instruments were parallel. The scan platform was mounted so that the scanning direction was in the plane of the spin axis of the spacecraft. The scanning range was 210°. The despun section of the spacecraft could be rotated to any angle about the spin axis. Assuming spin axis along the vertical, the instruments could be pointed to any angle in the azimuth plane and through 210° in the elevation plane.

Solid-State Imaging Camera

The solid-state imaging (SSI) camera was one of the four remote sensing instruments carried on the scan platform. The camera was an 800 line by 800 column solid-state imager that used a charge-coupled device (CCD) detector element. The camera employed a Cassegrain telescope with an aperture of 176 mm and a lens with 1500 mm focal length. The field of view was 0.46° circular. The camera was designed and assembled by JPL.

The camera had an eight-position filter wheel and a shutter in the optical path just before the focal point of the telescope. One of the eight filter positions was clear. The center wavelengths of the other filters were 413 nm (violet), 559 nm (green), 665 nm (red), 731 nm (methane 1), 756 nm (near infrared), 887 nm (methane 2), and 986 nm. The 731 nm filter was at a weak methane absorption band, and the 887 nm filter was at a strong methane absorption band. Filters from 734 nm to 986 nm were in the infrared range of the light spectrum. The CCD detector responded to the amplitude of light at each pixel. Color images were synthesized on the ground by combining images taken by the cameras through various filters.

Exposure time of the camera was controlled by a shutter. The exposure time could be set at 28 selectable times from 0.004 to 51 s. The camera was pre-flashed to remove residual images after the image had been read out. This was done by flooding the camera with light from a 930 nm source several times and then reading out the CCD to clear it.

After exposure, the amplitude of the charge of each of the 640,000 pixels in the CCD was read out as an analog video signal. The video signal was amplified and then digitized by an 8-bit analog-to-digital converter. Each full image frame was represented by 5.12 million bits of data. The time allocated to expose the image and transfer a full frame of data was 8.66 s. The bit stream was transferred to the tape recorder for temporary storage. The transfer rate to the recorder was 8.064×10^5 bps.

Data was read out of the tape recorder and underwent data compression to reduce the bit rate. Data compression was necessary to transfer image data to Earth because the high-gain antenna could not be deployed and the communication link could only accommodate low bit rates with the low-gain antenna. The camera had a compression feature called block adaptive rate compression built into the camera system. Some of the data was compressed using that system, but much of the data was compressed by using an integer cosine transform (ICT) compression system that was loaded into the system software. The ICT used 8 × 8-pixel blocks.

Because of failure of the high-gain antenna and limited capacity of the tape recorder, imaging was restricted to the most important items around Jupiter and its many moons. During the 2-year primary mission, 1645 images were taken of Jupiter and its moons. Additional images were taken during the 4-year extended mission.

An image of the moon, Io, taken on 3 July 1999 from a distance of 130,000 km is given in Fig. 7.13. The red splotches in the image were thought to be active volcanos. Close-up images showing eruption and lava flow from the Tvashtar Catena on Io at two different times are given in Fig. 7.14.

Fig. 7.13 Image of moon, Io, taken by Galileo in 1999 (credit: NASA/JPL/University of Arizona)

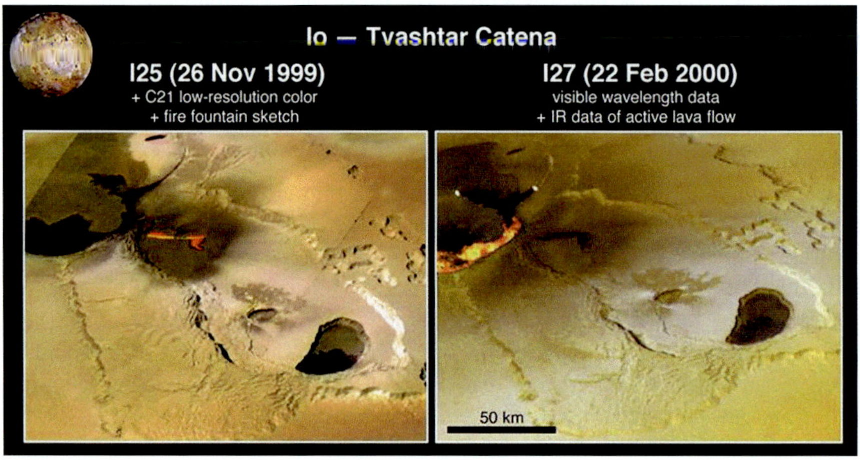

Fig. 7.14 Lava flow at Tvashtar Catena on Io at two different times (credit: NASA/JPL/University of Arizona)

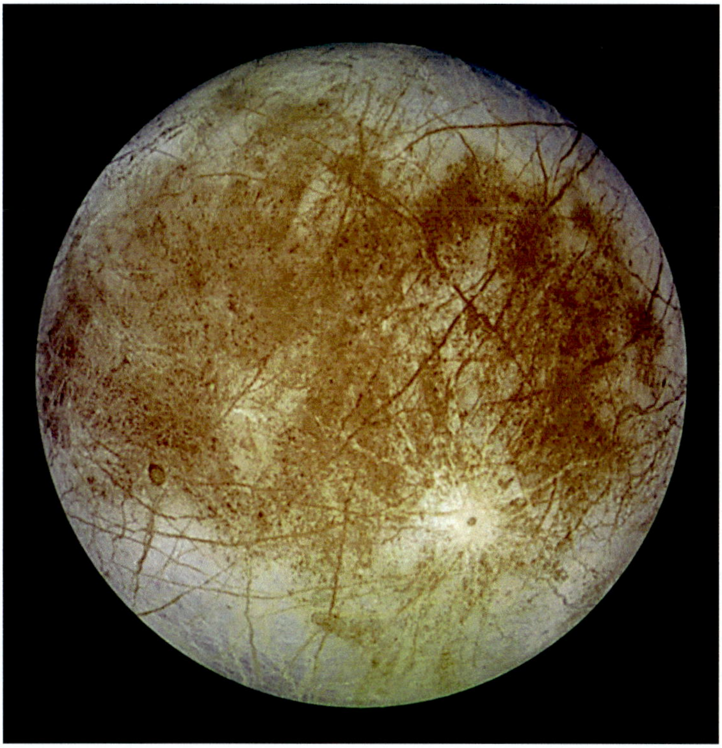

Fig. 7.15 Image of Europa taken during primary Cassini mission (credit: NASA/JPL/German Aerospace Center)

A picture taken of the moon, Europa, by Galileo's solid-state imaging instrument is shown in Fig. 7.15. The picture was taken from a distance of 677,000 km during the primary mission. Long cracks appear in the surface, and refrozen ice creates a jumble on the surface.

Near-Infrared Mapping Spectrometer (NIMS)

The near-infrared mapping spectrometer (NIMS) was composed of an imager and a spectrometer that made simultaneous measurements. It mapped mineral deposits on the surface of the moons and determined composition of the deposits. It investigated distribution cloud layers and determined constituents in the atmosphere and temperature as a function of altitude.

The NIMS used a Ritchey-Chrétien-type telescope with a scanning secondary mirror. The focal length was 800 mm, and the diameter of the input aperture was

Fig. 7.16 Photograph of near-infrared mapping spectrometer (NASA image)

22.8 cm. The output of the telescope was directed at a plane grating spectrometer that disbursed the light across a field of focal plane detectors. The detectors were cooled to about 64 K by a radiator that viewed cold space. The spectral wavelength range of the instrument was 700 nm to 5200 nm.

A photograph of the NIMS instrument from a NASA/JPL document is shown in Fig. 7.16. The telescope is at the right in the photograph, and the radiation cooler is at the left. A schematic diagram of the instrument from the same NASA/JPL document is shown in Fig. 7.17.

The NIMS was mounted on the scan platform on the despun section of the orbiter. The boresight was parallel to that of the solid-state imager. The field of view of the telescope was scanned in a plane parallel to the spin axis of the orbiter by rotating the secondary mirror in the telescope.

The spatial scan consisted of 20 steps of 0.5 milliradian (mR) increments for a total scan extent of 10 mR. A full scan cycle was a scan upwards for 20 steps and then a scan downwards for 20 steps.

A slit 400 µm wide in the plane of the mirror scan was located at the focal point of the telescope to act as a field stop. It set the field of view in the plane perpendicular to the scan plane of 0.5 mR. A tuning fork-type light chopper was located in the light path just before the field stop. The frequency of the chopper was 63 Hz. The duty ratio of light gated on by the chopper was 50%. The dark time was used to measure the dark current of the detectors. Positioning of the scanning mirror and positioning of the diffraction grating were made during the dark time. The scanning mirror was advanced to the next step position at each cycle of the chopper.

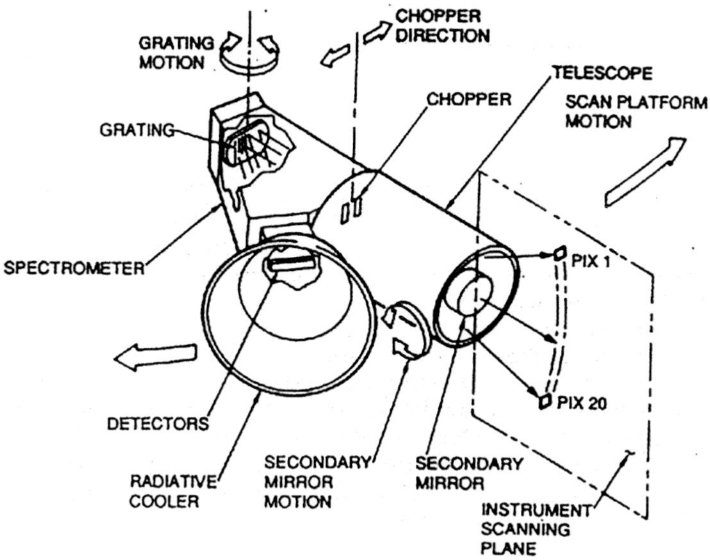

Fig. 7.17 Schematic diagram of near-infrared mapping spectrometer (NASA graphic)

A collimator following the field stop illuminated the diffraction grating. The diffraction grating could be rotated through 24 grating positions to generate a full wavelength spectrum. The disbursed light from the grating was collected by a flat-field camera that focused the image at the field stop onto a focal plane assembly. The focal plane assembly contained 17 detectors. Fifteen were indium antimonide (InSb) detectors, and two were silicon (Si) detectors.

During the 20-step scan of the secondary mirror in one direction, the diffraction grating was fixed at one rotation value. At the end of the scan, the diffraction grating was changed to the next angular position, and the mirror was scanned in the opposite direction. This process continued until all of the rotation settings of the diffraction grating had been made. The size and number of grating steps could be commanded from the ground.

There were 11 different NIMS modes labeled 1 through 11. Some of the modes, with the word "MAP," in the title had the spatial mirror scanning. Other modes, with the word "SPECTRUM," in the title held the spatial mirror in one position. One commonly used mode was Mode 1, "Full Map." That mode consisted of 12 diffraction grating positions with a gap of one grating position between each active position. There were 20 steps of the spatial mirror during each grating position. The time required for one cycle for Mode 1 was 4.33 s. Mode 3, referred to as "Long Map," used all 24 rotation positions of the diffraction grating. The time for a complete cycle for Mode 3 was 8.67 s.

Signals from each of the 17 detectors were processed for each step of the scanning mirror and each position of the diffraction grating. The focal plane assembly, which held the detectors, also held a field effect transistor preamplifier for each detector. An electronics assembly processed the signals from the 17 detector channels. It was mounted on the scan platform near the optical units.

The electronics assembly contained a variable gain amplifier followed by an integrator for each channel. The integration time was the dwell time for a particular reading. The output of the integrator was applied to an analog multiplexer. The multiplexer selected each channel in turn and applied the signal to a 10-bit analog-to-digital converter.

The operating modes of the NIMS and signal processing tasks were controlled by an RCA 1802 microprocessor in the electronic assembly. The microprocessor formatted science and engineering data for telemetry and communicated with the spacecraft command and data subsystem via the spacecraft data bus.

Results of NIMS Experiment

A global thermal map of the moon, Io, produced by the NIMS is shown on the right side of Fig. 7.18. The image on the left is of the moon in visible light taken by the SSI camera. The NIMS image, which was taken on 13 October 2001, is shown in false color to bring out temperature differences. The many hot spots on Io show up as white, yellow, and red. Colder areas are blue. The left half of the NIMS image was in darkness when the image was made. The two hot spots at the lower left-hand side of the image are the volcanos Pele and Pillan.

Scientists were able to use images from NIMS at different wavelengths to determine the type of material on the surface of the moons and the grain sizes. Images of Ganymede showing different materials are given in Fig. 7.19. The left image was

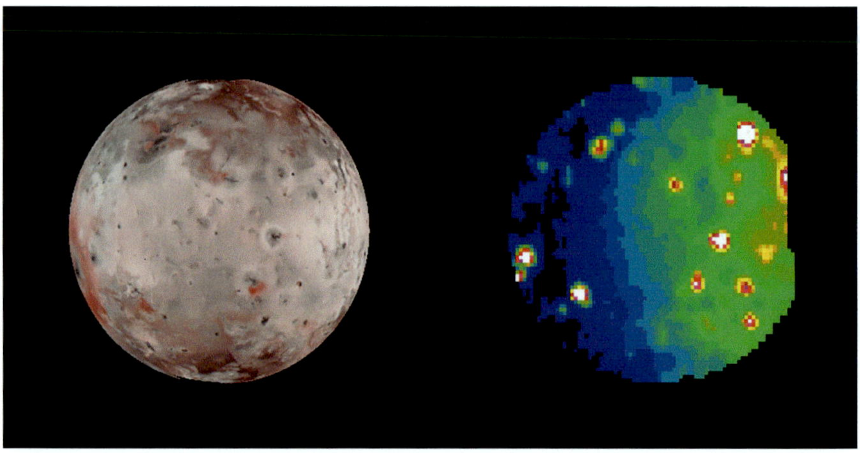

Fig. 7.18 Image of Io taken by SSI in visible light and by NIMS in infrared light. The infrared image is in false colors to show temperature difference. (NASA image)

Fig. 7.19 Middle image by NIMS shows abundance of water ice on Ganymede. Right image shows abundance of minerals (NASA image)

taken by the Voyager spacecraft in visible light. The middle image was taken by NIMS at a wavelength to distinguish water ice. The dark areas have less water ice, and the bright areas have more. The right image was taken at a wavelength to discern minerals on the surface. Minerals are shown in red, and size of ice grains is in shades of blue.

Photopolarimeter-Radiometer (PPR)

The photopolarimeter-radiometer (PPR) measured the distribution of particles in the clouds and haze of Jupiter and determined the thermal structure of the atmosphere. The thermal structure included absorbed solar energy. The PPR was also used to measure and map photometric, polarimetric, and thermal radiometric properties of Jupiter's moons.

A photograph of the photopolarimeter-radiometer from a NASA/JPL document is shown in Fig. 7.20. The instrument was mounted on the scan platform on the despun section of the orbiter. The boresight of the PPR was parallel to that of the other instruments on the scan platform.

The PPR used a series of filters to cover wavelength spectral bands for polarimetry, photometry, and radiometry. Polarimetry used filters with center wavelengths of 410, 678.5, and 944.6 nm. Photometry used seven filters spanning wavelengths from 618.7 nm to 891.8 nm. Radiometry used seven filters spanning wavelengths from 0.3 μm to 110 μm. One of the radiometer channels, referred to as "solar–thermal," had a broad filter covering 0.3–110 μm. Another channel, referred to as "solar," covered 0.3–0.4 μm. Taking the difference between the latter two channels was used to determine radiative energy balance for Jupiter.

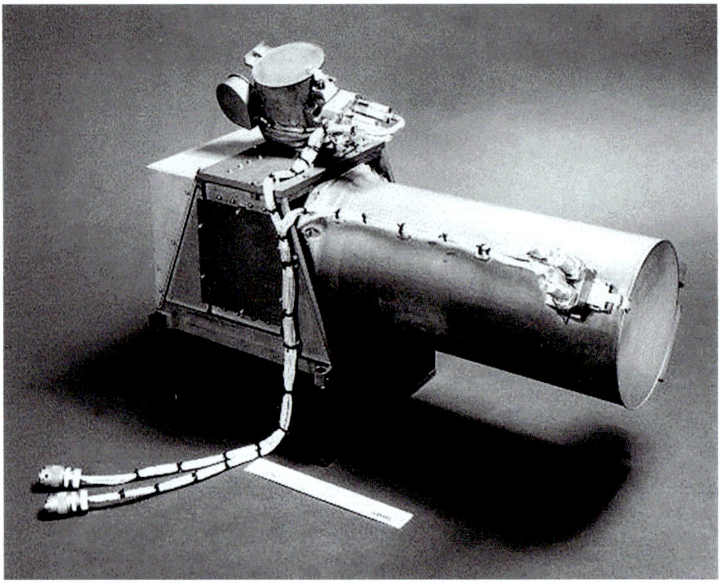

Fig. 7.20 Photopolarimeter-radiometer for Galileo orbiter (NASA image)

The photopolarimeter-radiometer was equipped with a telescope to examine narrow angular regions. The telescope was a Cassegrain type with an input aperture of 10 cm and focal length of 500 mm. The field of view was set by a field stop to 2.5 mR circular. An optical chopper was placed just before the field stop to alternately present the output of the telescope or a view of cold space to a filter in a filter wheel during radiometric measurements. The chopper was in the pass-through position for photometric and polarimetric measurements.

A 32-position filter wheel was positioned just behind the field stop. During photometric and polarimetric measurements, light at the output of the field stop traveled through a selected filter to a Wollaston prism. That prism separated polarized light into two orthogonally polarized beams. The two beams were detected by separate silicon photodetectors. The ratio of the outputs of the two detectors was used to determine polarization angle. The two detectors were also used for photometry measurements.

The filter wheel could be stepped through 32 positions. The positions, which were numbered 0 through 31, were separated by 11.25° in angle around the wheel. The filter wheel contained three groups of three filters each for polarimetric measurements. Half-wave retarders that shifted the polarization by 90° were part of these filter units. The three groups of filters had center frequencies of 945 nm, 678 nm, and 410 nm. The 945 nm channel had three filters separated by a blank space such that the filters were at positions 1, 3, and 5 on the wheel. The 678 nm channel had filters at positions 7, 9, and 11, and the 410 nm channel had filters at positions 13, 15, and 17.

Filter wheel positions 18 through 24 were used for radiometry. The filters for the seven radiometer channels were fitted with small mirrors that directed the output of each filter towards a lens system that focused the light onto a lithium tantalite pyroelectric detector. The range of wavelengths extended from 16.8 μm for filter 18–35.5 μm for filter 20. Filter 21 was centered at 27.5 μm. Filter 22 had a wavelength span of 45–110 μm. Filter 23 had a span of 0.30–4 μm. Filter 24 had a span of 0.3–110 μm.

One of the objectives of the radiometric measurements was to determine the radiative energy balance of Jupiter. Measurements in the 0.3–4 μm channel (filter wheel position 23) represent solar contribution, whereas measurements in the 0.3–110 μm channel (filter wheel position 24) represent solar contribution plus thermal energy from Jupiter. The difference between the two measurements was used to determine the radiative energy balance.

Filters 25 through 31 were used for photometry. The range of wavelengths extended from 618.7 nm for filter 25 to 891.8 nm for filter 31. Filters at wavelengths of 618.7, 840.3, and 891.8 nm were located in absorption bands of methane (CH_4). Filters at wavelengths of 648.0 and 788.7 nm were located in absorption bands of ammonia (NH_3).

Preamplifiers were mounted near the silicon detectors for the polarimetry and photometric channels and near the lithium tantalite detector for the radiometric channel. The outputs of the three preamplifiers were applied to gain controlled amplifiers. The gain was controlled by microprocessor-based electronics to keep the amplifiers in linear operating ranges. The amplified and conditioned signals from the three detector channels were multiplexed into a 12-bit analog-to-digital converter. The digital signals were applied to a CDS interface adapter and on to the spacecraft command data system. Commands from the CDS were received by the CDS interface adapter and passed on to the microprocessor-based electronics to control the PPR.

Results of Photopolarimeter-Radiometer Experiment

The PPR was used to measure and map thermal radiation from Jupiter and its moons. The moon, Io, which has a molten subsurface crust due to immense tidal forces from Jupiter and influence of the moons, Europa and Ganymede, was thermally mapped in some detail. Io was found to have several active volcanos and over 100 localized hot spots. The volcano Loki was the hottest and largest of the active volcanos.

A contour map of temperatures on a sector of the southern hemisphere of Io at night is shown in Fig. 7.21. The contours are 2.5 K apart. The contours are superimposed on an image taken in visible light by Galileo's camera. Contours at temperatures of 110 K and above were deleted for clarity. The map includes volcanos Pele (Pe), Pillan (Pi), and Babbar (Ba).

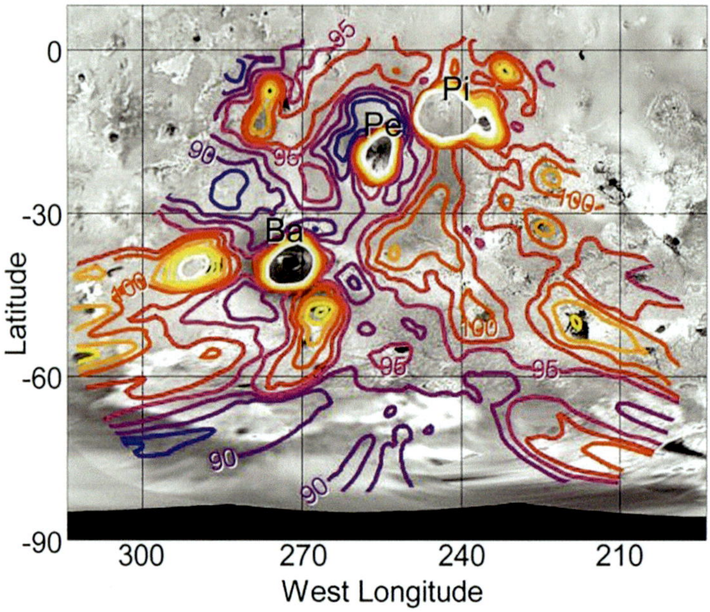

Fig. 7.21 Contour map of temperatures on nightside of Io (Credit NASA/JPL)

A global thermal map of the nightside temperature of Io is given on the left side of Fig. 7.22. A dayside image taken by Galileo's camera is shown on the right side. Volcanos in the image include Loki (L), Pele (Pe), Pillan (Pi), and Marduk (M).

Ultraviolet Spectrometer

The ultraviolet spectrometer consisted of two units: an ultraviolet spectrometer (UVS) and an extreme ultraviolet spectrometer (EUVS). The UVS was mounted on the scan platform on the despun section of the Galileo orbiter. The EUVS was mounted on the spinning section of the orbiter. The ultraviolet spectrometer covered the wavelength band 113–432 nm. The extreme ultraviolet spectrometer covered the wavelength band 54–128 nm. The purpose of the two instruments was to investigate the upper atmosphere of Jupiter, investigate volatile gases emanating from the moons, and investigate the Io plasma torus. The instruments probed constituents by generating spectra of emission, scattering, and absorption in the Jupiter system at ultraviolet and extreme ultraviolet wavelengths.

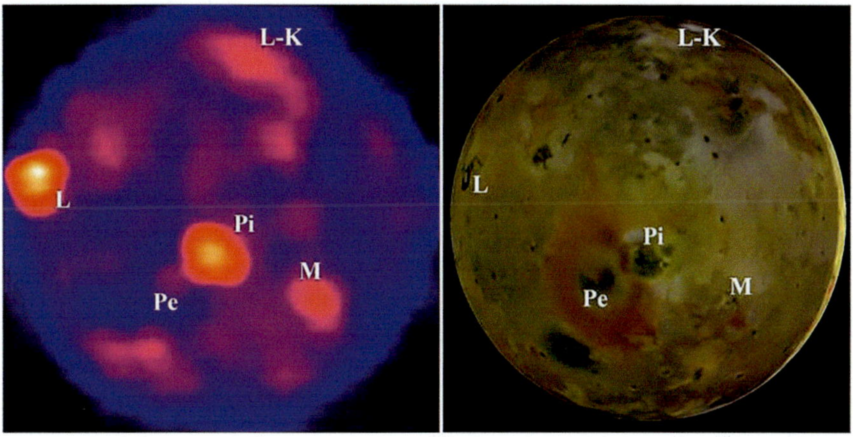

Fig. 7.22 Thermal map of nightside of the moon, Io, taken by Galileo's PPR (Credit NASA/JPL)

Ultraviolet Spectrometer (UVS)

The ultraviolet spectrometer was developed and built by the Laboratory for Atmospheric and Space Physics (LASP) at the University of Colorado. The instrument contained a telescope, a diffraction grating, and three detectors to cover the wavelength spectrum from 113 to 432 nm. A radiation-hardened RCA 1802 microprocessor was employed to control most operations of the UVS, including response to external commands, rotation of the diffraction grating, and organization of output data.

The telescope was a Cassegrain type with an elliptical primary mirror and spherical secondary mirror. The focal length was 250 mm, and the input aperture was 5.03 by 5.28 cm. A cylindrical sunshade affixed to the front of the telescope shielded the optics from off-axis scattered light. The field of view for two of the detectors (G and N) was set at 1.0° by 0.10° by a slit at the entrance to the spectrometer. A slit at the output of one of the spectrometers set the field of view for the third detector (F) to 0.40° by 0.10°.

The spectrometer was an Ebert-Fastie type that used a spherical mirror as a collimator and camera and a moveable plane diffraction grating. The focal length of the spectrometer was 125 mm. Three separate exit slits were located in the focal plane of the spectrometer. A detector was associated with each slit. Three detectors were used to cover the spectral range of the instrument. The characteristics of the detector and type of window before each detector established the wavelength range of each detector channel.

A far-ultraviolet detector (G channel) with a cesium iodide photocathode and magnesium fluoride window was used for the wavelength range 113–191 nm. A middle-ultraviolet detector (F channel) with a cesium telluride photocathode and quartz window was used for the wavelength range 162–323 nm. A near-ultraviolet detector (N channel) with a bi-alkali photocathode and a quartz window was used for the wavelength range 282–432 nm. The far-ultraviolet and near-ultraviolet detectors were mounted near the exit slits of the spectrometer. The middle-ultraviolet detector was illuminated by a periscope adjacent to the exit slit of that spectrometer.

The plane grating was rotated in steps to scan through the wavelengths of the instrument. During operation, the grating was stepped at each 7.6 ms for 528 steps within a 4.3-s science frame. The wavelength resolution of the G, F, and N detector channels was 0.67 nm, 1.36 nm, and 1.27 nm, respectively.

Two different scan modes could be commanded. They were referred to as “scan mode” and “fixed-wavelength scan mode.” The scan mode could be set to investigate one detector channel at a time. The grating was stepped up the wavelength scale during even-number science frames and stepped down during odd-numbered science frames. If two channels were to be investigated, such as the N and G channels, channel N would be examined during scan up of the grating during even-numbered science frames and channel G would be examined during scan down of the grating during odd-number science frames.

The fixed-wavelength scan mode was used to investigate a smaller span of wavelengths within a normal scan range. A particular detector channel would be selected along with the sector to be scanned within that channel. The scan would be set at the bottom of the selected scan interval, and then a programmed number of steps would be made up and down. The sawtooth scan pattern would be repeated during the 4.3-s scan frame.

Each detector was followed by an amplifier. The outputs of the amplifiers for the N and G channels were applied to a multiplexer, which in turn applied the selected signal to an integrator-counter. The counter output was buffered and applied to a data memory element. The F channel was not multiplexed, but rather the amplifier output was applied to an integrator-counter, buffered, and applied to the data memory element. The memory output was applied to a bus adapter that interfaced with the command and data subsystem of the spacecraft.

Extreme Ultraviolet Spectrometer (EUVS)

The extreme ultraviolet spectrometer was a flight spare from the Voyager program (Ultraviolet Spectrograph). It was developed and built by the Laboratory for Atmospheric and Space Physics (LASP) at the University of Colorado. The spare was modified for the Galileo mission. The extreme ultraviolet spectrometer covered the wavelength band 54–128 nm.

The instrument contained a collimator tube with 12 aperture plates. The input aperture was 4.0 cm by 6.0 cm. The collimator illuminated a diffraction grating set at an angle to the collimated light flow. The grating was concave in shape with a radius of 400 mm. The incoming light was disbursed by the grating onto a microchannel plate that fed a linear array of 128 detectors. The microchannel plate was made up of a large number of very fine hollow optical fibers. The fibers were short, and the microchannel plates were quite thin. The hollow fibers were coated on the inside with a material that liberated electrons when struck by another electron. When a photon of ultraviolet light struck the entrance to the plate, it liberated an electron. That electron in turn struck the coating on the inside of the fiber and liberated additional electrons. The microchannel plate acted as an electron multiplier. There were two microchannel plates in a series, and a photoelectron was amplified by a factor of about a million by the two plates.

Electrons at the output of the microchannel plates struck the anode of a particular detector located behind the plates. The anode accumulated the charge from the electrons until the detector was read out. There was an array of 128 detectors. Each individual detector could record photon event rates of about 300/s. The detectors were read out, and the amplitude of the charge pulse was converted to digital representation and placed in memory having 128 locations with 16-bit words. The process of reading out detectors, integrating, and accumulating data was retained from that in the Voyager spare. An interface between the Voyager-type output and Galileo data handling process was provided by an interface called PDS simulator.

A radiation-hardened RCA 1802 microprocessor was employed to control most operations of the EUVS, including response to external commands, scan of detector outputs, and organization of output data. The field of view of the EUVS was 0.17° in the dispersion direction and 0.87° in the cross-dispersion direction. The instrument was mounted on the spinning section of the spacecraft with the field of view perpendicular to the spin axis. As the spacecraft spun, the EUVS examined a ribbon in space 0.87° in extent perpendicular to the spin axis.

Results of Ultraviolet Experiment

The ultraviolet instruments measured emission, absorption, and reflective properties of atmospheres in the Jupiter system and surfaces of its moons. Measurements of the Jupiter's nightside polar auroral spectrum at wavelengths of 162–323 nm indicated the presence of a variant of molecular hydrogen. Spectra produced by the EUV observing the Io torus indicated a strong peak at 69 nm and other peaks at 83 nm (ionized atomic oxygen) and 102 nm (atomic oxygen).

Measurements of reflectivity of the surface of the moon, Callisto, by the UVS indicated the presence of carbon mixed with a redder material. Near the south pole of Callisto, reflectivity characteristics were thought to be produced by an organic species. Measurements of reflectance of Io by the UVS indicate a likely combination of SO_2 frost on the surface and SO_2 gas in the atmosphere. Measurements by the UVS around Ganymede indicated an ozone-like absorber that was strongest near the poles. The absorber was shifted about 20 nm to the red from the ozone line. Emission of hydrogen Lyman-α from Ganymede indicated a density of 1.5×10^4 atoms/cc.

Galileo received a gravitational assist from Venus on its journey to Jupiter. Emission measurements of the atmosphere of Venus were made by the EUV in the wavelength range of 55 nm to 125 nm. The emission spectrum showed peaks at 58.4 nm due to helium, 83.4 nm due to oxygen, and very strong peak at 121.8 nm due to hydrogen Lyman-α.

Atmospheric Probe

The atmospheric probe was designed to descend into the hostile atmosphere of Jupiter and determine its chemical composition, density profile, location and characteristics of clouds, profiles of temperature and pressure, and characteristics of electrical discharges (lightning). The atmospheric probe portion of the Galileo program was managed by NASA Ames Research Center. The probe was built by Hughes Aircraft Company.

The atmospheric probe was carried in the center bottom portion of the despun section. It was released from the Galileo orbiter on 13 July 1995 while 80 million km from Jupiter. Accelerated by the gravity of Jupiter, it entered the atmosphere on 7 December 1995 at a speed of about 170,550 km/h. The entry point was at latitude of 6.5° north and longitude of 4.4° west in the Jupiter coordinate system. After entry, the probe was slowed by a heat shield and then parachutes. It descended about 180 km into the atmosphere before radio transmission stopped, likely due to extreme temperature or pressure of the atmosphere.

Mechanical Configuration of Atmospheric Probe

The atmospheric probe included a heat shield, a descent module, and a back cover. A NASA drawing showing the configuration of the probe is given in Fig. 7.23. The conical heat shield was 1.25 m in diameter. It protected the descent module during the extreme high-speed entry into Jupiter's atmosphere. The overall height of the atmospheric probe was 86 cm. The overall weight of the probe was 337 kg.

The aeroshell portion of the deceleration module was a blunt cone with a half-angle of 45°. The widest portion of the aeroshell was 1.25 m in diameter. The aeroshell used carbon phenolic material for the front portion of the shield. Phenolic nylon was used for the afterbody cover. The atmospheric probe was designed to withstand temperatures up to 14,000 Kelvin expected during the entry of the probe into the atmosphere of Jupiter.

The deceleration module had mounting provisions for the descent module and to separate the two modules upon command. The weight of the deceleration module was 213.4 kg.

A NASA photograph of the descent module separated from the deceleration module is shown in Fig. 7.24. The main body of the descent module was about 0.8 m diameter and 0.7 m tall. The module weighed 117.6 kg.

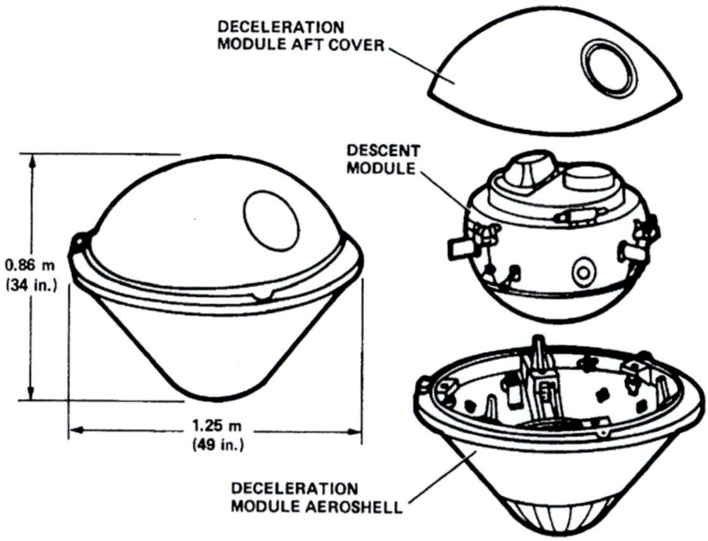

Fig. 7.23 Configuration of atmospheric probe (NASA graphic)

Fig. 7.24 Photograph of descent module and deceleration module of atmospheric probe (NASA image)

The descent module included seven scientific instruments as well as probe systems. Probe systems included power subsystem, command and data processing subsystem, and communications subsystem. A structural shelf within the module provided mounting for equipment. subsystem equipment were mounted on the top side of the shelf, and most of the science instruments were mounted on the bottom side.

Systems in Atmospheric Probe

Power Subsystem

Power for descent module systems and scientific instruments was obtained from three lithium/sulfur dioxide (Li-SO_2) battery modules. There were 13 D-size cells in each module, and the three modules were connected in parallel to supply power to the descent module. The total capacity of the batteries was 21 Ah. Power required for the mission was about 16 Ah. The terminal voltage of the battery assembly was about 32 V under load and about 39 V with no load. The batteries were built by Honeywell Power Sources Center.

A second battery system was provided to supply a few large current pulses needed to fire pyrotechnic pressure cartridges used to jettison the deceleration module and deploy the parachutes. That battery was a thermal type. The thermal battery was unique in that squibs were fired to melt an electrolyte and activate the battery.

The squibs raised the temperature within the battery to about 400 °C. The temperature was well above the melting point of the lithium-chloride plus potassium-chloride eutectic electrolyte. The squibs were fired using energy from the main Li-SO_2 battery.

The thermal battery had two groups of 14 cells connected in series, and the two groups were connected in parallel. The open-circuit voltage was 37 V. The battery was required to supply six current pulses ranging from 5.3 to 15.9 A. Each pulse was 22 ms in duration. The six pulses occurred within a time interval of 13 s. The thermal batteries were built by EaglePicher Technologies. Two thermal batteries were provided for redundancy.

Command and Data Handling Subsystem

The command and data handling subsystem provided system commands, organized the telemetry data, managed data storage, and provided timing functions. Timing functions included the coast timer that activated probe systems 6 h in advance of atmospheric entry. The command and data handling subsystem consisted of two redundant channels, each containing a data and command processor (DCP). Each channel also included a pyro-control unit and two acceleration switches. The entire sequence of operation of the probe from separation from the orbiter to the end of the probe's mission was contained in permanent memory.

The coast timer in the command and data handling subsystem was set before separation of the probe from the Galileo orbiter. The only item operating during the 5-month journey of the probe from separation from Galileo to Jupiter was the coast timer. The coast timer woke up the probe about 6 h before entry into Jupiter's atmosphere. Acceleration switches were used to sense entry into the atmosphere and start the entry sequence. That sequence included applying power to spacecraft systems and deploying parachutes. The acceleration switches also provided backup to the coast timer.

Communications Subsystem

The communications subsystem gathered data from the science instruments and spacecraft systems and transmitted the data to the Galileo orbiter for relay to Earth. Transmission to the orbiter was done at L-band. The transmission was picked up by an L-band parabolic antenna on the orbiter. The data was extracted by the orbiter and relayed to Earth.

The communications subsystem contained two redundant channels. Each channel included a frequency reference oscillator, an exciter, and a TWT power amplifier. Both channels operated during the mission.

One channel operated at a transmitted frequency of 1387.0 MHz and the other at a frequency of 1387.1 MHz. The frequency reference for the 1387.0 MHz channel was an ultra-stable oscillator. The 1387.1 MHz channel used a crystal-controlled oscillator for frequency reference. The channel with the ultra-stable oscillator allowed one-way Doppler tracking to measure probe velocity. The exciters, which drove the power amplifiers, were biphase modulated by data from the science instruments and engineering data. The data rate of both channels was 128 bps.

The power amplifiers were traveling wave tube (TWT) type that generated power output of 23 W at L-band. The outputs of the power amplifiers in the two channels were combined in a microwave hybrid. The two outputs of the hybrid connected to the two dipoles of a crossed dipole antenna. The antenna consisted of a crossed dipole in a cylindrical cup 25.4 cm in diameter. Feeds to the crossed dipoles were arranged so that one channel transmitted right-hand circular polarization and the other channel transmitted left-hand circular polarization. The beamwidth of the antenna was 56°, and the gain on boresight was 9.8 dB.

Relay Radio Equipment on Galileo Orbiter

An antenna and a receiver to gather signals from the atmospheric probe were mounted on the despun section of the orbiter. The antenna was a parabolic type, 1.1 m in diameter. The antenna had two feeds, one to receive right-hand and the other to receive left-hand circular polarized signals from the probe. The two outputs of the antenna were applied to separate receivers. One receiver was tuned to a frequency of 1387.0 MHz and the other to 1387.1 MHz. The outputs of the two receivers were processed by the command and data subsystem of the orbiter and applied to the tape recorder for later relay to Earth.

Flight of the Atmospheric Probe

The atmospheric probe was released from the Galileo orbiter on 13 July 1995 while 80 million km from Jupiter. The probe entered the atmosphere on 7 December 1995 at a speed of about 170,550 km/h. The entry point was at latitude 6.5° north and longitude of 4.4° west in the Jupiter coordinate system. The probe was slowed rapidly by its heat shield. The initial shock wave ahead of the shield reached a temperature of about 16,000 °C. Peak deceleration during entry was about 228 g's.

Since Jupiter is a gas giant planet, and it is unknown whether it has a solid surface, a reference altitude was established as being at an atmospheric pressure of 1 bar. One bar is the atmospheric pressure at the surface of the Earth. References to altitude at Jupiter are relative to the 1 bar atmospheric pressure level.

A NASA graphic showing events during the approach and descent into the atmosphere is given in Fig. 7.25.

The probe entered a tenuous region of Jupiter's atmosphere at an altitude of 450 km. The pressure was about 10^{-7} bars, and the temperature was 352 °C. A

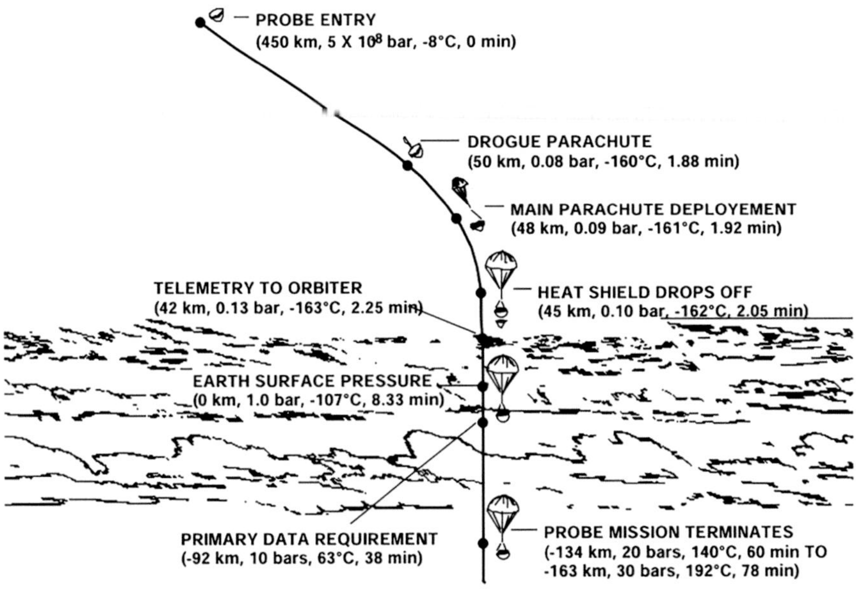

Fig. 7.25 Events during the descent of Galileo probe into the atmosphere of Jupiter (NASA graphic)

mortar was fired that deployed a drogue parachute at 2.86 min after entry. The altitude was 23 km, pressure was 0.4 bars, and temperature was −145 °C at the time. The drogue pulled off the back cover and deployed the main parachute a fraction of a second after drogue deployment. At 3 min after entry, the forward heat shield was jettisoned, and measurements by the science instruments began. The altitude was 21 km, the pressure was 0.45 bars, and the temperature was −145 °C at the time. At 3.8 min after entry and altitude of 16 km, the Galileo orbiter acquired signals from the probe. The pressure was 0.56 bars, and the temperature was −135 °C.

The probe reached 0 km altitude (defined as a pressure of 1 bar) 6.4 min after entry. The temperature was −107 °C. At 9.6 min after entry, the probe descended below a cloud layer at an altitude of −16 km, where the pressure was 1.6 bars and temperature was −60 °C. The signal from the probe ended at 61.4 min after entry. The altitude was −146 km, the pressure was 22 bars, and the temperature was 153 °C at the time. The probe likely succumbed due to the high pressure or high temperature of the atmosphere.

Data from the probe was received for 57.6 s during the descent. From the time of firing the drogue parachute at 2.86 s after entry to the end of transmission from the probe, it had descended about 170 km into the atmosphere.

Science Instruments on Atmospheric Probe

The atmospheric probe carried six scientific instruments on its descent into the atmosphere of Jupiter. The instruments all operated as intended and returned new information about the atmosphere. A list of experiments and a summary of purpose of each are given in Table 7.6.

Table 7.6 Experiments carried out by atmospheric probe

Experiment	Purpose	Principal investigator
Atmospheric structure instrument	Determine the temperature, pressure, and density of the atmosphere during descent of the probe	Alvin Seiff, San José State University
Neutral mass spectrometer	Analyze chemical and isotope composition of the atmosphere	Hasso Niemann, NASA Goddard
Nephelometer	Determine the nature of the clouds and particles in the atmosphere	Boris Ragent, NASA Ames
Helium abundance Detector	Measure the ratio of abundance of helium to hydrogen	Ulf von Zahn, University of Bonn
Net flux radiometer	Measure the temperature looking up and looking down to determine net heat flux	L. Sromovsky, University of Wisconsin
Lightning and radio emission detector/energetic particle instrument	Detect lightning and measure radio frequency emissions. Investigate electrons, protons, and heavy ions in the inner magnetosphere	Louis Lanzerotti, Bell Laboratories

Atmosphere Structure Instrument

The purpose of the atmospheric structure instrument (ASI) was to measure the physical properties of the atmosphere of Jupiter. The instrument included sensors for measuring temperature, pressure, and acceleration as the probe descended deep into the hostile environment of the gas giant planet. Accelerometer data was used to determine the density of the atmosphere. An electronics unit managed the operation of the sensors and gathered information from them. There were three different modes of operation of the ASI: calibration, entry, and descent.

Temperature Sensor

The temperature sensor was a platinum wire type where the resistance of a thin platinum wire changed with temperature. A constant current was applied to a circuit containing the platinum wire and a load resistor. The voltage across the load resistor became a function of temperature, and that voltage was digitized to become the output of the sensor.

The ASI contained two separate platinum wire temperature sensors. One, referred to as the primary sensor, used a 0.1-mm-diameter wire 1.2 m long. The resistance of the wire was about 14 ohms at a temperature of 273 K. The wire was wound around an open frame about 2.8 cm wide and 2.8 cm long. The frame was supported by a stem having low thermal conductivity that extended the temperature sensor outside of the skin of the probe and beyond the boundary layer. The atmosphere flowed through the wires as the probe descended.

The secondary temperature sensor used a 0.025-mm-dimeter wire 6 cm long. The resistance of the wire was 11.4 ohms at a temperature of 273 K. The wire was wound over a glass film on one of the tubes in the frame of the primary sensor.

The measurement range of the temperature sensors was 0 to 500 K (−273 to 227 °C). A constant current source was supplied to the wires at load resistors by the electronics unit. The electronics unit then sampled the resulting voltage across the load resistor and digitized it by a 10-bit analog-to-digital converter. The outputs of the two sensor elements were sampled alternately at 2-s intervals. The accuracy of the temperature sensor was expected to be within 1 K at a temperature of 500 K and within 0.1 K at a temperature of 100 K.

Temperature measured during the descent into the atmosphere was related to an altitude relative to zero altitude at a pressure level of 1 bar. A few measurements are given below:

Altitude, km	Temperature, °C
23	−145
16	−135
0	−107

Altitude, km	Temperature, °C
−16	−60
−146	153

Pressure Sensor

The pressure sensor was a diaphragm type with a vacuum cavity behind the diaphragm. The diaphragm was 1.5 cm in diameter. The deflection of the diaphragm was a function of pressure. The amount of deflection was measured by change in reluctance of a magnetic circuit. The output of the magnetic circuit was converted to a DC voltage with a range of 0 to +5 V.

There were three pressure sensors with different measurement ranges. The full-scale measurement ranges were 0.5 bars, 4 bars, and 28 bars. A bar is a convenient gauge for measuring atmospheric pressure since 1 bar is the pressure at sea level on Earth. The inlets to the three pressure sensors were connected to an inlet manifold. Tubing connected the inlet manifold to a pitot tube mounted outside the descent module and extending beyond the boundary layer.

The DC outputs of the three sensors were sampled by the electronics unit and converted to digital signals by a 10-bit analog-to-digital converter. Sampling of the pressure sensor outputs was made at the same time as sampling of the temperature sensor output, each 2 s. If the pressure was found to be <0.5 bars, the 0.5-bar and 4-bar sensors were sampled alternately. If the pressure was between 0.5 and 4 bars, the 4-bar and 28-bar sensors were sampled alternately. If the pressure was above 4 bars, only the 28-bar sensor was sampled every 2 s.

Pressure measured during the descent into the atmosphere was related to an altitude relative to zero at a pressure level of 1 bar. A few measurements are given below:

Altitude, km	Pressure, bars
23	0.40
16	0.45
0	1.0
−16	1.6
−146	22

Acceleration Sensor

The acceleration sensors were a flexure-supported mass type where movement of the mass due to acceleration was sensed. An electromagnetic coil could also position the mass. Should the mass move due to acceleration, a current was applied to the coil to restore the mass to its original position. The amount of current through the coil needed to maintain the mass in its original position was a measure of acceleration.

There were four acceleration sensors. Two redundant sensors were used to sense acceleration along the Z-axis (longitudinal), the third sensed acceleration along the X-axis, and the fourth sensed acceleration along the Y-axis. The Z-axis accelerometer operated in four ranges with full-scale accelerations of 0.012 g, 0.4 g, −4 g, and 410 g. The X-axis and Y-axis accelerometers operated in three ranges with full scale of ±0.0125 g, ±8 g, and ±12.8 g. The symbol "g" is the standard gravitation attraction of Earth. The peak deceleration measured during the entry of the probe into the atmosphere was 228 g's.

Electronics Unit

The electronics unit, which supported the various sensors in the atmospheric structure instrument, was microprocessor based. The microprocessor was a RCA CDP 1802 CPU with 8192 8-bit words of read-only memory (ROM) and 256 8-bit words of random access memory (RAM). The CPU was clocked at 2.2 MHz. Control and operating software for the ASI was contained in the ROM, and temporary storage was afforded by the RAM.

The electronics unit communicated with the command and data handling subsystem to receive input commands and transfer sensor data.

The electronics unit contained multiplexers and two analog-to-digital converters to convert analog outputs from the sensors to digital form. It also provided a constant current source for the temperature sensors.

The data rate available to the ASI was relatively low and dependent on the mode of the atmospheric probe. There were three operating modes: calibration, entry, and descent. Calibration was performed before reaching the atmosphere. The entry mode gave priority to deceleration measurements. The descent mode gave priority to temperature and pressure measurements. The data rate allocated to the ASI was 18 bps for the calibrate mode, 50 bpi for the entry mode, and 18 bpi for the descent mode.

Neutral Mass Spectrometer

The purpose of the neutral mass spectrometer (NMS) was to determine the chemical and isotopic composition of the atmosphere of Jupiter. The spectrometer was a quadrupole mass spectrometer type.

Atmospheric gas molecules passed through the inlet port of the mass spectrometer and into an ionization chamber. The gases were ionized by bombardment by energetic electrons generated by heated filaments. The electron bombardment broke up the molecules into fragments of positive ions. Those ions were accelerated and directed to the entrance of the mass analyzer by electric field lenses. The mass analyzer contained four narrow rods 15 cm long stacked close to each other in a square pattern. Ions were directed into the cavity at the center of the rod pattern.

Diagonally opposite pairs of rods were connected together electrically, and a combined DC voltage and RF voltage was imposed between the pairs. A particular combination of DC and RF voltage caused an ion of a particular mass-to-charge ratio (m/z) to set up a resonant oscillation in space without touching the rods and travel through the center of the rod arrangement to strike the detector. Ions with a different m/z struck a rod or escaped between the rods. The DC and RF voltages were swept to selectively pass ions with different m/z ratios. Ions that passed through the mass spectrometer were applied to a detector.

The measurement range of the spectrometer was 2–150 atomic mass unit (amu). An RF frequency of 2.83 MHz was used for the 2–19 amu range, and a frequency of 1.13 MHz was used for the 20 to 150 amu range. The amplitude of the RF voltage was changed in steps. The dwell time at each scan position was 0.5 s, and the mass steps were 1 amu. There were about 150 scan steps, and the total time for one scan was about 75 s.

The inlet port for the mass spectrometer was located near the bottom of the descent module near the stagnation point during descent. The outlet port was located near the minimum pressure point on top of the module. The pressure of the inlet sample was reduced by a series of capillary tubes before being applied to the spectrometer. There were two inlet sampling elements with associated valves that allowed taking two samples at different times.

The detector for the spectrometer was an electron multiplier type. One output of the spectrometer was the number of pulses in the time interval for each m/z ratio as set by DC and RF voltages applied to the rods. The second output was the measurement of the amplitude of the pulses.

The spectrometer was controlled by an electronics unit that included a read-only memory (ROM) with 8192-word 16-bit capacity. Each 0.5 s, the ROM was incremented to configure the spectrometer for the next measurement interval by setting the DC and RF voltages. The number of counts from the detector was read at the end of each 0.5-s interval by a 13-bit register. The pulse amplitude was also measured and digitized by an 8-bit analog-to-digital converter. The count data and amplitude data were applied to a digital multiplexer that formatted spectrometer output data for the telemetry system.

Molecular hydrogen (H_2) is the most abundant gas in the atmosphere of Jupiter. It makes up nearly 90% of the total. Measurements of a few constituents in the atmosphere by the mass spectrometer, expressed as ratios to molecular hydrogen, were as follows:

Ammonia (NH_3)/H_2 = 6.64×10^{-4}
Hydrogen sulfide (H_2S)/H_2 = 8.0×10^{-5}
Water vapor (H_2O)/H_2 = 4.9×10^{-4}
Methane (CH_4)/H_2 = 2.37×10^{-3}

Nephelometer

The purpose of the nephelometer was to measure the physical characteristics of particles making up the clouds and haze in the atmosphere of Jupiter. The instrument contained two nephelometers. The first was referred to as the forward-scatter nephelometer, and the second was referred to as a backscatter nephelometer. An electronics unit managed and gathered data from both nephelometers.

Both nephelometers contained a pulsed laser source operating at a wavelength of 904 nanometers (infrared). The pulsed laser light was scattered when it struck a particle. In the case of the forward-scatter nephelometer, a deployable arm allowed extending a vented sensor head 13 cm beyond the skin of the descent module so that the sampled volume was in undisturbed air. Mirrors in the sensor head were positioned to reflect scattered signals from particles into detectors at angles of 5.8°, 16°, 40°, and 70°.

The backscatter nephelometer was mounted to the base of the instrument, and the forward-scatter nephelometer was mounted on it. The vented sensor head of the backscatter nephelometer extended out just beyond the skin of the descent module. A mirror reflected the scattered signal at an angle of 178° (backscatter) into a detector.

The laser diode light source generated light pulses 200 ns in duration at a 2000 Hz rate and peak power output of 2 W. In case of the forward-scatter nephelometer, the light was collimated by baffles in a 40 cm long tube before it was used to illuminate substances in the vented sensor head. The detectors that received the scattered light pulses were p-i-n photodiodes. The effective scattering volume was 1.25 L for the 5.8° scattering path, 0.63 L for the 16° path, 0.65 L for the 40° path, 0.40 L for the 70° path, and 16.4 L for the 178° path.

The forward-scatter sensor assembly included four detectors. Each detector was followed by a charge amplifier and then a bandpass filter amplifier. The outputs of the four bandpass amplifiers were applied to the electronics unit. The electronics unit amplified each of the four input signals with a variable gain amplifier. The amplified signal in each channel was applied to an integrate-and-hold function and then on to an analog multiplexer. The amplified and filtered output of the backscatter nephelometer was also applied to the analog multiplexer. The output of the multiplexer was fed to an 8-bit analog-to-digital converter.

Data sampling was controlled by the electronics unit. Each of the five channels was sampled in turn by a burst of 64 pulses. The nephelometer was scheduled to collect and report data each km of descent. Early in the descent, the velocity was high and the data interval for 64 bursts in each of the five channels was 3 s. The 3-s sample period was repeated ten times. The sample period was increased as the spacecraft slowed. At 12.5 min after entry, it was set at 8 s, and that value was used for the remainder of the mission.

The nephelometer data frame was 800 bits long. The frame consisted of a synchronization word, engineering data, and data from ten scattering data records. The data was sent to the probe telemetry system.

The instrument was calibrated by introducing a known gas (ammonia for example) into the vented sensor heads and noting the proportions of light scattered for each of the operating angles of the instrument. Measurements of amplitudes of scattered signals at each angle during the mission could be used to identify a particular substance.

Results of the nephelometer experiment during the descent of the probe were reported by B. Ragent et al. (1998). The probe happened to descend in a relatively hot and dry region of the atmosphere. Measurements were made by the nephelometer over an atmospheric pressure range of 0.46 bars to 12 bars as the probe descended. Results indicated that the probe encountered one thin cloud during its descent. The cloud extended over an atmospheric pressure range of 0.76 bars to 1.34 bars. The corresponding altitude range, relative to zero altitude at 1.0 bar, was 6.5 km to −7.6 km. The peak particulate count occurred at an atmospheric pressure of about 1.25 bars. There was a sharp cutoff of particles below the base at 1.34 bars. Another slight rise in particle count occurred around the atmospheric pressure of 1.65 bars.

The size of the particles appeared to be in the range of submicron to a few microns. Ragent characterized the cloud encountered as having modest optical depth and very low mass loading.

Helium Abundance Detector

The atmosphere of Jupiter consists predominantly of hydrogen and helium. The mole fraction of molecular hydrogen (H_2) is about 86%, and the mole fraction of helium (He) is about 14%. Together, they constitute 99.5% of the atmosphere. Trace amounts of other gases make up the remainder. Accurate determination of the mole fraction of helium to hydrogen is important to researchers seeking the history of the He/H_2 fraction since those elements were first formed in the primordial fireball (big bang). The helium abundance detector was designed to make accurate measurements of the abundance ratio.

The helium abundance detector (HAD) instrument contained an interferometer to measure the refractive index of the atmosphere as the Galileo probe descended. Measurements were made during descent from an atmospheric pressure of 2.3 –11.8 bars.

The interferometer contained a light-emitting diode (LED) operating at a wavelength of 901.4 nm. A lens created a light beam parallel to the reference plane of the instrument. The light beam was directed at a Jamin plate oriented 45° from the beam. The Jamin plate, which consisted of a thick mirror that reflected light from the front and back sides, formed two parallel and coherent light beams. One beam was passed through a cell holding a reference gas, and the other beam passed through a cell holding a sample of the atmosphere of Jupiter.

The outputs of the reference cells and the atmospheric sample cells were combined by reflection by the Jamin plate. The speed of light through the reference cells was slightly different from the speed of light through the atmospheric sample cells. When combined, an interference pattern of bright and dark fringes was formed. The resulting interference pattern of bright and dark fringes was sensed by a linear array of nine photodetectors.

The reference gas was a mixture of 27.7% argon and 63% neon, which had the same refractive index as a mixture of 11% helium and 89% hydrogen. The helium and hydrogen mix chosen was close to that expected in the atmosphere of Jupiter. The reference gas was stored in a reservoir with a volume of 20 cc and fed into the reference cells as required.

The cells for the reference gas and for the atmospheric sample were sealed during the long journey to Jupiter. The seal on the atmospheric sample cell was designed to rupture at an atmospheric pressure of 2 bars. The resulting atmospheric pressure in the gas line to the cells pushed a sharp pin through a seal for reference gas that allowed reference gas to enter the reference gas cells. The reference gas was released into the referenced cells through a membrane valve that maintained the pressure of the reference gas in reference cells within 0.075 bars of that in the atmospheric sample cells.

The atmosphere entering the inlet of the instrument was passed through a chemical absorber that absorbed ammonia, water vapor, and methane before being fed to the atmospheric sample cells. The atmospheric gas and the reference gas were passed through separate heat exchangers to bring the gases to the same temperature as active portion of the instrument.

First fringes from the instrument were detected at 20 min after entry, and the contrast of fringes had stabilized by about 27 min after entry. The final dataset for helium abundance was taken at about 36 min after entry.

A data frame from the helium abundance detector contained 256 bits. Data in the frame included contents of the fringe counter, readings from three pressure sensors, readings from four temperature sensors, analog data from one of the photodetectors, and housekeeping data. The frame of 256 bits was transmitted from the HAD every 64 s.

Data from the first sign of fringes was contained in major data frame 19, and the final measurement was contained in major data frame 34. Using the measured data, U. von Zaln and colleagues determined that the mole fraction of helium to the atmosphere of Jupiter was 0.1359. The corresponding mass fraction of helium to the atmosphere was 0.234.

Net Flux Radiometer

The net flux radiometer (NFR) measured the net flux of solar energy and thermal emission from the planet as the Galileo probe descended through the atmosphere. One of the byproducts was identification of cloud layers. Five separate channels were used for the flux measurement. Each channel responded to a different wavelength of radiation. The wavelength span of the five channels was 0.3–500 μm (near infrared to far infrared).

The atmosphere was observed in the upward and downward directions to determine net flux as a function of altitude as the probe descended. The conical field of view of the instrument was 40°. The center of the field of view was alternately centered 45° above the horizontal and 45° below the horizontal.

The net flux radiometer consisted of two major subassemblies: the optical head and the electronics module. The optical head was 8.5 × 8 × 10.5 cm in size, and the electronics module was 13 × 19.5 × 16 cm in size. The optical head was mounted to the electronics module with a housing that contained a rotating portion extending through the skin of the probe. The housing had cutouts, so the rotating portion could view the atmosphere in the upward and downward direction.

The rotating portion of the optical head contained a diamond window through which the atmosphere was viewed. Radiation passing through the window struck a plane mirror that reflected the radiation towards a toroidal mirror. The toroidal mirror reflected the radiation through a condensing cone mirror to a series of six spectral filters in front of six detectors. One filter was ahead of each detector. The detectors were lithium tantalite pyroelectric type. Each detector was followed by a preamplifier. The channel designations and the filter bandpass wavelengths are given in Table 7.7. The blind channel was used for a zero-analog measurement.

The diamond window, mirrors, condensing cone, filters, detectors, and preamplifiers rotated together as a unit. Rotation was driven by a stepping motor and gear arrangement. Each step of the motor resulted in a rotation of the optics and detectors by 45°. If one imagines looking outwards along the rotation axis, 0° would be in the upward direction and 180° in the downward direction. Rotation of 45° would set the upward-looking angle for flux, and 135° would set the downward-looking angle. A calibration location at 90° allowed the sensor to view a black body at instrument ambient temperature. A calibration source located at 270° allowed the sensor to view a hot black body at a closely controlled temperature of 107 °C.

In the net flux mode of the instrument, the sensor was rotated between the 45-degree and 135-degree positions. This put the viewing angle alternately 45° above and then 45° below the horizontal. In the calibrate mode, the sensor was rotated between the 90-degree and 270-degree positions to alternately view the ambient temperature black body and the hot black body. In the upflux mode, the sensor was rotated between the downward-viewing angle (135°) and the ambient black body. A collection of measurements was 2 min in duration. In that time, there were 17 cycles of net flux measurements, one upflux measurement, one calibration measurement, and one zero analog measurement. In total, there were 20 integration periods of 6 s each in the 2-min data collection interval.

Signal processing in the electronics unit was controlled by an RCA CDP 1802 microprocessor. The microprocessor was supported by 256 8-bit words of random access memory (RAM) and 6144 8-bit words of programmable read-only memory

Table 7.7 Filter channels for net flux radiometer

Channel	Bandpass
A	3–500 μm
B	0.3–3.5 μm
C	3.5–5.8 μm
D	14–35 μm
E	0.6–3.5 μm
F	Blind channel

Table 7.8 Values of net flux vs. atmospheric pressure and sensor channels

Atmospheric pressure	Sensor channel	Net flux, W/m^2
0.44 bars	A (thermal)	6.5
	B (solar)	−3
	C (thermal)	0.2
	D (thermal)	3.3
	E (solar)	−2.2
1 bar	A (thermal)	5
	B (solar)	−2.2
	C (thermal)	0.2
	D (thermal)	2.3
	E (solar)	−1.9
8 bars	A (thermal)	2.6
	B (solar)	−0.1
	C (thermal)	2.6
	D (thermal)	0.1
	E (solar)	−0.1

(PROM). The outputs of the six preamplifiers in the rotating unit were applied to six post-amplifiers followed by six demodulators and six integrators. The integrators were controlled by the microprocessor. The six integrator outputs were applied to a multiplexer along with housekeeping signals. The output of the multiplexer was applied to an adjustable gain amplifier. The output of the amplifier was applied to a 12-bit analog-to-digital converter. The variable gain amplifiers maintained the signal level at the input to the analog-to-digital converter within operating range. The microprocessor controlled the stepper motor that determined the viewing direction. It also managed timing and sequencing for the instrument and formatting of the digital data output.

Flux measurements were made during the descent of the atmospheric probe from atmospheric pressure levels of 0.44 bars to about 14 bars. Plots of values of corrected net flux as a function of atmospheric pressure and sensor channel were generated by L. A. Sromovsky et al. (1998). According to the plots, approximate values for net flux for the various wavelength channels are given in Table 7.8. The dimension for net flux was W/m^2. The net flux values in all channels converged to low values at pressures approaching 14 bars.

Analysis of the data from the instrument indicated evidence of a weak cloud layer at altitudes above 0.45 bar atmospheric pressure at an altitude of about 21 km. There was clear evidence of a cloud layer near a pressure level of 1.35 bars at an altitude of about −13 km.

Lightning and Radio Emission Detector/Energetic Particle Instrument

The purpose of the lightning and radio emission detector was to detect lightning in the Jupiter system and to measure radio frequency emissions. The energetic particle instrument investigated electrons, protons, and heavy ions in the inner magnetosphere. The lightning and radio emission detector (LRD) and the energetic particle instrument (EPD) were separate instruments, but they shared a common electronics unit.

The Galileo probe was released from the Galileo orbiter spacecraft while 80 million km from Jupiter. Radio frequency measurements were made at distances of 5, 4, 3, and 2 Jupiter radii from the center of Jupiter. The corresponding distances from the center of Jupiter were about 357,460 km, 285,968 km, 214,476 km, and 142,984 km. Results of the measurements were stored, read out, and transmitted to the orbiter during descent of the probe into the atmosphere.

Lightning and Radio Emission Detector

The lightning and radio emission detector included a radio frequency antenna that intercepted radio frequency signals in the frequency range of 100 Hz to 100 kHz. The antenna was a ferrite core type. A sensing coil of wire was wound around the long core. The wound core was mounted inside of a cylindrical housing with insulating ceramic material around it. A radio frequency preamplifier was included in the antenna package. The dimension of the overall antenna package was 32 cm long and 3.3 cm in diameter. The back cover of the probe had a special window over the antenna that passed radio frequency signals to allow radio frequency measurements to be made while the heat shield and back cover of the probe were still in place.

The signal at the output of the antenna was applied to three different analysis elements: spectrum analyzer, waveform analyzer, and magnetic field measurement. The spectrum analyzer included three narrowband filter channels. Center frequencies of the channels were 3 kHz, 5 kHz, and 90 kHz. The radio frequency signal at the output of each channel was rectified and filtered. It represented the average noise level of that channel. Each filter output was also applied to a pulse height analyzer that measured the amplitude of a pulse.

The waveform analyzer accepted the wideband signal from the antenna. The signal was amplified, and the wideband noise level was measured. A pulse-height analysis was also performed. The amplified signal was sampled and digitized every 4 μs and applied to a waveform snapshot element. The snapshot element generated a 1-ms sample gate, each measuring a period of 256 s. The 1-ms gate sampled the digitized signal, which had been taken every 4 μs. The peak amplitude of a pulse provided a trigger point. The 64 samples that occurred before the trigger were stored along with 64 samples taken after the trigger and 61 samples taken at every other one of the 4 μs samples taken in the 1-ms interval. The 61 samples were used to observe the tailing edge of the pulse where the rate of change was lower. The 64 samples had been stored before the trigger provided details of the shape of the leading edge of the pulse.

The signal at the output of the antenna was also used for magnetic field measurements. The changing level as the descent module rotated provided a measure of the magnetic field component. The varying signal allowed determination of the spin period of the descent module as well. The lightning and radio emission detector also included two optical detectors. Coincidence between the optical detector outputs and the radio frequency signals was used to verify that lightning was the source of the radio frequency signals. The optical sensors had wide-angle lenses (fisheye) with fields of views looking perpendicular to the spin axis on opposite sides of the descent module. Together, the optical sensors provided nearly complete angular coverage in the space around the probe. The signals at the outputs of the two optical detectors were added together. The summed signal was used as a measure of brightness of the atmosphere and of pulses due to lightning.

The LRD was controlled by a radiation-hardened RCA 1802 microprocessor. The microprocessor controlled the mode of operation, generated statistical waveform data, determined the coincidence between optical and radio frequency information, and organized and formatted output data for telemetry.

The output data frame consisted of 256 bytes of data gathered during four periods of 64 s each. The first 64 bytes of the frame contained statistical and waveform data, average signal levels, and gin settings. The remaining 192 bytes in the frame contained waveform sampled data and a small amount of housekeeping data. The total frame occupied 256 s of time.

Results of the lightning and radio emission detector experiment were published by K. Rinnert et al. (1998). One waveform, thought to be from lightning, was recorded by the LRD during the 57 s that data was received from the probe. The event, with a magnetic field amplitude of 30 nT, happened at an atmospheric pressure of 16 bars during the descent.

Energetic Particle Instrument

The energetic particle instrument (EPI) investigated populations of energetic particles in the magnetosphere of Jupiter from a distance of five Jupiter radii to the top of the atmosphere. Particles investigated included electrons, protons, alpha particles, and heavy ions.

The instrument used a telescope shielded by a tungsten housing that contained two detectors. The first detector was mounted within a tungsten baffle that allowed a field of view of 73°. The second detector was mounted behind a brass shield and in line with the first detector such that particles needed to penetrate the first detector and the brass shield before striking the second detector. The field of view of the second detector was 44°. The center of the fields of view of the detectors was displaced 41° from the spin axis of the descent module with the fields of view passing through the rear cover. The EPI operated while the heat shield and rear cover were still attached to the descent module, and so particles had to pass through the rear cover before striking the telescope.

The detectors were silicon surface barrier type. We will call the upper detector “A” and the lower detector “B.” The diameter of each was 2.8 mm and the sensitive area of 6.2 mm^2. The thickness of the detectors was 0.5 mm. A sensor box contained electronics to read out the detectors and process the resulting data. Separate electronic channels were used for the two detectors. Final data processing and formatting were performed by the computer-controlled electronics unit that was also used by the lightning and radio emission detector.

Table 7.9 Energy channels of EPI instrument

Channel	Particle species	Detector	Energy range, MeV/nucleon
E1	Electron	A (0.1 MeV)	>3.2
E2	Electron	AB (0.1 MeV and 0.1 MeV)	>8
E3	Electron	B (0.1 MeV)	>8
P1	Proton	A (0.6 MeV)	42–131
P2	Proton	AB (0.6 MeV and 0.6 MeV)	62–131
P3	Proton	AB (0.6 MeV and 1.1 MeV)	62–92
He	Alpha particle	AB (2.3 MeV and 2.3 MeV)	62–136
HV	Heavy ion	AB (2.3 MeV and 24.5 MeV)	110–910

The output of detector A was amplified and applied to three discriminators having different energy level thresholds. The output of detector B was amplified and applied to five discriminators having different energy level thresholds. A series of coincidence circuits passed outputs only if particular discriminators in channel A corresponded in time with the output of channel B. In total, there were eight output channels fed to the microprocessor and digital processing sections of the electronics unit used by both the LRD and EPI. The energy ranges of the eight channels for particular species were given by H. M. Fischer et al. (1992). The energy ranges are summarized in Table 7.9.

The outputs of the eight measurement channels were sent to the electronics unit that was shared between the LRD and RPI. The computer-controlled electronics in that unit performed final data processing and formatting for telemetry.

Energetic particle measurements were made as the probe approached Jupiter at discrete distances of 5, 4, and 3 Jupiter radii from the center of Jupiter. Continuous measurements were made between distances of 2.4 and 1.25 Jupiter radii. Results of the measurements were stored, read out, and transmitted to the orbiter during the time of descent of the probe into the atmosphere.

The EPI experiment revealed that the inner radiation region of Jupiter had high electron and proton fluxes. The fluxes peaked at a distance of 2.2 R_J where R_J is the radius of Jupiter. The electron flux was 10^6 to 10^7 electrons/cm^2/steradian/s, and the proton flux was 10^5 to 10^6 protons/cm^2/steradian/s. The instrument also measured high fluxes of helium ions and high fluxes of heavy ions. Those fluxes peaked at distances of 2.2 R_J and 1.5 R_J. All of the fluxes decreased sharply at distances <1.35 R_J.

References

Bennett, Gary L., et al, *The General-Purpose Heat Source Radioisotope Thermoelectric Generator*, Proceedings of 4th International Energy Conversion Engineering Conference and Exhibit, June 2008

Carlson, R. W., et al., *Near-Infrared Mapping Spectrometer Experiment on Galileo*, Space Science Reviews, Vol 60, 1992

Crary, F. J., et al., *Galileo plasma spectrometer measurements of composition and temperature in the Io plasma torus*, Journal of Geophysical Research, Vol 103, No. A12, 1998

Frank, L. A., et al., *Galileo Plasma Subsystem Instrumentation (PLS) Description*, Space Science Reviews, Vol 60, 1992

Fischer, H. M., *Energetic Particle Investigation*, Space Science Review, Vol. 60, No. 1–4, 1992

Garrett, H. R. and Jun, I., *first Adiabatic Invariants and Phase Space Densities for the Jovian Electron and Proton Radiation Belts–Galileo and GIRE3 Estimates*, Journal of Geophysical Research: Space Physics 10:1029/2020JA028593, 2020

Givens, J. J., et al., *Galileo Atmospheric Entry Probe System: Design, Development, and Test*, Paper number AIAA-83-0098, AIAA 21st Aerospace Sciences Meeting, January 1983

Grun, Eberhard, et al., *The Galileo Dust Detector*, Space Science Reviews, Vol. 60, 1992

Jet Propulsion Laboratory, *Functional Requirement Galileo Orbiter Flight Equipment*, JPL Report GLL-3-180C, 1989

Johnson, Michael R., *The Galileo High Gain Antenna deployment Anomaly*, https://ntrs.nasa.gov/citations/19940028813, 1994

Johnson, T. V., Yeats, C. M., Young, R., *Space Science reviews Volume on Galileo Mission Overview*, Space Science Reviews, Vol. 60, 1992

Kruger, Harald, et al., *Jovian Dust: Streams, Clouds and Rings*, In: Jupiter. The Planet, Satellites and Magnetosphere. Edited by Fran Bagenal, et al., Cambridge University Press, 2004

Lanzerotti, L. J., et al., *The Lightning and Radio Emission Detector (LRD) Instrument*, Space Science reviews Vol. 60, 1992

Meltzer, Michael, *Mission to Jupiter*, NASA Report SP-207-4231, 2007

Meltzer, Michael, *Mission to Jupiter: A History of the Galileo Project*, NASA History Division, book version published in 2023

NASA, *Galleries | Galileo – NASA Solar System Exploration*, https://solarsystem.nasa.gove/missions/galileo/galleries

NASA, *In Depth | Galileo – NASA Solar System Exploration*, https://solarsystem.nasa.gove/missions/galileo/in-depth

NASA, *Overview | Galileo – NASA Solar System Exploration*, https://solarsystem.nasa.gove/missions/galileo/overview

NASA, *Timeline | Galileo – NASA Solar System Exploration*, https://solarsystem.nasa.gove/missions/galileo/timelne

Niemann, H. B., et al., *Galileo Probe Mass Spectrometer*, Space Science Reviews, Volume 60, 1992

O'Neil, W. J., et al., *Project Galileo at Jupiter*, Acto Astronautica, Vol. 40, Issues 2–8 January-April 1997

Ragent, B., et al., *Galileo Probe Nephelometer Experiment*, Space Science Reviews Vol. 60, 1992

Ragent, Boris, et al., *The clouds of Jupiter: Results of the Galileo Jupiter Mission Probe Nephelometer Experiment*, Journal of Geophysical Research, Vol 103, No. E10, 1998

Rinnert, K., et al., *Measurements of radio frequency signals from lightning in Jupiter's atmosphere*, Journal of Geophysical research, Vol. 103, No. E10, 1998

Seiff, Alvin, and Knight, T. D. C., *The Galileo Probe Atmosphere Structure Instrument*, Space Science Reviews, Vol 60, Issues 1-4, 1982

Sromovsky, L. A., et al., *Galileo Probe Measurements of Thermal and solar Radiation Fluxes in the Jovian atmosphere*, Journal of geophysical research, Vol. 103, E3, 1998

Statman, J. and Deutsch, L., *Galileo's Telcom Using the Low-Gain Spacecraft Antenna*, https://ntrs.nasa.gov/citations/20210005162, 1996

Von Zahn, U. et al., *Helium in Jupiter's Atmosphere: Results from the Galileo Probe Helium Interferometer Experiment*, Journal of Geophysical Research, Vol. 103, No. E10, 1998

von Zahn, U., and Hunten, D.M., *The Jupiter Helium Interferometer Experiment on the Galileo Entry Probe*, Space Science Reviews, Vol. 60, 1992

Yeates, C. M., et al., *Galileo: Exploration of the Jupiter System*, NASA Report SP-479, 1985

Chapter 8
New Horizons Spacecraft

New Horizons was the first spacecraft to conduct a close examination of Pluto, and it went on to investigate a large object in the Kuiper Belt named Arrokoth. The very capable spacecraft returned hundreds of images to Earth and a wealth of scientific information about Pluto, its large moon Charon, and the Kuiper Belt Object Arrokoth.

New Horizons was launched on 19 January 2006 and embarked on a 9.5-year journey to Pluto. Along the way, it received a gravitational boost in velocity from the flyby of Jupiter on 28 February 2007. The spacecraft made a close approach flyby of Pluto on 14 July 2015. It went on to make a close approach flyby of Arrokoth on 1 January 2019. At the time of this writing in 2024, 18.13 years after launch, New Horizons was 8.81 billion km from Earth traveling at 13.69 km/s through the Kuiper Belt.

An artist's rendering of New Horizons near Pluto is given in Fig. 8.1. Pluto was still classified as a planet when New Horizons was launched. It has since been downgraded to a dwarf planet.

Background of New Horizons Program

Pluto was the only planet that had not been explored by 2000 when a mission to Pluto was seriously considered. Missions to Pluto had been studied for several years but had not reached priority for funding. A mission to Pluto had strong support from a group of scientists that called themselves the "Pluto Underground." The eclectic group was led by Dr. Alan Stern. They vigorously lobbied scientific communities and NASA members for a Pluto mission and finally succeeded. There was reason to act promptly since it was necessary to obtain a gravitational velocity boost from

T. Lund, *Spacecraft Exploring Outer Planets*, Springer Praxis Books,
https://doi.org/10.1007/978-3-031-96544-9_8

Fig. 8.1 Painting of New Horizons near Pluto (credit: NASA/JHU-APL/SwRI)

Jupiter to reach Pluto. Launch would have to be before the end of January 2006 or Jupiter would be too far out of alignment to provide the needed boost.

NASA issued a request for proposals in January 2001 for a Pluto mission that would cost <$500 million.

Two of the five teams responding to the request for proposals were selected by NASA for Phase A studies. The two teams selected were a team headed by the Jet Propulsion Laboratory (JPL) of the California Institute of Technology and a team headed by the Applied Physics Laboratory of Johns Hopkins University

(JHU-APL). The principal investigator for the JHU-APL team was Dr. Alan Stern of Southwest Research Institute. The spacecraft proposed by APL was named New Horizons.

Phase A studies lasted for 3 months, after which final proposals were submitted by both teams. NASA selected the Applied Physics Laboratory-Johns Hopkins University team for the Pluto mission on 29 November 2001.

Development of the spacecraft began soon after the contract award, but work stopped after 2 months because of government budgetary issues. Low-level funding for the program was arranged by Congress through earmarks in spending bills to allow the work to continue. The Pluto mission then became part of the New Frontiers probe program. The New Frontiers program was begun by NASA in 2003 to encourage medium-size spacecraft missions. The missions were capped at one billion dollars in development and launch costs. New Horizons was the first spacecraft selected for the New Frontiers program.

Basic design of the spacecraft and identification of science instruments were completed by March 2003, and development and construction began. The finished spacecraft was sent to Johnson Space Center for environmental testing in March 2005. Following testing, it was sent to the Kennedy Space Center in September 2005 for final testing and propellant loading. It was mounted on top of a Star-48 upper stage, and the Star-48 stage with New Horizons attached was mounted on top of the Atlas V/Centaur launch vehicle.

New Horizons was launched from Pad SLC-41 at Cape Canaveral Air Force Station on 19 January 2006. The launch and subsequent burn of the Star-48 upper stage were successful. New Horizons sped away from earth at a velocity of 58,524 km/h, the highest speed ever attained for man-made objects in the vicinity of Earth. The journey to the distant dwarf planet Pluto took 9.5 years. New Horizons arrived at Pluto on 14 July 2015.

Mechanical Configuration of New Horizons

New Horizons was a relatively small, lightweight spacecraft. The launch weight was 478 kg including 77 kg of propellant. A photograph of the spacecraft taken at the Kennedy Space Center Payload Hazardous Servicing Facility is given in Fig. 8.2. The spacecraft is resting on a stand. The size of the spacecraft can be sensed by comparing with technicians. Another view of the spacecraft at Kennedy is shown in Fig. 8.3. The spacecraft was covered with thermal blankets at the time of the photograph.

The large black cylindrical object at the left of the photograph is the radioisotope thermoelectric generator.

A sketch of the spacecraft showing the roughly triangular shape of the frame is shown in Fig. 8.4. The primary structure was 0.7 m high, 2.7 m wide, and 2.1 m long. The high-gain antenna, which was mounted above the frame, was 2.1 m in diameter.

Fig. 8.2 New Horizons spacecraft at Kennedy Space Center Payload Hazardous Servicing Facility (NASA/KSC photograph)

Fig. 8.3 New Horizons spacecraft being serviced at Kennedy Space Center Payload Hazardous Servicing Facility (NASA/KSC photograph)

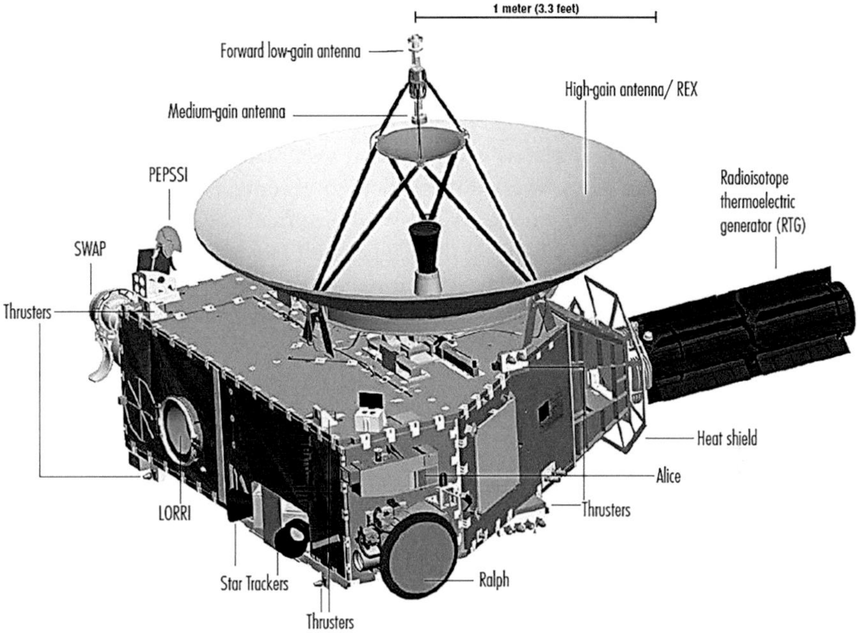

Fig. 8.4 Sketch of New Horizons spacecraft (NASA image)

The spacecraft was constructed with a central cylindrical thrust tube supporting top deck and lower deck panels of the frame. The cylinder was the main load-bearing structure of the spacecraft, and it interfaced with the upper launch stage. All of the panels of the frame were aluminum honeycomb material. A tank holding propellant for the thrusters was mounted inside of the thrust tube. Science instruments were mounted to the frame. Locations of several instruments, identified by their acronyms, are shown in Fig. 8.4.

The coordinate system of the spacecraft sited the +Y-axis along the boresight of the high-gain antenna. The +X-axis was through the radioisotope thermoelectric generator. The +Z-axis was 90° clockwise to the +X-axis.

A total of 16 small monopropellant thrusters were mounted in eight separate groups on the upper and lower decks of the spacecraft. Twelve of the thrusters had a thrust of 0.8 newtons (0.18 pounds) and were used to control the attitude and spin rate of the spacecraft. Four thrusters had a thrust of 4.4 newtons (1.0 pound) and were used for course corrections. Hydrazine fuel reacted with a catalyst in the thrusters. The hydrazine decomposed into very hot gases that exited the thruster nozzle to generate thrust.

The spacecraft was spin stabilized during long periods of cruise during the mission. The nominal spin rate was 5 rpm. The spin axis was along the Y-axis and through the boresight of the high-gain antenna. That allowed the antenna to be pointed at Earth for communications. The spin was stopped, and the spacecraft was stabilized in three axes during the flyby of bodies of interest.

Spacecraft Systems

Electrical Power

Electrical power for New Horizons was produced by a radioisotope thermoelectric generator (RTG). Solar arrays were not practical to generate power at Pluto's distance from the sun. The RTG produced 246 W of electrical power at the beginning of the mission. At the time of flyby of the Kuiper Belt Object, Arrokoth, the available power had decreased to 190 W. At that reduced power level, science instruments had to be selectively powered rather than having them all on at once.

The RTG was cylindrical in shape, 42.2 cm in diameter, and 114 cm long. It weighed 56 kg. The RTG was mounted horizontally at the narrow end of the spacecraft frame as shown in Figs. 8.3 and 8.4. RTGs generate electrical power by applying heat from radioactive decay of a material such as plutonium-238 to a series of thermoelectric couples. Plutonium-238 decays into uranium-234 by emitting alpha particles. The half-life of the decay process is 87.7 years. The kinetic energy of the alpha particles is converted to heat when the particles strike the surrounding material. The RTG contained a total of 10.9 kg of plutonium oxide.

The RTG used pellets of plutonium-238 oxide pressed into cylindrical shape 2.76 cm in diameter and 2.76 cm long. Each pellet was clad with a welded shell of iridium alloy. There were 72 such cladded pellets within a cylindrical region at the center of the RTG. A graphite cylindrical shell enclosed the pellets. A total of 572 silicon-germanium thermoelectric couples were mounted around the graphite shell. The thermocouples were mounted such that one end was in contact with the hot surface and the other end was in contact with the much cooler outer case of the RTG. The RTG had initial thermal output of about 3948 W, which resulted in initial electrical power output from the field of thermocouples of 246 W. The thermal to electrical power efficiency was 6.2%.

The voltage at the output of the RTG was somewhat over 30 V. A regulated 30-V power bus was formed by a shunt regulator unit that generated a 30-V power bus for the spacecraft. Excess power from the regulator was dissipated in resistors attached to radiator plates that radiated excess heat into space.

Propulsion Subsystem

Propulsion for the New Horizons spacecraft consisted of 12 small monopropellant thrusters for attitude and spin control and four larger monopropellant thrusters for trajectory correction maneuvers (TCMs). Hydrazine fuel for the thrusters was carried in a tank located within the central cylindrical thrust tube of the spacecraft structure. The small thrusters were Aerojet MR-103H type that provided 0.8 newtons (0.18 pounds) of thrust. The large thrusters were Aerojet MR-111C type that provided 4.4 newtons (1.0 pounds) of thrust.

The thrusters contained a solenoid valve to admit hydrazine, a low-power heater to keep the valve warm, and a low-power heater to warm the catalyst. The freezing point of hydrazine is about 2 °C.

The reaction chamber in the thrusters contained a bed of type S-405 catalyst. That catalyst was composed of granular aluminum oxide (Al_2O_3), liberally coated with indium. Indium made up about 33% of the catalyst. When hydrazine (N_2H_4) was sprayed on the catalyst, a spontaneous exothermic reaction took place with the hydrazine decomposing into very hot hydrogen, nitrogen, and ammonia gas. The hot gases exiting the thruster nozzle generated thrust.

The thrusters were arranged in eight locations on the spacecraft. Two of the 4.4 newton thrusters were mounted on the bottom deck and thrusted in the –Y direction. The other two 4.4 newton thrusters were mounted on the upper deck and were angled to thrust at a 45-degree angle to the +Y-axis to avoid impinging on the high-gain antenna. The 0.8 newton thrusters were mounted on the upper and lower decks in such a way that pairs of them could be fired to induce rotation about one of the three spacecraft axes. The thrusters were fired in a pulse-width modulation mode. The pulse width and number of pulses were controlled to achieve a required impulse.

Hydrazine fuel for the thrusters was contained in a "blowdown"-type tank where a diaphragm separated pressurized helium gas and liquid hydrazine. The helium forced fuel out of the tank when valves to thrusters were opened. Helium gas initially occupied about 30% of the volume of the tank, and initial pressure was 300 psi at 20 °C. The tank pressure decreased as hydrazine was used, the flow rate decreased, and the thrust decreased. About half of the propellant was expected to have been used by the time New Horizons reached Pluto, and the thrust from the thrusters was expected to be about half of the initial value. The amount of hydrazine used and the temperature were known, so the control system could compensate by firing the thrusters longer for a given impulse.

Guidance and Control

The guidance and control (G&C) system controlled the orientation of the spacecraft. The system included two star trackers, two inertial measurement units, two sun sensors, and two G&C processors. The redundant G&C processors were contained in two redundant integrated electronics modules.

The spacecraft had three attitude control modes: three-axis stabilized, passive spin, and active spin. The three-axis stabilized mode allowed the spacecraft to be rotated about any of the three principal axes to point the spacecraft in a particular direction. This mode was used during the flyby of an object to point particular science instruments at an object or to scan an object.

The passive spin submodes included passive spin hibernation where attitude was not controlled and electrical power demands were minimized. This mode was used during long periods of cruise between Earth and Jupiter and between Jupiter and Pluto. A second passive spin submode was passive spin normal where attitude was not controlled but power was available to other systems of the spacecraft. It was used when checking the health of the spacecraft. The active spin mode was used when attitude control of the spacecraft was required and to adjust the spin rate. The nominal spin rate of the spacecraft was 5 rpm.

The primary sensors for attitude control were two star trackers. Wide-angle pictures of the star field were taken by cameras ten times per second and compared with a stored map of 3000 stars. The comparison was used to calculate the orientation of the spacecraft. Two redundant inertial measurement units were used to determine attitude rates, and this data was combined with star tracker data to provide continuous spacecraft orientation. Two sun sensors were used to track the sun and provide a backup to the star trackers.

Information from the star trackers and inertial measurement units was fed to the guidance and control processor. The processor was a Mongoose V, radiation-hardened, 32-bit processor. The processor was run at a clock speed of 12 MHz. Should spacecraft orientation need adjustment, select thrusters were fired by signals from the G&C processor to adjust the orientation.

Command and Data Handling

The command and data handling (C&DH) system was the brains of the spacecraft. It used a Mongoose V, radiation-hardened, 32-bit processor along with comprehensive flight software to control spacecraft systems. The processor was clocked at 12 MHz.

The C&DH system included two redundant integrated electronics modules (IEMs). Each IEM contained the following cards: C&DH processor, solid-state recorder, instrument interface, downlink card, and uplink card. The C&DH processor was a Mongoose V, radiation-hardened, 32-bit processor that ran at a clock speed of 12 MHz. Only one of the C&DH processors operated at a time, and the other was a redundant backup.

Commands received from Earth were applied to the uplink card. The C&DH software extracted command packets that were either used by the processor or delivered to other spacecraft subsystems. The instrument interface card communicated with each of the scientific instruments. That function included sending commands to each instrument, collecting housekeeping and low-rate science data, collecting and formatting high-rate science data, and sending time markers to each instrument.

The solid-state recorder was a nonvolatile memory type with 64 Gb (Gbit) capacity. Science data could be streamed into the memory at rates up to 13 Mbps. The C&DH software had the ability to perform lossy and non-lossy data compression on the stored data. Raw sensor data was read out of memory, compressed, formed into data packets, and read back into memory for later transmission to Earth.

Telecommunications System

The telecommunications system in New Horizons received and demodulated command data uplinked from the very large antennas of the Deep Space Network (DSN) on Earth. The spacecraft transmitted scientific and engineering data on the downlink to the DSN. At the time of this writing in 2024, communication is still maintained with New Horizons at distances of 8.8 billion km from Earth. Communication with the DSN was carried out in the X-band frequency range. The uplink carrier frequency was 7182 MHz, and the downlink carrier frequency was about 8400 MHz.

Data rates during downlink to Earth depended on signal-to-noise ratio and hence distance to the spacecraft. The data rate was 38 kbps (kbits) during flyby of Jupiter (2.3 million km from Earth), 2 kb during flyby of Pluto (five billion km from Earth), and 1–2 kbps after flyby of Arrokoth 6.6 billion km from Earth.

Antennas

The telecommunication system included four different antennas: high-gain antenna, median-gain antenna, forward low-gain antenna, and aft low-gain antenna. All of the antennas had two feeds, one for right-hand circular polarization and the other for left-hand circular polarization. The high-gain antenna was a Cassegrain reflector type, 2.1 m in diameter with half-power beamwidth of 1.0° and gain of 44 dB at the downlink frequency. It had a specially shaped reflector to enhance performance. The antenna was mounted on the upper deck of the spacecraft with boresight along the +Y-axis. It was the primary antenna for downlinking data at long distances from Earth.

The medium-gain antenna was a parabolic reflector type located at the front of the secondary reflector for the high-gain antenna. The diameter of the reflector was 30 cm. The beamwidth was about 8°, and gain was about 25 dB. The boresight of the medium-gain antenna was along the +Y-axis.

The forward low-gain antenna was mounted at the forward tip of the feed for the medium-gain antenna. The antenna was an open-ended circular waveguide type with a plate with concentric groves attached to the outer wall to shape the antenna pattern. The half-power beamwidth was about 130°, and the gain was 6 dB. The boresight was along the +Y-axis of the spacecraft. The aft low-gain antenna was similar to the forward low-gain antenna. It was mounted at the bottom of the lower deck with boresight along the –Y-axis. The half-power beamwidth of the antenna was about 88°, and the gain was 10 dB.

The antennas were connected to a switch assembly that allowed the right-hand circularly polarized antenna port of any antenna to be connected to the receiver in integrated electronics module A and the left-hand circularly polarized antenna port of any antenna to be connected to the receiver in integrated electronics module B. The transmitter function included two 12-watt traveling wave tube amplifiers. One amplifier fed the right-hand circular polarization selector switch, and the other fed the left-hand circular polarization selector switch such that any antenna could be connected to either amplifier.

Diplexers were connected to the receiver/transmitter end of each of the two antenna selector switches. There were two diplexers. The diplexer allowed the transmit function and receive function to use the same antenna. The receive side of the diplexers was directed to a filter followed by a low-noise amplifier. The amplified received signals from the diplexers were applied to digital receivers in integrated electronics modules A and B. The transmit side of the diplexer was connected to one of the two traveling wave tube amplifiers. The two diplexers were paired with the two traveling wave tube amplifiers.

Receiver and Transmitter

There were two redundant receivers, one located in electronics module A and the other located in electronics module B. Received signals from the low-noise amplifiers were applied to receiver cards in the integrated electronics modules. The receiver cards performed X-band carrier tracking, command detection and demodulation, command decoding, and ranging tone demodulation and provided signals used in the radio science experiment. The receivers were double-conversion superheterodyne type. After conversion of the X-band received signal to a second intermediate frequency of 2.5 MHz, most of the rest of the receiver was implemented by digital circuitry. The amount of regulated power consumed by each receiver was only 2.3 W because of the digital implementation.

The first local oscillator frequency for the receiver and the transmitter frequency was derived from an ultra-stable oscillator (USO). The USO had an output frequency of 30 MHz. The frequency stability with temperature was 1×10^{-12}/°C. There were two USOs for redundancy. Each oscillator had two outputs to allow cross strapping of the feed to the two receivers and the two transmitters.

The transmit card in each of the two integrated electronics modules contained a frequency synthesizer that used the output of the ultra-stable oscillator as reference to develop a frequency that was a factor of four lower than the transmit frequency of 8438 MHz. The synthesized output frequency was multiplied by two and applied to two phase modulators in series. The first biphase modulator imposed the ranging signal. The second biphase modulator was driven by encoded telemetry data. Telemetry data was encoded using turbo code error-correcting code. The frequency of the phase-modulated signal was multiplied by two and amplified to form the X-band exciter output. The downlink data rate was controlled by a 16-bit register that allowed selecting 65,532 different data rates from 6.35 to 104.16 bps.

The X-band exciter outputs from the two integrated electronics modules were applied to a microwave hybrid device that combined the two outputs and applied them to two 12-watt traveling wave tube amplifiers. Either exciter could be used to drive either amplifier. The hookup also allowed both amplifiers to be operated together to increase radiated power since each amplifier had a separate path to the antenna. Each traveling wave tube amplifier fed a separate section of the antenna switch assembly described previously.

There were two unique features of the New Horizons telecommunications system. The first was reconstituting the ranging signal. Rather than simply retransmitting an offset copy of the uplink carrier with noisy range tones embedded, the ranging signal was extracted and reconstituted before it was used to phase modulate the downlink carrier. The clean ranging signal transmitted improved range measurement accuracy on the ground.

The second unique feature was that the downlink carrier was not coherent with the uplink carrier signal. Previous spacecraft generated a downlink carrier that was offset but coherent with the uplink carrier. The two-way Doppler was extracted on the ground and used to compute very accurate relative velocity of the spacecraft. New Horizons compared a reference frequency related to the transmitted frequency with uplink carrier frequency. The difference frequency was measured, and the measurement was included in the telemetry data. Knowing the difference frequency allowed determining two-way Doppler as before.

Flight of New Horizons

New Horizons was launched from pad SLC-41 at Cape Canaveral Air Force Station on 19 January 2006. The launch vehicle was an Atlas V-551 with a Centaur upper stage. Atlas V-551 was a powerful launch vehicle with five strap-on solid-fuel booster rockets attached to the central core first stage. The total thrust of the RD-180 engine in the common core stage plus the five solid-fuel boosters was 12,141 kN (2.73 million pounds). The weight of the launch vehicle plus spacecraft at the time of launch was 571,300 kg, which was easily lofted by Atlas.

The five booster rockets burned for 95 s and then were jettisoned 1 min 50 s after launch. The protective shroud around New Horizons was jettisoned 3 min 50 s after launch. The RD-180 engine of the common core booster continued to burn for 4 min 30 s after launch. The common core stage was then separated from Centaur and dropped away.

The first burn of the Centaur engine lasted for 5 min 30 s and placed the spacecraft into a low Earth parking orbit. The spacecraft coasted in the parking orbit for about 20 min, and then Centaur was fired again 29 min 59 s after launch. The burn lasted for 9 min27 s. Centaur shutdown occurred 39 min 26 s after launch. The second burn increased the velocity to about 12.4 km/s, which was sufficient to escape Earth's gravity. Centaur was then separated from the Star upper stage and dropped away.

The Star-48B upper stage began spin up at 40 min15 s after launch. A spin rate of about 68 rpm was achieved, and then the motor was fired. The burn lasted for 86 s and accelerated New Horizons to a velocity of 16.34 km/s. Star-48B was separated from New Horizons at 44 min55 s after launch. The velocity of New Horizons was the fastest ever achieved for a man-made object leaving Earth.

One day after launch, New Horizons fired its thrusters to reduce the spin rate from 68 rpm to 20 rpm. Three days after launch, the thrusters were fired again to reduce the spin rate to about 5 rpm, a value that would be nominal for the flight. The spacecraft was put into the nominal spin active mode, and the star trackers and attitude control system were used to place the spacecraft in the nominal cruise attitude. Two trajectory correction maneuvers (TCMs) were made to refine the trajectory. The spacecraft had to pass through a small window near Jupiter to achieve gravity boost and course correction to Pluto. The first TCM was on 28 January, and the second was on 30 January 2006. The first and second TCMs changed spacecraft velocity by 5 m/s and 13.3 m/s, respectively. A third TCM was made on 9 March 2006 to fine-tune the trajectory. A velocity change of 1.16 m/s was achieved.

Flyby of Jupiter occurred on 28 February 2007 with the closest approach distance of about 2.3 million km. Gravity assist during the encounter increased spacecraft velocity by 14,000 km/h. New Horizons conducted detailed observations of Jupiter for about 4 months including approach and departure from Jupiter. A total of 36 Gb of data was gathered by its scientific instruments.

New Horizons was placed in the hibernation mode on 28 June 2007 for the long cruise to Pluto. The spacecraft was taken out of hibernation for about 2 months' duration several times during the 8-year cruise to Pluto. Wake-up events occurred on 16 December 2008, 27 August 2009, 29 August 2014, and 6 December 2014.

New Horizons began its approach phase to Pluto on 15 January 2015. A trajectory correction maneuver was made on 10 March 2015 to refine the trajectory. A final 0.25-m/s tweak was made on 29 June 2015. Finally, on 14 July 2015, New Horizons made flyby of Pluto and of its large moon, Charon. Closest approach distance to Pluto was 7800 km, and closest approach to Charon was 28,800 km. A total of 6.25 GB of data was recorded during observation of the Pluto system. That large amount of data required 15 months to download. A photograph of Pluto taken by New Horizons during the flyby is shown in Fig. 8.5.

After surveying Pluto, New Horizons continued on into the Kuiper Belt. The Kuiper Belt is a population of icy bodies that extends from 30 AU to 55 AU from the sun. It is thought to be leftover debris from the formation of the solar system. Two potential Kuiper Belt Objects (KBOs) that could be reached by New Horizons were identified by the Hubble telescope. The New Horizons team selected KBO 2014 MU69 for exploration, which was later named Arrokoth.

Arrokoth was about 1.6 billion km farther out in the Kuiper Belt from Pluto. Five TCMs were performed between the time of flyby of Pluto and the last correction on 1 February 2017. New Horizons was put into hibernation mode on 10 April 2017 when about halfway between Pluto and Arrokoth.

The approach to Arrokoth began in the fall of 2018 by taking increasingly higher definition photographs as the distance decreased. Flyby of Arrokoth occurred on 1 January 2019. The closest approach distance was 3500 km. Detailed science observations began 24 h before the closest approach and continued until well after the flyby. The speed of the spacecraft relative to the sun was 51,948 km/h at Arrokoth flyby. The distance to Earth was 6.62 billion km. An enhanced image of Arrokoth taken during the flyby is given in Fig. 8.6.

Fig. 8.5 Image of Pluto taken by New Horizons in 2015. Credit: NASA/JHU-APL/SwRI

New Horizons continued to travel through the Kuiper Belt. In April of 2022, the mission was extended for 2 years to explore the Kuiper Belt out to 63 AU from Earth. The mission was recently extended to the end of 2025. At the time of this writing in 2024, 18.13 years after launch, New Horizons was 8.81 billion km (58.9 AU) from Earth traveling at 13.69 km/s through the Kuiper Belt.

Scientific Experiments on New Horizons

New Horizons carried seven scientific instruments to investigate Pluto, Arrokoth, and other items of interest in its travels. A list of scientific instruments and summary of purpose of each are given in Table 8.1. A sketch of New Horizons showing locations of instruments is given in Fig. 8.7.

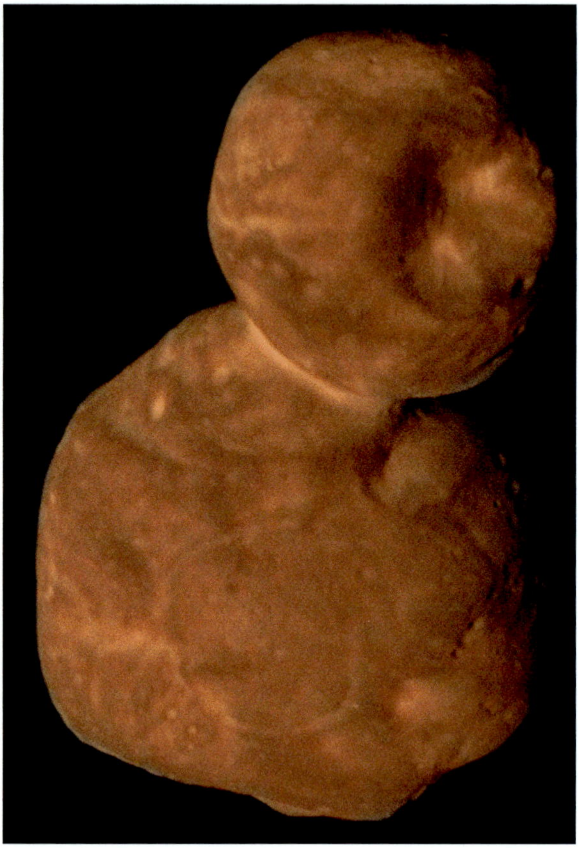

Fig. 8.6 Enhanced image of Arrokoth taken by New Horizons in 2019. Credit: NASA/JHU-APL/SwRI/Roman Tkachenko

Long-Range Reconnaissance Imager (LORRI)

The long-range reconnaissance imager (LORRI) was a high-resolution, narrow field-of-view camera that produced high-quality images of Pluto, its moons, and the Kuiper Belt Object Arrokoth from afar. Its high angular resolution also produced detailed images of the surfaces of Pluto and Arrokoth during flybys.

LORRI consisted of an optical telescope, a focal plane unit, associated support electronics, and a door over the input aperture. The telescope was a Ritchey-Chrétien type with hyperbolic primary and secondary mirrors. The input aperture was 20.8 cm in diameter, and the focal length was 263 cm. The telescope was mounted within a cylindrical baffle tube to minimize stray light. A moveable cover (door) was attached to the structure of the spacecraft and could be closed over the aperture to protect the camera until the time for use. The camera was installed on the –X side of the spacecraft with the field of view parallel to the –X-axis. A photograph of the camera being installed in the spacecraft is shown in Fig. 8.8.

Table 8.1 Scientific instruments carried by New Horizons spacecraft

Experiment	Purpose	Principal investigator
Long-range reconnaissance imager (LORRI)	Generate high-resolution hemispheric panchromatic maps of Pluto and Arrokoth and high-resolution select images	A. Chung, JHU-APL
Ralph: Visible color imager (MVIC) and IR spectral imager (LEISA)	MVIC imaged objects through clear and four color filters. LEISA mapped constituents of the atmosphere using near-IR wavelengths	S. A. Stern, SwRI
Ultraviolet imaging spectrograph (Alice)	Conduct spectroscopic investigation of atmospheres and surfaces at extreme ultraviolet (EUV) and far-ultraviolet (FUV) wavelengths	S. A. Stern, SwRI
Pluto energetic particle spectrometer science investigation (PEPSSI)	Investigate neutral atoms that escape Pluto's atmosphere and become charged by interaction with the solar wind. Measure mass, energy, and spatial distribution of charged particles	R. McNutt, JHU-APL
Solar wind around pluto (SWAP)	Investigate interaction of particles around Pluto with the solar wind	D. McComas, SwRI
Venetia Burney student dust collector (VB-SDC)	Map the size and density of dust particles along New Horizons' path through the solar system	M. Horanyi, University of Colorado
Radio science experiment (REX)	Use occultation to determine the temperature and pressure of the atmosphere near the surface of Pluto. Measure surface temperatures by radiometer	L. Tyler, Stanford University

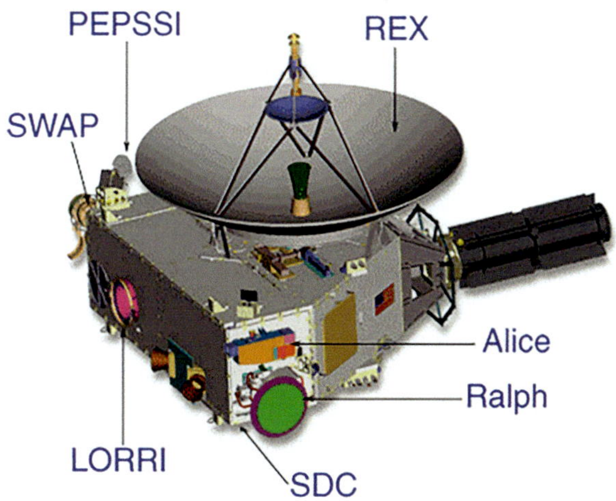

Fig. 8.7 Location of instruments on New Horizons (NASA sketch)

The field of view of the camera was 0.29 by 0.29°. The focal plane unit, which was mounted to the telescope structure, employed a CCD with an active field of 1024 by 1024 pixels. The charge on each pixel of the CCD was sampled and digitized by a 12-bit analog-to-digital converter. The resolution of each pixel was 4.94 μrad (2.82×10^{-4} degrees). Typical exposure time by the electronic shutter was

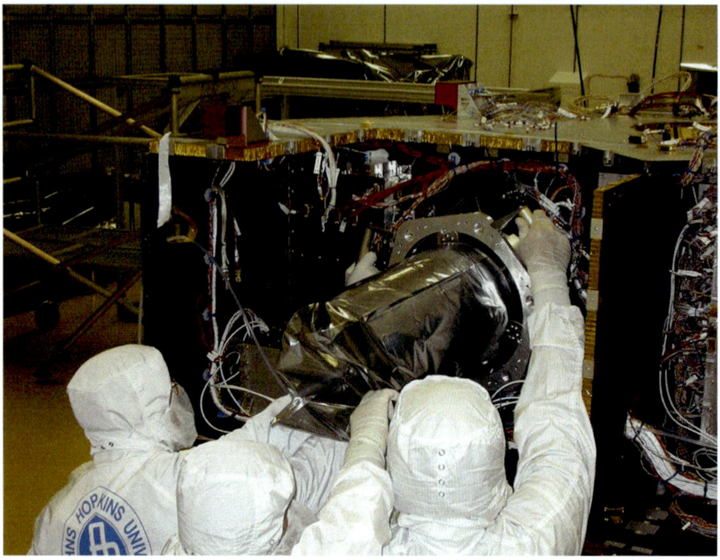

Fig. 8.8 Installation of LORRI on spacecraft (NASA photograph)

between 50 and 200 ms. The image frame transfer time was 13 ms. Timing and control of the CCD and image data were managed by circuitry within a field programmable gate array. A serial image data stream was sent by a low-voltage differential signaling (LVDS) data bus to the associated support electronics unit.

The associated support electronics (ASE) processed the data and sent it via an LVDS bus to instrument interface cards in redundant integrated electronics modules. The image data was formatted and put into memory for later transmission to Earth. The ASE also received commands from the instrument interface cards to control the LORRI.

Development of LORRI was managed by the Johns Hopkins University Applied Physics Laboratory. The telescope was built by SSG Precision Optronics of Wilmington, Massachusetts, and the CCD was furnished by e2v of Chelmsford, England.

A few images produced by LORRI reveal its capability. Figure 8.9 is an image taken at eight million km from Pluto a week before flyby. The heart-shaped, icy Tombaugh Regio plain is clearly visible.

Images became more distinct each day during approach. A close-up image produced at a distance of 17,000 km, just before the closest approach, is shown in Fig. 8.10. The image shows a section of the shoreline between the icy surface of the Sputnik Planitia basin and mountains of great blocks of water ice over 2 km high. An image taken by LORRI of Pluto's large moon, Charon, is given in Fig. 8.11. The image was made during the 14 July 2015 flyby of the moon at a range of 74,176 km.

LORRI also brought the Kuiper Belt Object named Arrokoth into sharp view during the 1 January 2017 flyby at a distance of 6640 km. The image is shown in

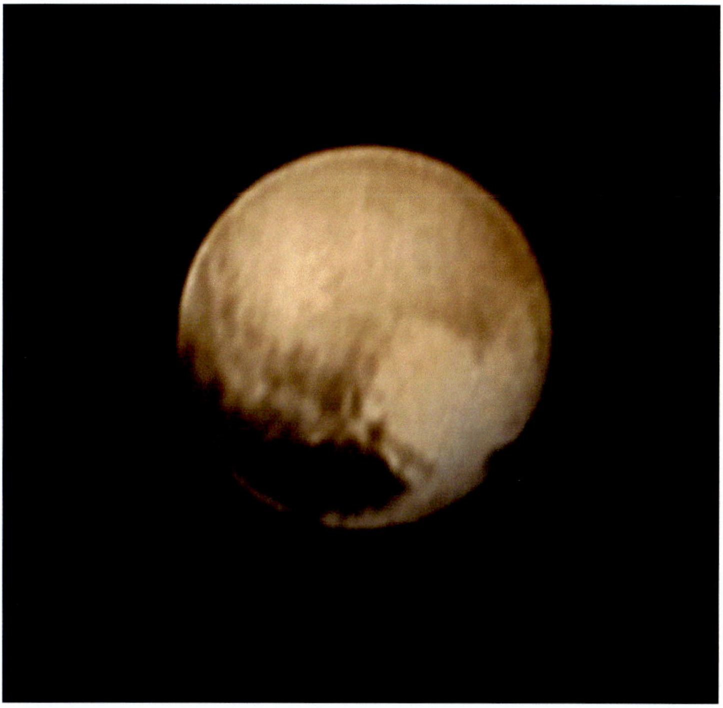

Fig. 8.9 Image of Pluto from a distance of eight million km taken by New Horizons LORRI camera. (Credit: NASA/JHU-APL/SwRI)

Fig. 8.12. The resolution of the image is 33 m/pixel. Arrokoth is about 35 km long. LORRI imaged in grayscale. A composite image with color added from a lower-resolution image taken by MVIC was shown in Fig. 8.6 previously.

Ralph: Visible Color Imager (MVIC) and IR Spectral Imager (LEISA)

The Ralph instrument contained two separate channels: multispectral visible imaging camera (MVIC) and linear etalon imaging spectral array (LEISA). Development of Ralph was a joint effort involving the Southwest Research Institute (SwRI), Ball Aerospace & Technologies Corp., and the NASA Goddard Space Flight Center (GSFC).

The name "Ralph" may have been chosen to add a little levity to the otherwise serious effort of developing the instrument on time. Ralph was operated in conjunction with an ultraviolet spectrometer named Alice. Ralph and Alice were an inseparable pair in a popular television show at the time, "The Honeymooners."

Fig. 8.10 Image of shoreline between Sputnik Planitia basin and ice mountains taken by New Horizons LORRI. (Credit: NASA/JHU-APL/SwRI)

Ralph contained a telescope made up of three off-axis mirrors that concentrated light onto two different focal planes. Light was applied to a dichroic beam splitter before the focal planes. The beam splitter transmitted infrared wavelengths longer than 1.1 μm through to the LEISA focal plane, and it reflected shorter wavelength light to the MVIC focal plane.

The telescope had an input aperture of 75 mm diameter and focal length of 658 mm.

Multispectral Visible Imaging Camera (MVIC)

Images from the telescope were reflected by the dichroic beam splitter onto a customized CCD detector located at the focal plane of the multispectral visible imaging camera (MVIC). The CCD was furnished by the e2v Corporation of Chelmsford, England. The custom CCD had six time delay and integration (TDI) 5024 × 32 pixel CCDs and one 5024 × 264 pixel frame transfer CCD on a single device.

Fig. 8.11 Image of Pluto's moon, Charon, taken by New Horizons LORRI from a distance of 74,176 km. (Credit: NASA/JHU-APL/SwRI/Alex Parker)

The TDI implementation avoided smearing of the image by motion of the platform. The registers of the CCD were clocked in synchronization with platform motion, and the charge was coupled from row to row in the multirow CCD. The rotation rate of the spinning spacecraft was known from internal measurements. The frame transfer array was divided up into a 5024 × 128 pixel image gathering area and a 5024 × 136 pixel image storage area.

A filter with five segments was placed in the light path over the six CCDs. Two CCD arrays had panchromatic filters (400–975 nm wavelength). Each of the other four CCD arrays had individual color filters. The filters were blue (400–550 nm), red (540–700 nm), near IR (780–975 nm), and methane (860–910 nm). Individual color images were used to made composite color images during processing on Earth. The field of view of each pixel was 19.8 by 19.8 μrad. The field of view of the framing camera was 5.7 by 0.146°. The frame transfer integration time could be selected between 0.25 s and 10 s.

A composite image of Pluto and its moon, Charon, assembled from images gathered by MVIC is shown in Fig. 8.13. A section of the icy surface of Pluto's Sputnik Planum as imaged by MVIC is shown in Fig. 8.14. That image is about 400 km wide.

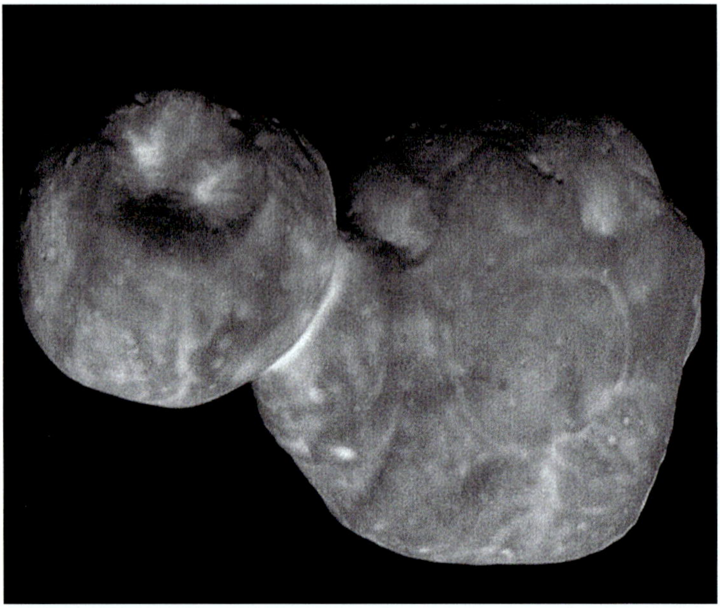

Fig. 8.12 Image of Kuiper Belt Object Arrokoth taken by New Horizons LORRI (Credit: NASA/JHU-APL/SwRI/National Optical Astronomy Observatory)

Linear Etaton Imaging Spectral Array (LEISA)

The LEISA instrument was developed by Goddard Space Flight Center in Greenbelt, Maryland.

Infrared light from the telescope with wavelengths longer than 1.1 μm passed through the dichroic beam splitter and was applied to the LEISA focal plane. Central to the focal plane was an etalon-type spectrometer. The etalon element consisted of several thin layers of sapphire material that dispersed light in angle as a function of wavelength (similar to a prism). The disbursed light spectrum was applied to the detector with the dispersion direction across rows. This, in effect, created 256 separate wavelength filters.

The detector was a 256 by 256 pixel mercury cadmium telluride (HgCdTe) Picnic-type device made by Rockwell Scientific Corp. The Picnic device was structured with four independent quadrants. Each quadrant of 126 by 128 pixels had a horizontal shift register and a vertical shift register to clock out the array. Each row of the detector represented a unique wavelength. The target was scanned across the rows by rotation (spin) of the spacecraft. The field of view of each pixel was 60.8 by 60.8 μrad.

LEISA scanned across the entire illuminated face of Pluto during the flyby. Spectral data from LEISA allowed generating a map of spatial distribution of the main molecules of nitrogen (N_2), methane (CH_4), carbon monoxide (CO), and water

Fig. 8.13 Composite image of Pluto and its moon, Charon, from images gathered by New Horizons MVIC (Credit: NASA/JHU-APL/SwRI)

(H_2O). A study by B. Schmitt et al. (2017) presents a set of curves of reflectance as a function of wavelength for various sites on Pluto. That information indicated the predominant molecule for various sites as follows:

Site	Predominant molecule
North polar region	H_2O
Venera Terra	CH_4
Sputnik Planum	H_2O
Pulfrich crater	CH_4

Solar Illumination Assembly (SLA)

The solar illumination assembly included a small optical lens with an aperture 4 mm in diameter. The lens was coupled by an optical fiber to project a diffuse image of the sun onto both the MVIC and LEISA focal planes. The field of view of the SLA was displaced 90° from the telescope in the X–Z plane. The SLA was

Fig. 8.14 Image of portion of Pluto's Sputnik Planum taken by New Horizons' MVIC (Credit: NASA/JHU-APL/SwRI)

also used to examine the atmosphere by viewing the limb of Pluto during solar occlusion.

Electronics for Ralph

Electronics subassemblies for Ralph were contained in an electronics box mounted to the spacecraft under the telescope detector assembly. The electronics provided biases and clocking to both focal planes, amplified signals from the CCDs, and performed analog-to-digital conversion of the signals. Analog CCD data was converted to digital form at 12 bits per pixel. The electronics also interpreted commands from the spacecraft and formatted and transferred science data to the spacecraft.

Ultraviolet Imaging Spectrograph (Alice)

Alice was an adjunct to Ralph, extending spectroscopic coverage to ultraviolet wavelengths. The names Ralph and Alice came from a popular television show at the time called the "Honeymooners." The purpose of Alice was to conduct spectroscopic investigation of atmospheres and surfaces at extreme-ultraviolet (EUV) and far-ultraviolet (FUV) wavelengths. The wavelength range covered was 87–187 nm (520–1870 Angstroms).

Alice equipment included a telescope with two inputs and a Rowland circle-type spectrograph. The main path through the telescope, called the airglow channel (AGC), had a rectangular input aperture 40 mm by 40 mm in size. Light through the rectangular aperture struck an off-axis paraboloidal primary mirror located at the far end of the instrument. The second input to the telescope was the solar occultation channel (SOC), which was designed to view the sun through the atmosphere at the time of occultation by Pluto. SOC had a circular input aperture 1 mm in diameter. Light from the SOC aperture was reflected into the telescope path by a small mirror. The boresight of SOC was 90° displaced in the Y–Z plane from that of AGC.

Light at the entrance to the primary mirror in the telescope was focused onto the entrance slit of a Rowland circle-type spectrograph 15 cm in diameter. The entrance slit was composed of a separate slit for the airglow and occlusion channels. The airglow slit was 1° wide in the spectral axis and 4° long in the spatial axis. The SOC channel had an opening 2° by 2° adjacent to the airglow slit. The entrance slits were at the periphery of the circle, and a diffraction grating was located on the other side of the circle.

A diffraction grating was located at one side of the circle with the center of the grating along the diameter. The entrance slits were at the periphery of the circle but located away from the centerline of the grating so that light struck the grating at an angle. The dispersion direction of light at various wavelengths from grating was along the periphery of the circle. Microchannel plate detectors were curved to fit around the periphery of the circle to intercept the dispersed wavelengths of light.

Electronics following the detectors included preamplifiers, time-charge digital converters, and pulse-pair analyzer circuits. Only those signals above a threshold were digitized by the time-charge digital converters. Those digital signals were identified by pixel address locations. A 10-bit spectral address and a 5-bit spatial address in the Y-axis accompanied the signal amplitude data sent to command and data handling circuits for processing and data handling.

Results of solar occlusion measurements of the atmosphere of Pluto by Alice were reported by S. A. Stern et al. (2015). The data indicated column densities of molecules of nitrogen (N_2) = $9 \times 10^{10}/cm^2$, methane (CH_4) = $5.6 \times 10^{15}/cm^2$, and higher hydrocarbons = $2.6 \times 10^{15}/cm^2$. Spectral data had not been downloaded at the time of the study.

Pluto Energetic Particle Spectrometer Science Investigation (PEPSSI)

The Pluto energetic particle spectrometer science investigation (PEPSSI) instrument investigated neutral atoms that escaped Pluto's atmosphere and became charged by interaction with the solar wind. The instrument measured the mass, energy, and spatial distribution of those charged particles.

The PEPSSI instrument had six sectors arranged in a ring. Half of the ring contained six input sections, and the other half contained six measurement sections. Each section pair consisted of a collimator, entrance foil, a drift space for time-of-flight measurements, stop foil, and solid-state detector. The start and stop foils were biased at −2600 V. Electrons and ions passed through the collimator and passed through the entrance foil. Secondary electrons generated by the impact were gathered by a start anode to generate a start pulse. Incoming ions then passed through a 6-cm-long drift space before striking and passing through a stop foil. Secondary electrons were again generated that were gathered to form a stop pulse. The particle then struck the solid-state detector where its energy was measured. The time between start and stop pulses was measured and combined with the drift distance to computed particle velocity. Knowing energy and velocity of the ion, its mass could be determined.

The energy of electrons was measured over a range of 25 keV to 500 keV. Energy and time of flight of protons were measured over a range of 25 keV to 1 MeV. Energy and time of flight of molecular ions such as nitrogen (N_2) and oxygen (O_2) were measured over a range of 100 keV to 1 MeV.

The PEPSSI instrument was mounted at the corner of the upper deck of the spacecraft opposite to the side holding Ralph and Alice instruments. The total field of view of the six sections of the instrument was 160 by 12°. The angular resolution was 25 by 12°.

A study by P. Kollmann et al. (2019) gives some measured results for He^+ ions. Curves were given of intensity vs. energy in keV. Intensity had dimensions of 1/(keV cm^2 sr s). The curves at one instance of time show nearly constant intensity of 2000 from energy levels of 2–7 keV. The intensity dropped off with increasing energy to 1000 at 10 keV, 220 at 2 keV, 50 at 5 keV, and 10 at 100 keV.

Solar Wind Around Pluto (SWAP)

The SWAP instrument investigated interaction of particles escaping from the atmosphere of Pluto with the solar wind. It measured velocity and density of solar wind particles to determine the amount of slowing by local particles. The solar wind at Pluto is about a factor of 1000 less dense than around Earth since Pluto orbits about 32 astronomical units (AU) from the sun. The instrument had to be very sensitive to measure that tenuous solar wind.

The instrument was mounted to the –Z side of the spacecraft near the top and near the –X side. The field of view was about 200° in the X–Z plane and 10° in the Y–Z plane.

The electro-optical section of SWAP included a retarding potential analyzer (RPA) and an electrostatic analyzer (ESA). The ESA consisted of a path for ions between two curved teardrop-shaped elements. A particular voltage across the two curved plates allowed an ion with particular energy-to-charge ratio to follow the curved path and strike a detector at the narrow end of the teardrop shape. The voltage across the curved elements was stepped to accept a spectrum of ions with matching energy-to-charge ratios.

There were two input ports to the device in a line that was tangent to the broad curvature of the teardrop-shaped gauntlet. A retarding potential analyzer (RPA) consisted of a screen at a retarding voltage potential that was placed across each of the two entrance ports to the instrument. Only ions with energy-to-charge ratios higher than the threshold set by the retarding potential passed through the RPA. The retarding potential could be changed to reject ions with lower energy-to-charge ratios. Ions with larger energy-to-charge ratios passed through the RPA to the ESA. The two openings into the RPAs were covered by long, narrow doors to protect the instrument during ground handling and launch. The doors were opened after launch.

Ions selected by the ESA traveled through the gauntlet, struck a very thin carbon foil at the narrow end of the gauntlet, and passed through to a primary detector. Secondary electrons generated by striking the foil were detected by a secondary detector. Triggering on noise was minimized by requiring coincidence between the primary detector and secondary detector outputs to record an ion.

Electronics to control the instrument, generate high voltages needed for bias of the elements, and process science data was contained on printed circuit cards within the SWAP unit. SWAP communicated with the spacecraft for command input and science and housekeeping data output through a redundant RS-422 data bus.

SWAP measurements of solar wind velocity in the vicinity of Pluto were undertaken to see if the velocity was being slowed by interaction with molecules escaping from the atmosphere of Pluto. The measurements indicated that the solar wind velocity was nearly constant during the approach, suggesting that very few molecules were escaping from the atmosphere.

Venetia Burney Student Dust Collector (VBSDC)

The Venetia Burney student dust counter (VBSDC) was designed to map the size and density of dust particles along New Horizons' path through the solar system. The VBSDC instrument was included in New Horizons as part of NASA's Education and Public Outreach program. The instrument was designed, built, and operated by students at the University of Colorado with guidance from New Horizons scientists.

The name was chosen to honor Venetia Burney who was an 11-year-old student who suggested the name "Pluto" for the ninth planet in 1930.

The dust collector consisted of 12 separate sensors, each 14.2 by 6.5 cm in size, mounted adjacent to one another on a flat detector support panel. The material used was polyvinylidene fluoride plastic film. That material generated an electrical signal when penetrated by a dust particle. Larger particles produced larger signals. Two sensors were mounted on the underneath side of the support panel to provide a noise reference. The support panel was mounted above the thermal blankets on the –Y side of the spacecraft.

The VBSDC detected one impact during the period between 5 days before the closest approach to Pluto and 5 days after the closest approach. From that one measurement, F. Bagenal et al. (2016) estimated an upper limit of dust density in the Pluto system of 4.6 /km^3.

Radio Science Experiment (REX)

The radio science experiment (REX) was designed to investigate the temperature and pressure of the atmosphere near the surface of Pluto by occultation. Radiometer measurements were also made of the thermal emission temperature of the surface of Pluto on the day- and nightsides. A secondary objective of the experiment was to search for an ionosphere of Pluto. The experiment made use of the high-gain antenna, receiver, and transmitter of the communications system on New Horizons. Existing equipment was augmented by a small amount of special signal processing circuitry.

The experiment received highly stable signals at X-band transmitted from the ground. Four antennas of the Deep Space Network each transmitted 20 kW of power at a wavelength of 4.2 cm to the spacecraft. Two of the antennas transmitted right circular polarization, and the other two transmitted left circular polarization. Two separate receivers in the spacecraft received the refracted signal after it passed through the atmosphere during occultation. The received signal was sorted by polarization, and signals of each polarization were applied to a separate receiver.

Separate ultra-stable oscillators (USOs) were used as reference for the two receivers and to set the transmitted frequency. The difference frequency between that received from the ground and a reference from the USO was measured and transmitted to the ground. Knowing the transmitted frequency from the ground, the frequency of the USO, and the difference frequency allowed two-way Doppler shift to be computed on the ground for an accurate measure of relative velocity of the spacecraft.

A portion of the 4.5 MHz intermediate frequency of the receiver was taped off, filtered, and converted to 12-bit digital format. A digital down-converter converted the digital signal to baseband (0 Hz). That digital signal was applied to two paths: a digital integrator and a digital filter. The integrator generated 16-bit data proportional to the amplitude of the 4.5 MHz IF signal. The filter passed perturbations in the occluded received signal having frequencies lower than 1000 Hz. The filtered signal was digitized to 16 bits. Both the filtered signal and the amplitude signals were sent to the solid-state recorder for later playback to Earth.

Combined phase measurements in the two receivers allowed determining pressure and temperature of the atmosphere during occultation of the radio signal. The occultation event happened soon after the closest approach. Curves of altitude vs. temperature and pressure were presented in the work by S. A. Stern et al. (2018) for lower, middle, and upper atmospheres. A few data points from those curves are given below:

Altitude, km	Temperature, K	Pressure, millibars
0	40	10.2
10	97	7.8
20	106	6.5
40	105	4.3
100	90	1.2
200	81	0.2
300	70	0.032

Radiometer measurements were made by scanning the high-gain antenna across the surface of Pluto on the nightside and dayside. Detailed data was reported by Linscott et al. (2021) for two scans made shortly after passing the closest approach. Both scans were made largely on the nightside of Pluto. The first scan was initiated 5.5 min after the closest approach. The spacecraft was about 15,000 km from the center of Pluto (13,812 km above the surface) at the time of the scan. The scan, which took 41 s, traced a path diagonally across the planet from Pluto coordinates of about 38° south, 336 east to 7.1° north, 179 east. Brightness temperatures computed from the measurements indicated temperature of about 18 K at the scan entrance limb, a broad peak of 30 K about a quarter of the way through the scan, and 21 K at the terminator into daylight and at the exit limb.

A second scan, which passed close to the south pole, was made from coordinates of 11.9° south, 146° east to 55.9° south, 31.3° east. The most southern point in that scan occurred at latitude of 84.6° south. The scan lasted for 25 s. Brightness temperatures were about 17 K at the entrance limb, 19 K at the terminator into night, 33 K near the south pole, and 38 K at the exit limb.

References

Bagenal, F., et al., *Pluto's interaction with its space environment: Solar wind, energetic particles, and dust*, Science, Vol. 351, Issue 6279, March 2016

DeBoy, Christopher C., *The RF Telecommunications System for the New Horizons Mission to Pluto*, IEEE Aerospace Conference Proceedings 3:1478 Vol. 3, 2004

Fountain, Glen H., et al., *The New Horizons Instrument Suite*, Johns Hopkins APL Technical Digest, Vol. 37, Number 1, 2023

Fountain, Glen H., et al., *The New Horizons Spacecraft*, https://www.boulder.swri.deu/pkb/ssr/ssr-fountain.pdf, 2007
Kollmann, P., et al., *Pluto's Interaction with Energetic Heliospheric Ions*, Journal of Geophysical Research: Space Physics/Vol 124, Issue 9, 2019
Kusnierkiewicz, David Y., *A description of the Pluto-bound New Horizons spacecraft*, Acta Astronautica Vol 57, 2005
Linscott, I. R., et al., *High-resolution radiometry of Pluto at 4.2 cm with New Horizons*, Icarus, Vol. 363, July 2021
McComas, D., et al., *The Solar Wind Around Pluto (SWAP) Instrument Aboard New Horizons*, Space Science Reviews, Vol 140, Issue 1-4, October 2008
McNutt, Ralph L., et al., *The Pluto Energetic Particle spectrometer Science Investigation (PEPSSI) on the New Horizons Mission*, Space Science Reviews, Vol 140, Issue 1-4, October 2008
NASA, NSSDCA/COSPAR ID: 2006-001A, *New Horizons Kuiper Belt Flyby*, 01 July 2025
Reuter, Denis C., et al., *Ralph: A Visible/Infrared Imager for the New Horizons Pluto/Kuiper Belt Mission*, Space Science Reviews, Vol. 140, 2008
Schmitt, B., et al., *Physical state and distribution of materials at the surface of Pluto from New Horizons LEISA imaging spectrometer*, Icarus, Vol 287, May 2017
Siddiqi, Astif A., *New Horizons*, A Chronicle of Deep Space Exploration, 1958–2016, NASA History Program Office, 2018
Spaceflight 101, *Project History & Mission Profile*, https://spaceflight101.com/newhorizons/project-history-mission-profile, 2024
Stern, S. A., et al., *The Pluto system: Initial results from its exploration by New Horizons*, Science, Vol. 350, Issue 6258, 2015
Stern, S. Alan, et al., *The Pluto System After New Horizons*, Annual Review of Astronomy and Astrophysics, Vol 56:357–392. 2018
Stern, S. Allan, et al., *New Horizons and Planetary Exploration*, Johns Hopkins APL Technical Digest, Vol. 37, Number 1, 2023
Weaver, Harold A., et al., *Science Highlights from NASA's New Horizon Mission*, Johns Hopkins APL Technical Digest, Vol. 37, Number 1, 2023
Zastrow, Mark, *Pluto Revealed: New Horizons' Historic Flyby, Five Years Later*, Astronomy, July 2020

Chapter 9
Juno Spacecraft

The Juno mission to Jupiter was a great success. The very capable spacecraft transmitted hundreds of images to Earth and a wealth of scientific information about Jupiter and its large moons. The spacecraft was still operating and in polar orbit around Jupiter at the time of this writing in 2024. Juno entered orbit in July 2016, and it has completed the prime mission, which was a 5-year, 35-orbit information-collecting endeavor. Juno is presently on an extended mission, which is planned to last until 2025. An artist's rendering of Juno in orbit around Jupiter is given in Fig. 9.1. Juno, in Roman mythology, was the long-suffering, feisty wife of the lord of the gods, Jupiter.

The main goal of the Juno mission was to gather detailed information about Jupiter to further understand its formation. Jupiter, by far the largest planet in the solar system, had a profound influence on the formation of other planets and ordering of their orbits. Understanding the formation of Jupiter, its makeup under the thick clouds, and what generates the strong magnetic field are questions important to those striving to learn about the formation of the solar system.

T. Lund, *Spacecraft Exploring Outer Planets*, Springer Praxis Books,
https://doi.org/10.1007/978-3-031-96544-9_9

Fig. 9.1 Artist's rendering of Juno in orbit around Jupiter (NASA/JPL image)

Background of Juno Program

The main goal of the Juno mission was to gather detailed information about Jupiter to further understand its formation. Jupiter, by far the largest planet in the solar system, had a profound influence on the formation of other planets and ordering of their orbits. Understanding the formation of Jupiter, its makeup under the thick clouds, and what generates the strong magnetic field are questions important to those striving to learn about the formation of the solar system.

Jupiter had been examined by the flyby spacecraft Pioneer 10, Pioneer 11, Voyager 1, Voyager 2, and the Cassini Saturn orbiter. More detailed exploration was undertaken by the equatorial orbiting Galileo spacecraft. Many questions about Jupiter remained, and so the polar orbiting Juno mission was planned. The goal was to help understand the Jupiter system and unlock mysteries of Jupiter's formation and perhaps the formation of the solar system.

The Juno mission was part of the New Frontiers probe program, which was begun by NASA in 2003 to encourage medium-size spacecraft missions. The missions were capped at one billion dollars in development and launch costs. The first spacecraft selected for the New Frontiers program was New Horizons, described in Chap. 8 of this book. Juno was the second spacecraft to be selected.

The National Research Council identified the Juno mission as a candidate for the New Frontiers program in 2003. The Juno mission was approved by Congress and received a go-ahead in June 2005.

The Juno program was managed by the Jet Propulsion Laboratory (JPL) of the California Institute of Technology. JPL is an affiliate of NASA as a Federally Funded Research and Development Center. The Juno spacecraft was built by Lockheed Martin Space Systems near Denver, Colorado. Scientific instruments were designed and built by various academic and scientific groups. The principal investigator for Juno was Scott Bolton of Southwest Research Institute. The project manager at JPL was Jan Chodas and the project manager at Lockheed Martin was Tim Halbrook.

Construction of the basic spacecraft was completed at Lockheed Martin, and it was transported by aircraft to Astrotech Space Operations (ASO) in Titusville, Florida, on 8 April 2011. ASO, which is operated by Lockheed Martin, has performed final assembly and checkout of spacecraft for over 30 years. It performed the final work on Juno including filling the liquid propellant tanks, installing the science instruments, and installing the solar arrays. After complete checkout of the finished spacecraft, ASO transported Juno to the Cape Canaveral launch facility on 27 July 2011. Juno was launched by an Atlas 551 first stage and Centaur upper stage launch vehicle from Pad SLC-41 at Cape Canaveral Air Force Station on 5 August 2011. After leaving Earth, Juno was placed in an orbit around the sun that returned the spacecraft to the vicinity of Earth. A gravitational assist from Earth (slingshot maneuver) increased spacecraft velocity and set Juno on a trajectory to Jupiter.

Juno arrived at Jupiter on 4 July 2016, and it was inserted into a polar orbit. It has completed its 5-year primary mission, and the orbiting spacecraft is presently on an extended orbiting mission.

Mechanical Configuration

Juno was a large spacecraft, measuring 4 m tall, and the core section was about 4.5 m wide. Three solar arrays were attached to the core. When unfurled, two of the arrays were 8.9 m long and 2.9 m wide. The third array was foreshortened to mount a magnetometer boom at the end. A photograph of Juno with solar panels folded being lifted at the Astrotech Space Operations facility before being taken to Cape Canaveral is shown in Fig. 9.2. The arrays were folded to fit into the launch shroud. The large triangular structure at the right side of the picture was a boom to hold magnetometers. The painting of Juno in Fig. 9.1 shows the solar arrays extended.

Another view of the spacecraft with only one of the three solar panels installed that shows some of the structure of the spacecraft is given in Fig. 9.3. A sketch of the spacecraft from NASA Space Science Data Coordinated Archive with labels identifying major elements is given in Fig. 9.4.

The spacecraft structure contained two octagon-shaped decks. The spacing across the flats of the octagons was about 4.5 m. The spacing between the two decks was about 1.7 m. A strong central cylinder with high tensile stiffness held the decks in place. Six spherical propellant tanks and three helium pressurization tanks were fit in the space around the cylinder between the decks.

Fig. 9.2 Juno being hoisted at Astrotech Space Operations (NASA photograph, cropped by author)

A boxlike structure, referred to as the vault, was mounted in the center on top of the upper deck. The vault protected spacecraft electronics from high-level radiation in the vicinity of Jupiter. The vault, which was about 1 m on a side, was made of titanium 1 cm thick. The vault is apparent on the upper shelf in the photograph in Fig. 9.3. The vault weighed about 182 kg empty. It protected 20 electronic assemblies mounted inside. The total weight with electronic assemblies was about 200 kg.

A high-gain antenna, which was a parabolic reflector type 2.5 m in diameter, was mounted by trusses above the vault. A medium-gain antenna and a low-gain antenna were mounted on struts just to the left of the high-gain antenna in Fig. 9.3. Two low–gain antennas were mounted to the bottom of the lower shelf. Several scientific

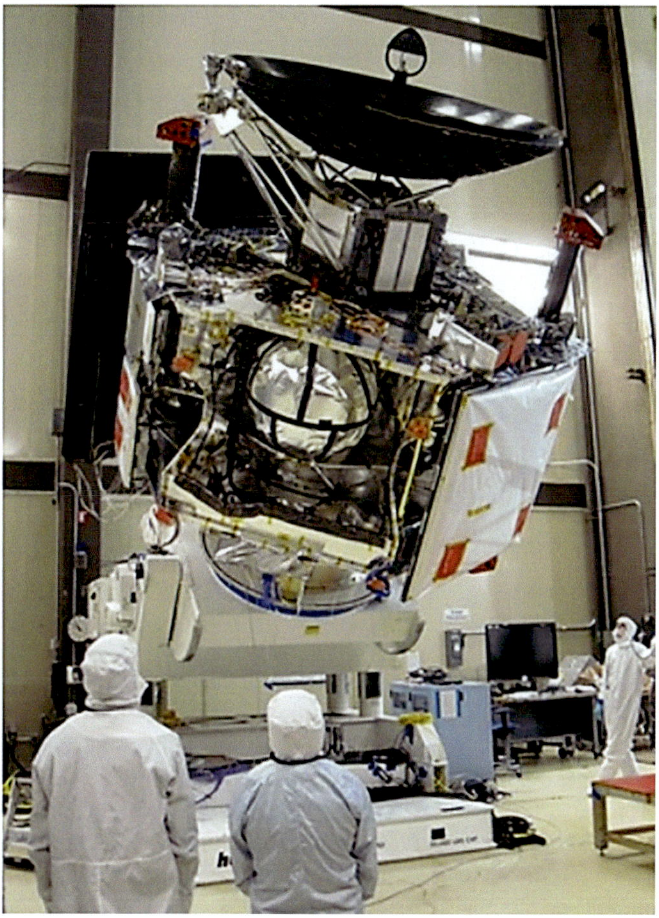

Fig. 9.3 View of Juno with two solar panels removed (NASA image)

instruments were mounted on the top of the upper shelf and on the bottom of the lower shelf.

The main rocket engine, which provided 635 newtons (143 pounds) of thrust, protruded through the center of the lower deck. The engine thrusted in the opposite direction to the boresight of the parabolic antenna.

The engine used hydrazine for fuel and nitrogen tetroxide for oxidizer. The fuel and oxidizer ignited upon contact in the engine.

A total of 12 thrusters, mounted in four groups of three, were used to control attitude and spin rate of the spacecraft. They were also used for small velocity changes. The thrusters used hydrazine fuel that reacted with a catalyst in the thrusters to develop four newtons (0.9 pounds) of thrust. Propellants were contained in six spherical tanks about 0.9 m in diameter. Four of the tanks held hydrazine and two held oxidizer. Hydrazine was used by both the main engine and the thrusters.

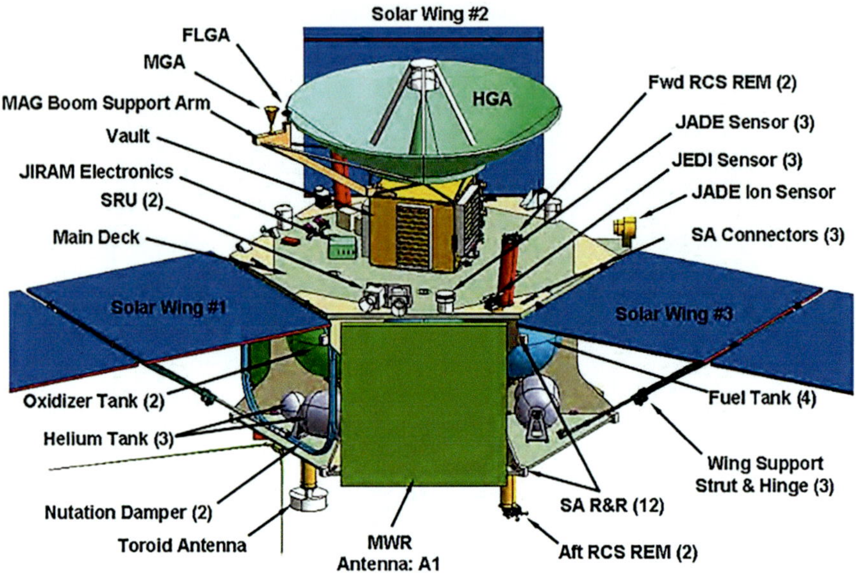

Fig. 9.4 Sketch of Juno spacecraft (NASA image)

Juno was a spin-stabilized spacecraft. The spin rate was 1 rpm during the cruise to Jupiter and 2 rpm during data gathering by the scientific instruments. The spin rate was increased to 5 rpm to maintain stability during burn of the main engine. The spin axis was along the longitudinal axis of the spacecraft.

The coordinate system of the spacecraft sited the Z-axis along the spin axis with +Z in the direction of boresight of the high-gain antenna. The X-axis was parallel to the solar array with magnetometer attached with +X in the direction of the boom. The +Y-axis was 90° clockwise to the +X-axis.

The spacecraft made extensive use of thermal blankets to ward off the extreme cold of space. The weight of the Juno spacecraft at launch was 3625 kg. Of that weight, 2032 kg was propellants.

Spacecraft Systems

Electrical Power

Power for the Juno spacecraft and its instruments was provided by three solar arrays. The foldable solar arrays are shown extended in Fig. 9.1. Two of the solar arrays have four panels that fold upon one another to fit within the launch shroud. The other array has three folding panels with structure holding magnetometers at the far end instead of the fourth solar panel. The weight of the magnetometer structure was the same as the solar panel that it replaced. The two full solar arrays were 8.9 m long and 2.7 m wide. The total area of the three arrays was about 71 m^2. Each array could be tilted up to 4.5° in the Z-axis and moved 1.9° in the X–Y plane.

The three arrays held a total of 13,598 triple-junction solar cells built by Spectrolab Inc. in California. The efficiency of the solar cells was about 28%, which was very good for production-type solar cells at the time. Three layered junctions in the solar cell resulted in the high efficiency. The layers were stacked with a top layer of gallium/indium/diphosphorus ($GaInP_2$), second layer of gallium/arsenic (GaAs), and third layer of germanium (Ge). The area of each solar cell was 26.6 cm^2, and the total active area for 13,598 solar cells was about 49 m^2.

The power available from the array was about 14,000 W in the vicinity of Earth where the solar irradiance is 1361 W/m^2. Jupiter orbits a factor of 5.2 farther from the sun than Earth. The solar irradiance at Jupiter is 50.26 W/m^2, about 3.7% of that of Earth. The predicted power available at Jupiter was about 480 W. The power available at the end of the mission was predicted to be about 420 W.

Two lithium-ion batteries with a capacity of 55 Ah each powered the spacecraft before the solar arrays deployed and when the arrays were not pointed at the sun. Since Juno was in a polar orbit, the spacecraft was always exposed to the sun during orbit. A central power distribution unit in the electrical power subsystem received power from the solar arrays, conditioned it, and distributed power forms to spacecraft systems and scientific instruments.

Propulsion Subsystem

Propulsion for Juno spacecraft consisted of a main engine for large velocity changes and 12 small thrusters for attitude control and small velocity changes. The main engine, which provided 635 newtons (143 pounds) thrust, was a LEROS-1b type built by AMPAC-ISP in Westcott, England. The minimum specific impulse was 317 s. Propellants for the engine were hydrazine fuel and nitrogen tetroxide oxidizer. The fuel and oxidizer ignited on contact in the engine (hypergolic). The motor contained two solenoid operated inlet valves, one for fuel and the other for oxidizer.

Hydrazine fuel was carried in four spherical tanks, and oxidizer was carried in two spherical tanks. The six tanks, which were all the same size, were covered by layers of insulation. According to the photographs, the diameter of the tanks, including insulation, was about 0.9 m. Three smaller tanks held helium pressurant to promote fuel and oxidizer flow from the tanks. The hydrazine fuel tanks also supplied fuel to the monopropellant hydrazine-powered thrusters.

A total of 12 monopropellant hydrazine thrusters were used for attitude control of the spacecraft and to make small velocity changes. The thrusters were Aerojet General MR-111C type, which provided four newtons (0.9 pounds) of thrust. The specific impulse was 224 s. The thrusters were 38 mm in diameter and 169 mm long. They contained a solenoid valve to admit hydrazine, a low-power heater to keep the valve warm and a low-power heater to warm the catalyst. The freezing point of hydrazine is about 2 °C.

The reaction chamber in the thruster contained a bed of type S-405 catalyst, which was composed of granular aluminum oxide (Al_2O_3), liberally coated with indium. Indium made up about 33% of the catalyst. When hydrazine (N_2H_4) was sprayed on the catalyst, a spontaneous exothermic reaction took place with the hydrazine decomposing into very hot hydrogen, nitrogen, and ammonia gas. The hot gases exiting the thruster nozzle generated thrust.

The thrusters were organized in four groups of three thrusters in each group. Each group was mounted on top of a reaction control system (RCS) tower about 1 m long. Two RCS towers were mounted on the upper shelf, and two were mounted on the lower shelf. The thrusters were used to rotate the spacecraft around the three coordinate axes of the spacecraft. They were located to minimize impingement of the plumes on the spacecraft. One thruster in each group, referred to as an axial thruster, was pointed 10° from the Z-axis in the Y direction of the spacecraft, and the other two in a group were pointed in lateral directions.

Command and Data Handling Subsystem

Juno contained two redundant command and data handling (C&DH) units. The C&DH was the brains of the spacecraft. It processed input commands from Earth, controlled major elements of the spacecraft, managed science instruments, and assembled scientific data for transmission to Earth. The C&DH also determined spacecraft attitude in inertial space and controlled thrusters to set attitude.

Each redundant C&DH contained a RAD750 computer on a single printed circuit card. The computer included a RAD750 radiation-hardened microprocessor with attendant memory and support circuits. The microprocessor and the single card computer were built by BAE Systems in the United States. The computer for Juno operated at 132 MHz and had 128 megabytes of dynamic random access memory (DRAM) and 256 megabytes of flash memory. The throughput capacity of the computer was 100 Mbps.

The spacecraft was spin stabilized. The spin rate was one revolution per minute (RPM) during cruise, 2 RPM during science gathering, and 5 RPM during main engine firing. The spin rate and the direction of the spin axis were controlled by firing select thrusters.

It was important for the spacecraft to know its orientation in inertial space. The C&DH obtained primary orientation information from two star trackers referred to as stellar reference units (SRUs). The SRU imaged the star field with a low-light-level, wide-angle camera and compared the image to a database of stars stored on the spacecraft. The basic accuracy would be about two arcseconds if the spacecraft were not spinning. However, image of stars was smeared by rotation of the spacecraft during the 0.25-s exposure time. Two cameras with boresights offset and their images combined were used to reduce the effect of smearing. The resulting orientation accuracy was about 20 arcseconds (0.0055°). The C&DH used attitude data from one of the two redundant inertial measurement units between updates from the star trackers.

Telecommunications System

The Juno spacecraft communicated with Earth through the very capable Deep Space Stations of NASA's Deep Space Network. Three main Deep Space Stations were used for the Juno mission. They were Goldstone located near Barstow California; Robledo near Madrid, Spain; and Tidbinbilla near Canberra, Australia. These stations all had a 70-m-diameter antenna for communications with Juno as well as one or more 34-m-diameter antennas. Juno's telecommunications equipment provided communication with Earth when Juno was up to 9.7×10^{11} km distant. A downlink data rate of 18,000 bps could be supported at that distance.

Uplink and downlink communications with Juno were conducted in X-band. The receive (uplink) frequency was 7153 MHz, and the transmit (downlink) frequency was 8404 MHz. The uplink carrier from the ground was modulated with command data and a ranging signal. Juno's receiver demodulated the command and ranging signals. Command data was sent to the command and data handling units, and the ranging signal was retransmitted to Earth.

Juno's transmitter generated an X-band carrier signal referenced to either an onboard crystal oscillator or a phased-locked, frequency-shifted version of the uplink carrier signal. If the uplinked carrier were used for reference, the ratio of transmitted to receive frequencies was 880/749. The downlink signal was bi-phase modulated by telemetry and/or ranging signals. The coherent downlinked carrier was used on the ground to track two-way Doppler shift and establish very accurate velocity of the spacecraft. The delay of the two-way ranging signal allowed determining the range to the spacecraft.

A Ka-band receiver and transmitter were provided for high-frequency Doppler tracking as part of a gravity science experiment. The equipment received a carrier at 34,365 MHz from Earth and retransmitted a coherent and frequency offset carrier at a frequency of 32,085 MHz to allow two-way Doppler tracking on Earth. The high-gain antenna had feeds for both X-band and Ka-band.

Antennas

Four different antennas were available on the spacecraft: high-gain antenna, medium-gain antenna, forward low-gain antenna, and toroidal low-gain antenna. The high-gain antenna was a parabolic reflector type 2.5 m in diameter with feeds for X-band and Ka-band. The antenna gain was 44.5 dB at the X-band transmit frequency and 47.5 dB at the Ka-band transmit frequency. The beamwidth was 0.5° at X-band and 0.5° at Ka-band. The polarization was right-hand circular for both transmit and receive. The high-gain antenna was used for communications when the spacecraft was in the vicinity of Jupiter.

The medium-gain antenna had a wider beamwidth and attendant lower gain than the high-gain antenna. It was a conical horn type mounted next to the high-gain antenna. Its boresight was in the same direction as that of the high-gain antenna. It was used during cruise and when in safe mode during the orbit of Jupiter. It only operated at X-band. The gain at the transmit frequency was 18.8 dB, and the beamwidth at the transmit frequency was 18°.

The low-gain antennas had a wide beamwidth and relatively low gain. There were two of them: forward low-gain antenna and aft low-gain antenna. The forward low-gain antenna was mounted next to the high-gain antenna with boresight in the same direction. The aft low-gain antenna was mounted to the bottom of the lower shelf of the spacecraft. The antennas were conical spiral type with a gain of 7.7 dB and beamwidth of 80° at the X-band transmit frequency. The low-gain antennas were used for near-Earth communications early in the cruise phase.

The toroidal low-gain antenna was a biconical type that provided a right-hand circular polarized radiation pattern symmetrical around the spin axis. It was mounted to the bottom of the lower shelf of the spacecraft. The antenna was used during burns of the main engine. The gain of the antenna was 6.5 dB, and the beamwidth normal to the spin axis was 20°.

Radio Frequency Subsystem

The radio frequency subsystem contained two redundant small deep space transponders. Both transponders operated at X-band, but one also has an exciter output that could be multiplied by a factor of four external to the transponder to provide a Ka-band carrier frequency for a Ka-band transmitter used for the gravity science experiment. That transponder was identified as X/X/Ka signifying that the uplink was at X-band, and it had outputs to drive either an X-band or a Ka-band downlink. The other transponder with X-band uplink and X-band downlink was identified as X/X. The X/X/Ka transponder was the primary transponder, and the X/X transponder was a redundant spare.

The transponders contained a receiver, exciter, and a digital processing module. The digital processing module performed convolutional encoding of data to be downlinked, and it provided baseband telemetry and ranging signals to the exciter. The exciter generated the downlink carrier frequency and biphase modulated it with telemetry data and ranging signals. The two transponders were cross-strapped to the two command and data handling units. The X-band RF outputs of the two transponders were cross-strapped to the inputs of two redundant traveling wave tube power amplifiers. Either transponder could be used with either power amplifier.

The radio frequency subsystem also contained a Ka-band frequency translator to support the gravity science experiment. The frequency translator was furnished to the Juno program by the Italian Space Agency.

The Ka-band frequency translator received a Ka-band carrier signal at a frequency of 34,364 MHz through the high-gain antenna from one of the Deep Space Network stations on Earth. The carrier was translated by a factor of 3360/3599 to a coherent downlink carrier signal at 32,085 MHz. The downlink signal was amplified by a solid-state Ka-band amplifier to a power level of 2.5 W. The amplifier output was sent to the Ka-band port of the high-gain antenna for transmission to Earth.

A second means of generating the downlink signal for the gravity science experiment used the reference signal from the X/X/Ka transponder. The reference signal was frequency multiplied by a factor of four to 32,088 MHz. That Ka-band signal could be switched to the input of the 2.5-W Ka-band amplifier in the Ka-band translator instead of the translated input.

Traveling Wave Tube Power Amplifier

The spacecraft contained two redundant X-band traveling wave tube amplifiers. Each traveling wave tube amplifier had an accompanying electronic power converter to power it. The output power at X-band was 25 W. The output of either amplifier could be switched to the high-gain antenna, the toroidal antenna, or the low-gain antennas.

Flight of Juno

Juno was launched from pad SLC-41 at Cape Canaveral Air Force Station on 5 July 2011. The launch vehicle was an Atlas V-551 with a Centaur upper stage. Atlas V-551 had five strap-on solid-fuel booster rockets attached to the central core booster. The total thrust of the RD-180 engine in the common core booster plus the five solid-fuel boosters was 12,141 kN (1.238 million kg or 2.73 million pounds). The weight of the launch vehicle plus spacecraft at the time of launch was 577,700 kg, which was easily lofted by Atlas. The launch vehicle did not have sufficient energy to place Juno on a direct trajectory to Jupiter, so a gravity assist from Earth was planned. The spacecraft was placed in a heliocentric orbit adjusted to return Juno to Earth for a gravity assist flyby to establish a trajectory to intercept Jupiter.

The five booster rockets burned for about 90 s and then were jettisoned about 1 min 44 s after launch. The protective shroud around Juno was jettisoned about 4 min 45 s after launch. The RD-180 engine of the common core booster continued to burn for 4 min 27 s after launch. It was separated from Centaur at 4 min 33 s after launch, and the common core booster dropped away. The first burn of the Centaur engine began at 4 min 43 s after launch. The burn lasted for 6 min and placed the spacecraft into a parking orbit around Earth at a velocity of 28,158 km/h. The spacecraft coasted in the parking orbit for about 31 min, and then Centaur was fired again.

The burn lasted for 9 min and was cutoff 50 min 34 s after launch. The second burn increased the velocity sufficiently to escape Earth's gravity and to cause the spacecraft to go into a heliocentric orbit. Centaur was separated from Juno about 54 min after launch.

Near aphelion of the heliocentric orbit, beyond the orbit of Mars, Juno fired its main engine twice to alter the orbit so that Juno would approach Earth on the return path. A gravitational assist from Earth was needed to gain sufficient velocity for Juno to reach the orbit of Jupiter. The first firing was 2 days before aphelion, and the second firing was 2 days after aphelion. Each firing lasted for about 30 min, and each changed velocity by about 344 m/s. Flyby of Earth occurred on 9 October 2013 with a closest approach distance of about 560 km. The resulting slingshot maneuver around Earth resulted in a boost in velocity of 7.3 km/s (26,289 km/h). Juno left the vicinity of Earth on a trajectory that would intercept Jupiter in its orbit.

A NASA/JPL graphic of the trajectory of Juno from launch on 5 July 2011 to orbit insertion around Jupiter on 4 July 2016 is shown in Fig. 9.5. The graphic shows orbits of the planets outward from the sun of Mercury, Venus, Earth, Mars, and lastly Jupiter for reference.

After the boost from Earth, Juno began a 32-month cruise to intercept Jupiter in its orbit. A trajectory correction maneuver was made to fine-tune the trajectory as the spacecraft approached Jupiter.

Finally, on 5 July 2016, Juno began a flyby of Jupiter at the remarkable velocity of 57.9 km/s (208,440 km/h). At a predetermined time, the main engine was ignited for the Juno Orbit Insertion (JOI) burn. The engine thrusted with a force of 65.8 kg. A series of operations were performed to ready Juno for the JOI burn and to configure the spacecraft after the burn. Those operations included the following:

Time from JOI	Event
-5 days	Turn off science instruments
−4 days	Juno computer begins running JOI commands
−125 min	Switch communications to medium-gain antenna
−122 min	Begin maneuvering to JOI attitude
−37 min	Switch to toroidal low-gain antenna
−28 min	Begin fine-tuning of JOI attitude
−22 min	Begin spacecraft spin-up from 2 rpm to 5 rpm
0	Start JOI burn of main engine
+35 min	End engine burn
+37 min	Begin spacecraft spin-down from 5 rpm to 2 rpm
+49 min	Begin changing attitude to point solar arrays at sun
+53 min	Switch communications to medium-gain antenna
+58 min	Begin transmitting telemetry

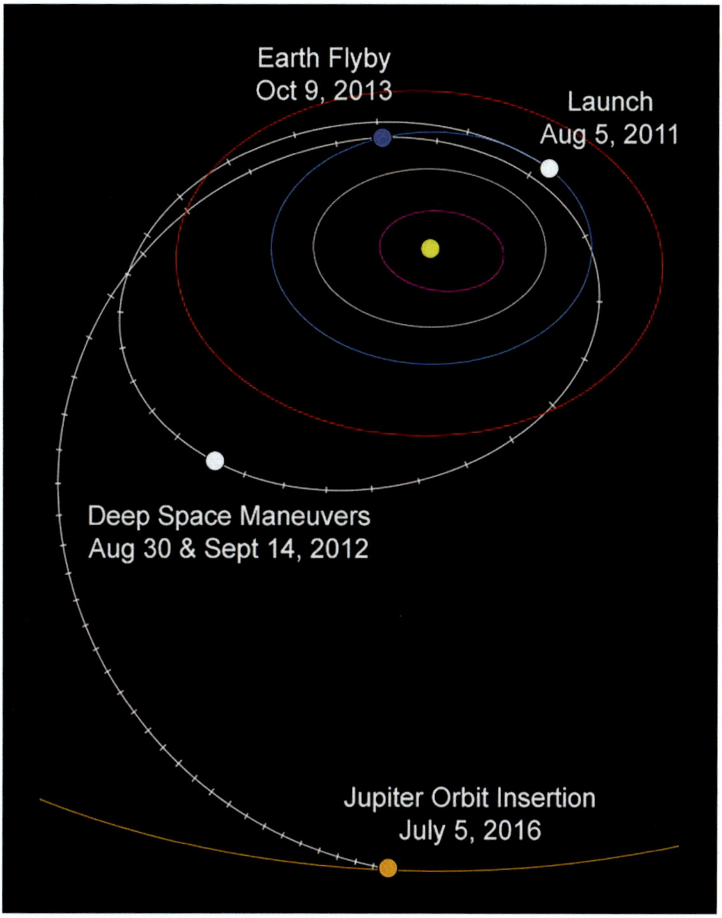

Fig. 9.5 Trajectory of Juno from launch to orbit insertion around Jupiter (Credit: NASA/JPL-Caltech)

The orbit insertion burn lasted for 35 min and changed velocity by 542 m/s (1951 km/h). The burn caused Juno to enter a highly elliptical polar orbit around Jupiter with a period of 53.5 days. The first orbit was referred to as a capture orbit. Aphelion of the orbit was about 8.1 million km from Jupiter, and perijove was about 4150 km above the cloud tops. Perijove of the orbit was reached on 27 August 2016.

It had been planned to reduce Juno's capture orbit of 53 days to orbits lasting 14 days for science gathering. However, during preparations for the orbit decreasing burn, it was noticed that two helium check valves took several minutes to open rather than opening immediately. A decision was made to retain the 53-day orbit rather than risk firing the engine again to reduce the orbit period.

The revised primary mission included 32 science orbits of 53 days each. Most of the science was performed in a period of 8 h centered at the point of closest approach. The polar orbits gave good views of the poles of Jupiter. One such image of the south pole of Jupiter is shown in Fig. 9.6. The image was taken from an altitude of 52,000 km over pole. The oval features in the image are cyclones, some up to 1000 km in diameter.

The latitude of perijove of the orbit slowly migrated northward during the mission. Perijove was at 2.7° north at the beginning of the mission, and it had increased to 28.8° north by the end of the primary mission.

This favored science coverage of the northern hemisphere. An extra orbit was included at the end of the primary mission for a flyby of the moon, Ganymede, on 7 June 2021. The encounter with Ganymede at a closest approach distance of 1050 km reduced the orbital period of Juno around Jupiter from 53 days to 43 days.

The very successful primary mission was completed in July 2021. The spacecraft was still performing well, so the mission was extended through September 2025. An additional 42 orbits were planned. The extended mission included investigation of the moons, Ganymede, Europa, and Io. A flyby of Europa on 29 September 2022 at a closest approach distance of 352 km reduced the period of Juno's orbit to 38 days. Flybys of Io on 30 December 2023 and on 3 February 2024 at a closest approach distance of 1500 km for both encounters reduced the orbital period to 32 days.

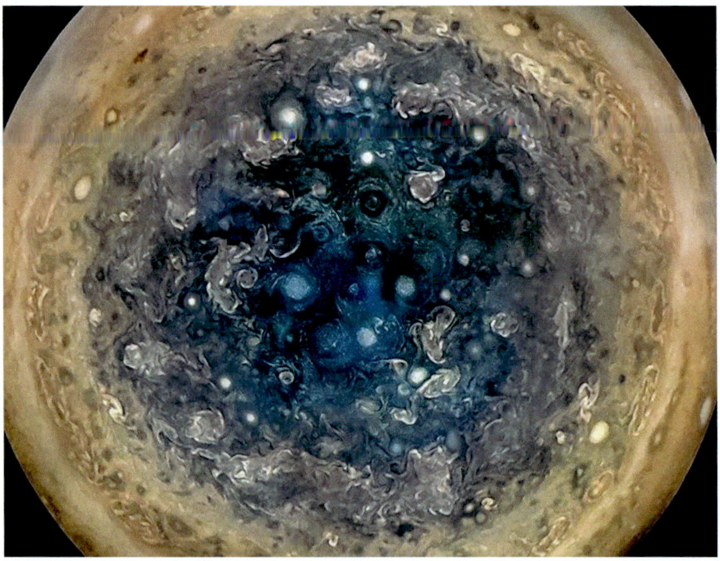

Fig. 9.6 Image of south pole of Jupiter taken by Juno. Credit NASA/JPL-Caltech/SwRI/MSSS/Betsy Asher Hall/Gervasio Robles

Scientific Experiments on Juno

Juno carried nine scientific instruments to investigate Jupiter and some of its moons. A list of scientific instruments and summary of purpose of each are given in Table 9.1.

JunoCam

JunoCam was a visible-light, color-imaging camera. It consisted of two units: a camera head and a Juno digital electronics assembly. The camera head was mounted on top of the upper deck between solar arrays #2 and #3 with the center of the field of view perpendicular to the spacecraft spin axis. The Juno digital electronics

Table 9.1 Scientific instruments carried by Juno spacecraft

Experiment	Purpose	Principal investigator
JunoCam	Take pictures of Jupiter's cloud tops, including the region over the north and south poles	Michael Ravine, Malin Space Science Systems
Microwave radiometer	Determine the atmospheric structure and chemical composition deep in the atmosphere of Jupiter	Michael Janssen, Jet Propulsion Laboratory
Magnetometer	Measure the 3-axis magnetic field strength to generate a 3-dimensional map of Juno's magnetic field	John Connerney, Goddard Space Flight Center
Jovian auroral distributions experiment	Identify particles and processes that produce Jupiter's auroras	Phil Valek, Southwest Research Institute
Jupiter energetic particle detector instrument	Characterize electron and ion environments in Jupiter's magnetosphere	Barry Mauk, Applied Physics Laboratory of Johns Hopkins University
Jovian infrared auroral mapper	Study Jupiter's atmosphere around the auroras to investigate interactions between the auroras and the magnetic field	Alberto Adriani, Italian National Institute for Astrophysics and Planetology
Ultraviolet imaging spectrograph	Measure far-infrared emissions from Jupiter's aurora to determine morphology, brightness, and spectral characteristics of the aurora	Randall Gladstone, Southwest Research Institute
Waves	Measure radio and plasma waves in Jupiter's magnetosphere. Radio waves associated with the auroras were of particular interest	Bill Kurth, University of Iowa
Gravity science	Use Doppler data on transmitted signal from Juno to map Juno's gravitational field	William Folkner, Jet Propulsion Laboratory

assembly was mounted near the camera head on the upper deck. A NASA picture of the camera head of JunoCam mounted to the spacecraft is given in Fig. 9.7.

JunoCam was developed by Malin Space Science Systems (MSSS) in California. The principal investigator was Michael Ravine.

Science objectives of JunoCam included imaging of the poles of Jupiter to investigate polar meteorological phenomena, investigate atmospheric phenomena as exhibited in the cloud tops, and serve as eyes for other instruments. A loftier objective was to produce striking pictures of Jupiter and its moons to reward faithful taxpayers with spectacular images and let them see that NASA was putting its share of tax dollars to good use. To that end, raw images were made public soon after they were obtained, and the public was invited to further process them. Some accounts refer to JunoCam as an "Outreach Camera." Jupiter is a very photogenic planet, and the pictures were impressive.

Some of the resulting images processed by citizen scientists, which appear to have used enhanced contrast and enhanced color saturation, were truly spectacular. As an example, a raw image of the moon, Europa, is shown on the left side of Fig. 9.8, and a processed image by citizen scientist Navaneeth Krishnan is shown on the right side. The image was taken by JunoCam on 29 September 2022 while 1521 km above Europa.

Another example of processing by a citizen scientist of Jupiter's belts and zones is shown in Fig. 9.9. The image was taken on 7 September 2023 from 3000 km above the cloud tops and processed by citizen scientist Tanya Oleksuik.

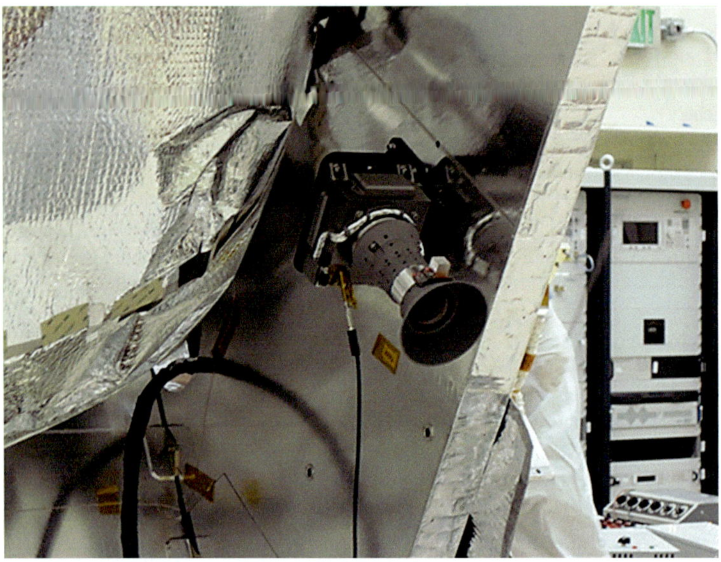

Fig. 9.7 Camera head of JunoCam mounted to the spacecraft. Image cropped by author (credit: NASA/JPL-Caltech/MSSS)

Fig. 9.8 Raw and processed image of Jupiter's moon Europa (NASA image processed by citizen scientist Navaneeth Krishnan)

Fig. 9.9 Jupiter's belts and zones imaged by JunoCam and processed by citizen scientist Tanya Oleksuik

JunoCam obtained best images of Jupiter between times of 1 h before and 1 h after the closest approach. Distances were 3000–5000 km above the cloud tops. The Juno spacecraft spun at 2 RPM. To avoid image smearing due to spinning motion, JunoCam took a series of snapshots at the same rate as spin of the spacecraft. The short duration exposures were integrated together by shifting the images to account for rotation. The image was made up in a push-frame fashion. When a series of exposures and time delay were complete, the frame was pushed into local memory. Thin strips of images were later stitched together to form an image.

The camera head of JunoCam used a 14-element refractive lens with a focal length of 11 mm and a field of view of 58°. The detector was a Kodak KAL-2020 charge-coupled device (CCD). The CCD had an active field of 1640 by 1200 pixels. The angular size of each pixel was 0.67 mrad. Near perijove of the orbit at about

5000 km, the resolution would be about 3 km. At the poles at a distance of about 70,000 km, the resolution would be about 50 km.

Four rectangular filters were bonded side by side to the front of the CCD. Three of the filters were color filters of blue, green, and red, and the fourth filter was at infrared and sensitive to the narrow absorption band of methane. It was referred to as the methane filter. The filters were rectangular shape and 1600 pixels long. The width of each of the color filters was 150 pixels. The width of the methane filter was somewhat larger.

The filter parameters included blue with center wavelength of 480.1 nm and bandwidth of 45.5 nm, green with center wavelength of 553.5 nm and bandwidth of 79.3 nm, red with center wavelength of 698.9 nm and bandwidth of 175.4 nm, and methane with center wavelength of 893.3 nm and bandwidth of 22.7 nm. The methane filter was useful in observing the fine structure of the clouds.

An image of Jupiter's cloud swirls is shown in Fig. 9.10. The image was taken by JunoCam on 5 August 2020 over Jupiter's northern midlatitude region. It was processed by citizen scientist Kevin M. Gill. An image of Jupiter's prominent Great Red Spot is given in Fig. 9.11. The image is a composite of two images stitched together by Kevin M. Gill.

A detailed image of the Great Red Spot taken on 10 July 2017 while Juno was directly over the spot at an altitude of 9866 km is given in Fig. 9.12. The image was processed by citizen scientist Gerald Eichstadt.

Fig. 9.10 Image of clouds at mid-north latitude taken by JunoCam. Credit: NASA/JPL-Caltech/SwRI/MSSS/Kevin M. Gill

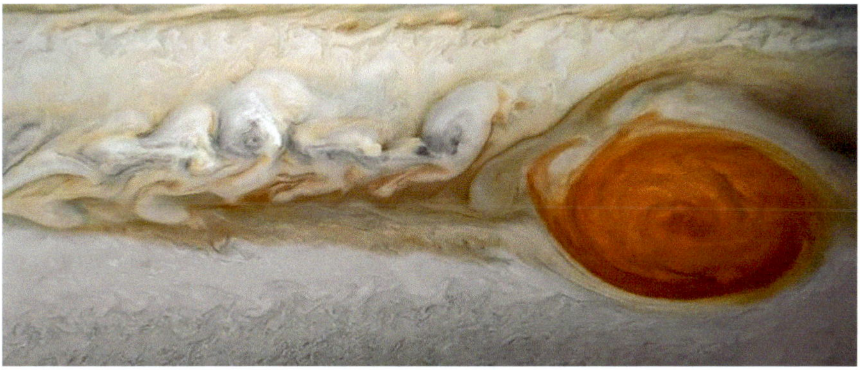

Fig. 9.11 JunoCam's rendition of Jupiter's Great Red Spot. Image credit: NASA/JPL-Caltech/SwRI/MSSS/Kevin M. Gill

Microwave Radiometer

The microwave radiometer used the penetrating properties of microwave frequencies to investigate atmospheric structure and chemical composition deep in the atmosphere of Jupiter.

Emissions from Jupiter at microwave frequencies were measured and converted to temperature using Planck's law. According to Planck's law, the spectral radiance per steradian emitted from one square meter of surface is a function of the wavelength of radiation and temperature as well as Planck's constant and several other constants.

The microwave radiometer used six different frequencies to probe different levels of the atmosphere. Low frequencies penetrated much deeper into the atmosphere and returned signals from great depths. The emission frequencies chosen spanned a range of 600 MHz to 21.900 GHz (wavelength from 50 cm to 1.37 cm). A separate antenna and a separate receiver were provided for each frequency channel. A listing of frequency channels and parameters of the antennas and receivers for each channel is given in Table 9.2. The parameters were obtained from a study by Janssen et al. (2017).

The depth of peak response in Table 9.2 was taken from a set of curves presented by M. A. Janssen. The values given were scaled from the curves and are approximate. Calculations for the curves assumed proportions of ammonia and water vapor in the atmosphere three times that of the sun. Like the sun, Jupiter's atmosphere is composed of mainly molecular hydrogen and helium.

Water vapor and ammonia are the main sources of microwave opacity in the Jupiter's atmosphere. The measurement of spectrum of microwave radiation by the MWR allowed determining the temperature and composition of the atmosphere.

The microwave radiometer measured both the thermal radiation and the dependence of emission angle on the radiation. The spinning spacecraft allowed measurement at emission angles up to at least 60° from the nadir. The ratio of the amount of

Fig. 9.12 Jupiter's Great Red Spot. Image taken by JunoCam and processed by citizen scientist Gerald Eichstadt

limb darkening at an angle of 60° to the value at the nadir was given as an example by Janssen; the data was given as a function of frequency of measurement and amounts of water vapor and ammonia in the atmosphere. Comparing measured emission levels at various emission angles with that at the nadir helped deciphering amounts of constituents in the atmosphere.

Table 9.2 Parameters of antennas and receivers for microwave radiometer

Channel	Frequency, GHz	Antenna beamwidth, degrees	Receiver bandwidth, MHz/%	Depth[a] of peak response, km
1	0.600	20.6	21/3.50	265
2	1.248	21.0	43.75/3.51	160
3	2.597	12.1	84.5/3.25	91
4	5.215	12.1	169/3.24	49
5	10.004	12.0	325/3.25	25
6	21.900	10.8	770/3.52	0

[a]Depth is below the center of the narrow ammonia cloud in the atmosphere. The ammonia cloud is at about the 1 bar pressure level in the atmosphere

Table 9.3 Characteristics of antennas for microwave radiometer

Channel	Frequency, GHz	Antenna beamwidth, degrees	Antenna type	Antenna size, cm
1	0.600	20.6	5 × 5 Patch array	160 × 160 × 13.1
2	1.248	21.0	5 × 5 Patch array	76.8 × 76.8 × 9.8
3	2.597	12.1	8 × 8 Slotted array	77.1 × 67.3 × 8.9
4	5.215	12.1	8 × 8 Slotted array	38.6 × 34.0 × 5.7
5	10.004	12.0	8 × 8 Slotted array	20.1 × 17.9 × 4.4
6	21.900	10.8	Horn	15.3 × 15.3 × 34

The microwave radiometer instrument was made up of three major functions: antenna subsystem, receiver subsystem, and electronics subsystem. The receiver subsystem and electronics subsystem were located inside the radiation-shielded vault to protect them from the fierce radiation in Jupiter's environment.

Antenna Subsystem

The antennas were quite large at the lower frequencies and became progressively smaller as the frequency increased. The beamwidth, type of antenna, and size are given in Table 9.3. The 0.6 GHz and 1.248 GHz antennas were patch type with patch elements supported above a metal backplane. The 2.587 GHz, 5.215 GHz, and 10.004 GHz antennas were slotted waveguide arrays. The 21.900 GHz antenna was a corrugated conical horn.

A photograph of Juno from the NASA Jupiter Orbit Insertion Press Kit that shows the large 0.60 GHz antenna on the side of the spacecraft is given in Fig. 9.13. The 5 × 5 array of patch elements can be seen on the antenna.

Another picture from the Press Kit shows a different side of the spacecraft that held five microwave radiometer antennas, which is given in Fig. 9.14. The lower antenna in the photograph is the 5 × 5 array of patch elements of the antenna for channel 2 of the radiometer.t to the left is the slotted waveguide array antenna for channel 3. The smaller slotted array antenna to the right is for channel 4. The very

Fig. 9.13 Juno spacecraft showing the antenna for the 0.60 GHz microwave radiometer channel (NASA image)

Fig. 9.14 Juno spacecraft showing five antennas for the microwave radiometer (NASA photograph cropped by author)

small waveguide array above the upper deck of the spacecraft is for channel 5. The circular object to the right of the channel 5 antenna is the horn antenna for channel 6.

Microwave Receivers

A separate receiver was used for each of the five radiometer channels. The receivers were Dicke type where a Dicke switch is chopped between a 60 ohm resistive load at known temperature and the input from the antenna. The low-level square wave output of the Dicke switch was amplified by up to 55 dB by a cascade of five low-noise amplifiers before being applied to a diode detector. Two filter stages were incorporated between amplifier stages to provide about 3.5% bandwidth for the six frequency channels.

A coupler at the input to the receiver was used to insert a known noise level from a diode noise source for calibration purposes. The noise diode, which was fed by a constant current source, could be activated upon command. Two other noise diodes, which were coupled into the system along the amplifier chain, could be activated for calibration and diagnostic purposes.

The output of the low-noise amplifier chain was applied to a diode detector. In the case of the two lowest frequency channels, channel 1 and channel 2, the signal levels were expected to be quite high, and two detectors were used. One detector was fed directly from the amplifier, and the other was fed from a 10 dB coupler to lower the signal level by 10 dB. The output of each diode detector was applied to a video amplifier, which in turn supplied an amplified video signal to a voltage-to-frequency converter. The outputs of the voltage-to-frequency converter for each channel were applied to the electronics subassembly.

Electronics Subassembly

The electronics subassembly contained three major items: power distribution unit, command and data handling unit (CDU), and housekeeping unit. The power conversion unit converted 28-volt DC power from the spacecraft power bus to the various DC voltage forms required by the microwave radiometer.

The command and data handling unit contained an 8051 microcontroller that ran software that controlled operation of the microwave radiometer. The microcontroller and associated circuitry executed incoming commands from the spacecraft. It used data from the receivers to produce output data proportional to the amplitude of microwave radiation received in each frequency channel. It also controlled the Dickie switches and gain states of the amplifiers in the receivers and application of calibration signals.

The output signal from each receiver was in the form of a variable frequency. The frequency was proportional to the amplitude of the thermal microwave radiation received. The variable frequency signals from the six receivers were converted to digital form in the CDU by counting for a 99-ms integration time. One millisecond was allocated for readout and transfer of data for a total time of 100 ms. The resulting digital data was read out and placed in memory by the 8051 microcontroller.

The housekeeping unit includes two voltage-to-frequency converters to accommodate 128 multiplexed housekeeping channels. Parameters measured include a large number of temperatures and 16 different voltages. Control signals from the CDU selected each of the multiplexed channels in turn. The variable frequency outputs of the two voltage-to-frequency converters were applied to the CDU where the frequencies were counted for 99 ms and placed in memory.

Communication between the spacecraft and the microwave radiometer was via dual-redundant RS-422 data buses. The data transfer rate was 57.6 kbps.

Results of MWR Experiment

Jupiter is adorned with distinctive belts and zones of white and reddish bands of clouds that encircle the planet. Narrow strips of strong winds moving in opposite directions separate the bands. The white zones are thought to be ammonia ice particles. The cause of the reddish bands is still being investigated. In addition, Jupiter has numerous cyclone storms. The best known is the Great Red Spot, a cyclone storm over 16,000 km wide (about 1.25 times as wide as Earth) that has persisted for centuries.

Measurements by the microwave radiometer were used to probe great depths under the Great Red Spot. Data from a pass over the cyclonic storm during orbit "Perijove 7" is shown in graphical form in Fig. 9.15. The graphic was presented in NASA/JPL Photojournal PIA22177. The upper trace in the graphic is an image taken by JunoCam for reference. The graphic of layers at different depths shows very warm temperatures near the top (0 km depth). The temperature decreases with depth, and warmth is still evident to at least 90 km below the spot.

Magnetometer (MAG)

Measurements by magnetometer were used to generate a detailed three-dimensional map of Juno's magnetic field. Two three-axis fluxgate magnetometers were mounted on a magnetometer boom to space them away from incidental magnetic fields of the spacecraft.

Two magnetometers were used for redundancy and to allow canceling small intrinsic magnetic fields of the spacecraft. The magnetometer boom was a substantial, nonmagnetic structure shown in stowed position in Fig. 9.16. When deployed, it extended about 12 m from the body of the spacecraft along the +X-axis. The boom was covered with a reflective thermal blanket.

Each magnetometer was attached to a stiff rectangular optical bench made of carbon silicon carbide. Two camera heads for an advanced stellar compass were attached to the optical bench near the magnetometer. The camera heads were

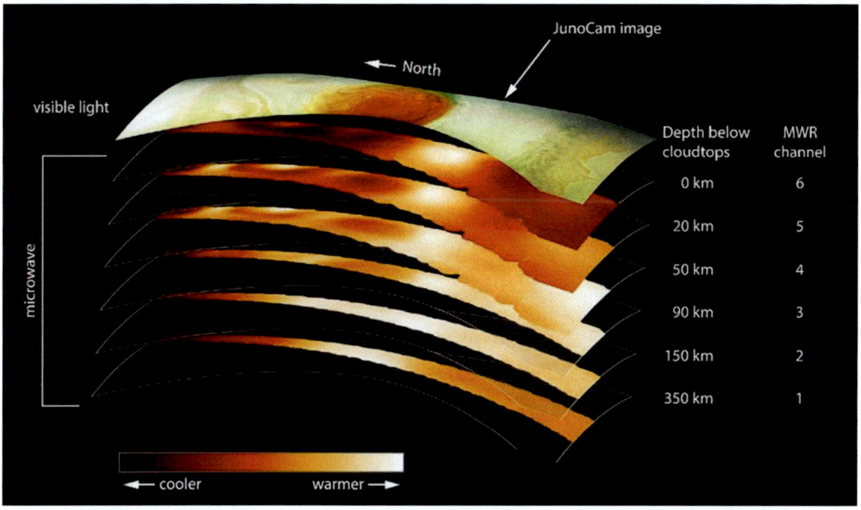

Fig. 9.15 Slices of temperature measurements by the six channels of MWR. (Image credit: NASA/JPL-Caltech/SwRI)

equipped with tapered rectangular baffles to shield them from stray light. One optical bench was mounted on the magnetometer boom near the end, placing the magnetometer 11.8 m from the attachment end of the boom. The other bench was mounted about 2 m inboard, placing the magnetometer 9.8 m from the attachment end of the boom. The camera heads were splayed out about 13° from the spin axis in the direction of the Y-axis. The field of view of the cameras was 14° by 20°. The baffles for the camera heads can be seen in the photograph of Fig. 9.16. The magnetometers are the rectangular packages near each pair of cameras.

The cameras of the advanced stellar compass imaged the star field and compared the image with a stored map of stars to determine the orientation of the optical bench to high precision. Orientation of the magnetometer boom would not have been accurate enough for the planned science to be conducted. The camera heads were designed and built by the Danish Technical University in Copenhagen, Denmark.

Each camera captured a new image of the star field every 4 s. The eight-bit grayscale image was 752 by 580 pixels in size. The exposure time was 0.25 s. Camera images were transferred to a data processing unit that was part of the advanced stellar compass subsystem. Each of the four images from the four cameras was stored in a read-only memory. The images were screened to eliminate large bright objects such as planets and then compared with a star catalog to update the attitude of the spacecraft.

The advanced stellar compass cameras on the spinning spacecraft were also used to observe the atmosphere of Jupiter at low angles encompassing the horizon. Haze layers and elevated clouds were investigated.

Fig. 9.16 Magnetometer boom in stowed position. (Image credit: NASA/JPL-Caltech/LMSS)

The magnetometers were designed and built by NASA Goddard Space Flight Center. The magnetometers were fluxgate type, long used in spacecraft and in aircraft. A description of fluxgate magnetometers is given in Chap. 4 of this book. The unit of magnetic field measurements in the International System of Units is the Tesla (T). For reference, the magnetic field at the surface of the Earth is about 40,000 nT where nT is 10^{-9} Tesla. The dynamic range of the magnetometers was up to 1.6 million nT per axis. The resolution at the most sensitive dynamic range of ±1600 nT was 0.05 nT. The three-axis magnetic field was sampled simultaneously by both magnetometers 64 times a second.

The 53-day polar orbit of Juno took it into the outer regions of the magnetosphere and across the bow shock where the solar wind is slowed by the strong magnetic field of Jupiter. The magnetic field was mapped in that region and down to perijove altitudes. A magnetic field strength of 7.8×10^5 nT was measured when passing over the magnetic equator.

Jovian Auroral Distributions Experiment (JADE)

The Jovian Auroral Distributions Experiment (JADE) measured properties of electrons and ions in the magnetosphere of Jupiter. Observations were made in the polar regions of Jupiter to investigate the polar magnetosphere and the aurora. An image of the northern aurora taken by Juno's ultraviolet spectrograph is given in Fig. 9.17. The displacement of the magnetic north pole from Jupiter's spin axis of 10.3° is apparent in the image.

The Jovian Auroral Distributions Experiment equipment was developed by the Southwest Research Institute in Texas. The principal investigator was Philip Valek.

JADE measured the angular distribution of electrons and ions along with the energy of electrons and mass-to-charge ratio of ions. The instrument included three identical electron sensors (JADE-E), one ion sensor (JADE-I), and an electronics box. The electronics box was located inside the radiation-shielded vault.

The three JADE-E sensors were located on top of the upper deck of the spacecraft and spaced 120° around the spacecraft. The ion sensor was located on top of the upper deck with the sensor portion hanging over the side. A sketch showing the location of the sensors is given in Fig. 9.18. The green item in the sketch is the high-gain antenna.

Fig. 9.17 Image of the northern aurora taken by Juno's ultraviolet spectrometer (NASA image)

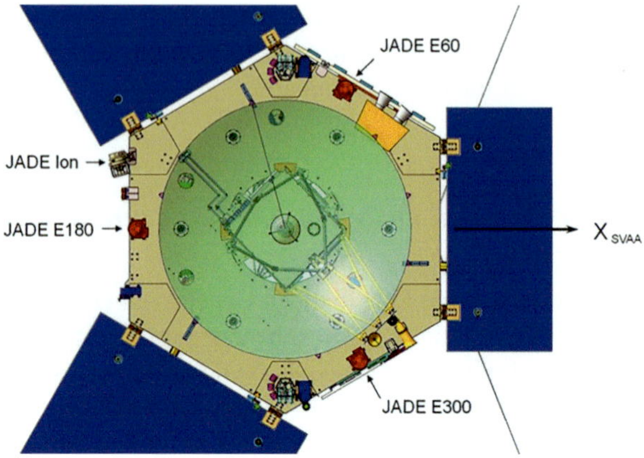

Fig. 9.18 Placement of JADE sensors (image credit: NASA/SwRI)

Table 9.4 Measured characteristics of JADE sensors

Parameter	Electron sensor	Ion sensor
Energy range	100–95 keV	5–43 keV
Energy resolution	11%	18–28%
Field of view	360 × 70°	270 by 90°
Mass range		1–40 amu
Mass resolution		2.5–11%
Time resolution	1 s	4 s

Measured characteristics of the electron sensors and ion sensor are given in Table 9.4.

Electron Sensor

The electron sensor consisted of a spherical top-hat electrostatic analyzer. The analyzer used curved upper and lower deflection plates followed by a microchannel plate electron multiplier with a ring of anodes underneath. The azimuth angular coverage of each sensor was 120°. A ring of 16 anodes under the microchannel plate covered an angular extent of 120°. A background anode was added to provide background reference for the signal anodes.

Electrons entered between the upper and lower deflection plates. The plates were biased with stepped positive DC voltages up to 10,000 volts. The biasing was limited to limit the elevation field of view of the instrument to ±35°. Electrons with a particular energy passed through the gap between plates at a particular voltage field. The voltage was varied in steps to develop a spectrum of electron energies.

Electrons that successfully passed through the electrostatic analyzer struck the microchannel plates. The microchannel elements generated an avalanche of secondary electrons. Those electrons were attracted to be impinged on a particular anode plate in a ring of 16 anode plates below the microchannel plates. The particular anode plate depended on the azimuth angle of arrival of the electron into the electron sensor. Each anode plate responded to electrons in a field of view of 7.5° in azimuth.

A charge amplifier was connected through a capacitor to the output of each of the 16 signal anodes and the background anode. The charge amplifier converted the charge signals and the background signal to digital form. Digital circuitry following the charge amplifiers gathered the digital outputs of the 17 charge amplifiers and converted them to a serial data stream. The serial data stream was sent to the electronics box located in the radiation-shielded vault.

Ion Sensor

The ion sensor was mounted on the side of the spacecraft at the top of the upper shelf. The mounting allowed a field of view of 270° in the elevation plane. The field of view in the azimuth direction was 9°. Viewed from the top of the spacecraft, the field of view was 15° clockwise from the −X-axis of the spacecraft.

The sensor included an electrostatic analyzer that contained upper and lower curved deflection plates. The voltage across the plates was varied in steps between 2.5 volts and 10,000 volts. A particular voltage allowed ions with a particular energy-to-charge ratio pass through the plates without striking the walls. Those ions were accelerated by a potential of 10,000 volts into a time-of-flight section. The time-of-flight section had a very thin carbon foil in the path of the ions. The ions passing through the carbon foil generated secondary electrons, which were attracted to the center of a microchannel plate and detected to serve as a start pulse. The ion then traveled towards a partial ring of 12 anode plates. A stop pulse was generated when the ion hit one of the plates. The angular extent of each plate was 22.5°. The 12 plates resulted in a total angular sector of 270°. Time of flight was determined from the difference between start and stop pulses. The known spacing between start and stop pulse items allowed the velocity of the ion to be determined. The charge deposited on an anode plate was a function of the energy of the ion. Knowing the velocity and energy, the mass of the ion could be computed.

Each of the anode plates was followed by a charge amplifier. A 13th plate was included to monitor the background level. The charge amplifier converted the charge signals to digital form. Digital circuitry following the charge amplifiers gathered the digital outputs of the 13 charge amplifiers and converted them to a serial data stream. The serial data stream was sent to the electronics box located in the radiation-shielded vault.

Electronics Box

The electronics box, which was mounted inside the radiation-shielded vault, contained an instrument processing board (IPB), sensor interface board (SIB), low-voltage power supply, and two high-voltage power supplies.

Instrument Processing Board (IPB)

The instrument processing board responded to commands from the spacecraft and organized low-rate science and housekeeping data and high-rate burst mode science data. It formatted the data into packets and transmitted the data to the spacecraft. During the burst mode, the data is placed in a buffer and then transmitted at a high rate over a dedicated high-speed serial bus.

The IPB contained an Atmel AT697E processor. That radiation hard device was a 32-bit reduced instruction set computer (RISC) processor based on Scalable Processor Architecture (SPARC) V8 architecture. It was capable of executing up to 86 million instructions per second. Memory provided for the processor on the IPB included 128 kbit PROM, 512 kbit EEPROM to hold flight software and lookup tables, and 4 Mbit SRAM.

Spacecraft commands were received by the IPB from the spacecraft over a universal asynchronous receive and transmit (UART) interface. Low-speed telemetry was written into memory by software and transmitted to the spacecraft over the UART interface. During the burst mode, telemetry data was placed in a buffer and then transmitted at a high rate over a dedicated high-speed serial data bus.

Sensor Interface Board (SIB)

The sensor interface board managed the functions of the Jovian Auroral Distributions Experiment. It regulated timing for all events, controlled the stepped high voltages for the electrostatic analyzer deflection plates, processed raw data from the electron sensors and ion sensor, and performed time-of-flight measurements for the ion sensor. It used an analog multiplexer and an analog-to-digital converter to read and store 80 different housekeeping analog signals.

Power Supplies

A low-voltage power supply used 28-volt spacecraft power to generate power forms of 3.3 volts, ±5 volts, and ±12 volts for electronic circuitry of the JADE. The high-voltage power supply used the ±12-volt power from the low-voltage power supply to generate high voltages needed for deflection plates and electrostatic analyzer of the electron sensor and ion sensor.

The three electron sensors were separately supplied stepped voltages in the range of either 0–300 volt or 0–10 kV. Each electron sensor was supplied with separate stepped voltages for the upper deflection plate, lower deflection plate, and electrostatic analyzer. Each was also supplied a voltage in the range of 0–3.8 kV for the microchannel plates.

The ion sensor was supplied stepped voltages in the range of either 0–70 volts or 0–10 kV for the upper deflection plate and the lower deflection plate. The ion sensor was also supplied with stepped voltages of either 0–160 volts or 0–10 kV for the electrostatic analyzer. A voltage form in the range of 0–3.8 kV was supplied for the microchannel plate.

Results of Jovian Auroral Distributions Experiment

Data taken during the Juno's first perijove pass indicated a broad band of high-energy electrons between latitudes of 30° and 60° north and altitudes between 10 R_J and 7 R_J (R_J is the radius of Jupiter). The energy level of electrons in that region was over a million electron volts. There was also a region in the southern hemisphere where energies of electrons were about a million eV. That region extended from south latitudes from about −30° to −80°. Ion intensity, given by the ratio of energy to charge (eV/q), was about 10,000 in those regions.

Jupiter Energetic Particle Detector Instrument (JEDI)

The Jupiter energetic particle detector instrument (JEDI) was used to characterize the electron and ion environment in Jupiter's magnetosphere. Particular interest was given to understanding energies leading to the powerful auroras. JEDI complemented the JAFE instrument by investigating higher-energy electrons and ions. It processed electrons in the energy range from 25 to 1000 keV. The energy range of ions processed depended on species: hydrogen 10 to 2000 keV, helium 25 to 2000 keV, and oxygen 45 to 10,000 keV. JEDI was designed and built by the Applied Physics Laboratory of Johns Hopkins University. The principal investigator was Barry Mauk.

JEDI was comprised of three separate, but essentially the same, assemblies. Each assembly contained a cylindrical sensor head and an electronics unit. The cylindrical sensor heads of two of the assemblies had a fanlike field of view of 160° by 12° with the wide angle in the plane of the mounting feet. The third assembly had a field of view of 148° by 12° with the wide angle perpendicular to the mounting feet. An illustration of two configurations of JEDI is given in Fig. 9.19. The sensor on the left was referred to as JEDI-A180, and the sensor on the right was referred to as JEDI-90 and JEDI-270. The different orientations of the cylindrical sensor heads are apparent in the images.

The three sensors were placed on the upper shelf of the spacecraft as shown in the sketch in Fig. 9.20. Sensor JEDI-A180 generated a 148° by 12° fan acceptance angle in the plane of the spin axis with the wide portion of the fan extending along the −X-axis of the spacecraft. The fan was interrupted by a shield that cut out 12° in the region of the +Z-axis to avoid looking at the sun. The fan swept out nearly the whole region except for the exclusion of the sun as the spacecraft rotated.

The JEDI 90 and the JEDI 270 sensors had fanlike fields of views with acceptance angle of 12° by 160° with the fans approximately normal to the spin axis. The center of the 160-degree fan of JDI 90 was along the +Y-axis of the spacecraft, and

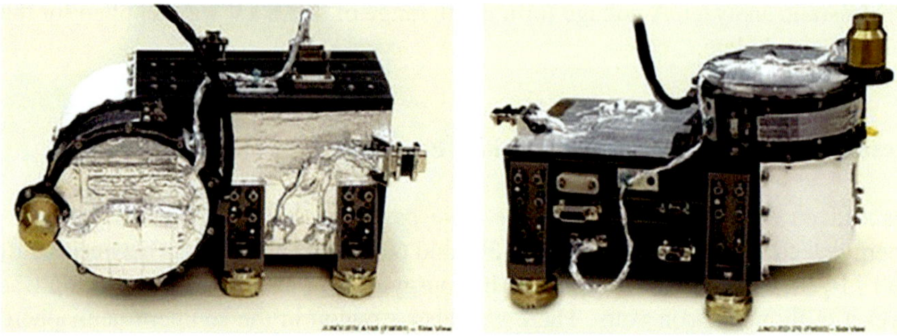

Fig. 9.19 Design of sensor head for Jupiter Energetic Particle Detector Instrument (NASA/JHU image cropped by author)

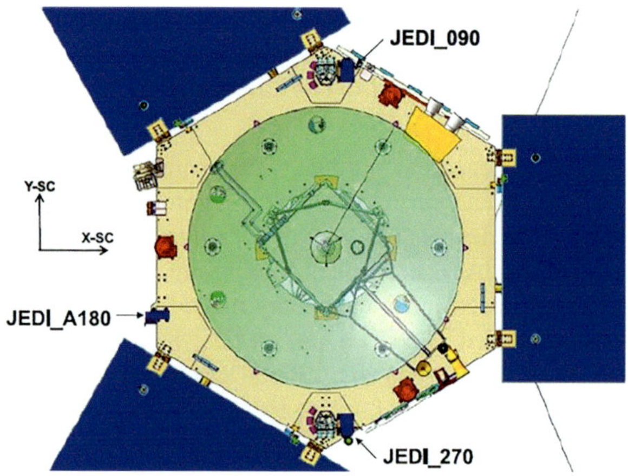

Fig. 9.20 Placement of JEDI sensors on spacecraft (image credit: NASA/JHU)

the center of the 160-degree fan of JDI 270 was along the −Y-axis. The fans were tilted upwards by 10° to provide clearance from the solar arrays.

The cylindrical sensor head had an open sector on the periphery that corresponded to the 160-degree field of view. A circular collimator consisting of a series of holes was positioned in the open sector. A very thin entrance foil was placed behind the collimator to exclude low-level particles. A circular thin carbon foil, referred to as the start foil, was positioned behind the collimator and entrance foil. Separate paths in the sensor head were provided for ions and electrons.

Ion Sensor

When an ion hit and penetrated the start foil, it generated secondary electrons. A potential of 500 volts between the start foil and the microchannel plate directed the secondary electrons towards the microchannel plate. The secondary electrons were amplified by the microchannel plate and applied to a set of six start anodes located along a semicircle near the microchannel plate. Time delay between the responses of the start anodes to the secondary electrons was used to determine the direction of arrival. For example, a start anode lying directly under the path of arrival would be first to generate an output pulse. The charge deposited on the start anode was processed as a "start pulse."

The ion passed through the thin start foil with little decrease in velocity. It traveled through the inside of the sensor head for 6 cm until striking a stop foil in the form of a semicircle on the opposite side of the sensor. Again, secondary electrons were generated as the ion passed through the stop foil. Those electrons were amplified by the microchannel plate and applied to a series of six stop anodes located along a semicircle. The charge deposited on the stop anode was processed as a "stop pulse."

The ion that passed through the stop foil struck one of the six solid-state silicon detectors that were arrayed in a semicircle on the opposite side of the sensor from the entrance aperture. The detector that was struck was in line with the direction of arrival of the ion. The voltage pulse output of the solid-state silicon detector was a function of energy of the ion.

Each of the six solid-state detectors (SSDs) had separate sections for ions and electrons. Each SSD had two electron pixels and two ion pixels. The ion section and the electron section were located side by side. Each section had a "large pixel" that covered half of the width of the assembly and a "small pixel" in the center of the large pixel. The electron section of the detector was covered with a flashing of aluminum two microns thick to reject low-energy protons.

The velocity of the ion could be determined by measuring the time between the start and stop pulses. The mass of the ion was computed from velocity and energy measurements.

Electron Sensor

Energetic electrons entering the instrument were first decelerated by a −2.6 kV potential that was part of the ion detection time-of-flight arrangement. The electron then passed though the start foil, through the drift chamber, and passed though the stop foil. It was then accelerated again by a +2.6 kV potential. The electron then struck one of the solid-state silicon detectors where the energy of the electron was measured. Time-of-flight elements were not used for electron detection since the mass of electrons was known and only energy needed to be measured. The particular SSD in the array of SSDs in the semicircular collection area of the instrument that responded gave the angle of arrival of the electron.

Electronics Unit

The electronics unit, to which the cylindrical sensor heads were mounted, contained an event board, support board, and power board. The event board processed the outputs of the solid-state detectors and the outputs of the preamplifiers following the anodes. It measured time of flight for ions and direction of arrival for both electrons and ions. The support board plugged into the event board, and together they provided command, control, data processing, and telemetry functions. The support board also contained a high-voltage power supply to generate very high voltages for the microchannel plates and optics for the instrument. The power board converted spacecraft primary power to various voltage forms required by the instrument.

Results of Jupiter Energetic Particle Detector Instrument Experiment

Some results of the JEDI experiment are given in a study by B.H. Mauk et al. (2017b). A few measurements from that source are given here. Data from a northern polar pass indicated electron rates of about 4000/s and ion rates of about 500/s at 46° latitude. Electron intensity was about 7×10^5 (1/cm^2.s.sr.keV) at that latitude.

Jovian Infrared Auroral Mapper (JIRAM)

The Jovian Infrared Auroral Mapper (JIRAM) generated high-resolution infrared images of Jupiter's auroral region and atmosphere. The instrument included an infrared imager and an infrared spectrometer. Both sensors operated in the wavelength range of 2.0–5.0 micrometers (μm). The optical head for the combined instrument was mounted under the lower shelf of the spacecraft. An electronics unit for the instrument was mounted on top of the upper shelf. The center of the field of view of JIRAM was in the same direction as the JunoCam camera.

JIRAM was developed by the Leonardo Company in Italy under the supervision of the Institute for Space Astrophysics and Planetology in Rome. It was funded by the Italian Space Agency. The principal investigator was Alberto Adriani of the Italian National Institute for Astrophysics and Planetology.

To avoid blurring of the image by the spinning spacecraft, a spinning mirror synchronized with spacecraft spin (2 rpm) was placed at the entrance to the optics of the instrument. The spinning mirror compensated for spacecraft spin motion in the along-track direction. The spinning mirror was followed by a telescope with equivalent aperture of 44 mm and focal length of 160 mm. A beam splitter at the output of the telescope split the beam into two paths. One path fed the imaging channel, and the other path fed the spectrometer channel. The beam splitter applied 30% of the light to the imaging channel and 70% to the spectrometer channel.

Light entering the imaging channel was directed to two bandpass filters. One filter, called L, was centered at 3.455 μm wavelength with 290 nanometers (nm) bandwidth. The other filter, called M, was centered at 4780 μm with 480 nm bandwidth. The L filter channel responded to positive H_3 ions in the polar aurora. The M channel was used to investigate hot spots in Jupiter's atmosphere. Both filter channels had a field of view of 1.75° along-track and 5.94° cross-track. The output of the two filter channels was applied to a 270 by 438 pixel imaging array of HgCdTe photodiodes. Each filter channel was allocated 128 by 432 pixels. A 10-pixel-wide opaque area separated the two array fields.

The spectrometer included a slit located at the focal plane of the telescope after the beam splitter. Light at the output of the slit was applied to a mirror that directed light through optics to a diffraction grating. The flat diffraction grating was 60 by 32 mm in size and had a ruling of 30 groves/mm. A second set of optics focused the diffracted light from the grating onto a 256 by 336 array of detectors. The 336-pixel direction was in the spectral dispersion plane. The resulting spectral resolution was 9 nm over the spectral range of 2.0–5.0 μm. The spectrometer had a field of view of 3.5° across-track and 50 arcseconds along-track.

The imager and the spectrometer were operated simultaneously. The optical head was passively cooled by two separate radiators facing deep space. One radiator was used to cool the detectors to about 80 K. The other radiator was used to cool the optical paths of the instrument to about 120 K.

Measurements gathered by the optical head were transferred to the electronics unit. The electronics unit included a power supply for the instrument along with electronics and instrument controller functions. The instrument controller received commands for JIRAM functions, managed collection of data from the imager and spectrometer, controlled the despin mirror, and controlled heater and internal calibration functions. It sent science data to the spacecraft command/data handling system via a RS-422 data bus.

Results of JIRAM Experiment

JIRAM gathered several thousand images and spectra of Jupiter's auroras and atmosphere. A composite image of Jupiter's north pole derived from data taken by JIRAM is given in Fig. 9.21. The infrared image shows densely packed cyclones and anticyclones in the north polar region. A composite image of the aurora at the south pole of Jupiter at infrared wavelengths is given in Fig. 9.22.

A collection of spectacular images of the polar regions of Jupiter from data taken by JIRAM is given by Alberto Adriani et al. in the study *Two-Year Observations of the Jupiter Polar Regions by JIRAM on Board Juno* (Adriani et al. 2020). Adriani was a scientist at the Italian National Institute for Astrophysics and Planetology and in Rome. He was the lead investigator for the JIRAM experiment.

Ultraviolet Imaging Spectrograph (Juno-UVS)

The purpose of the ultraviolet imaging spectrograph (Juno-UVS) experiment was to observe and characterize ultraviolet emissions from Jupiter's auroras. It was an imaging spectrometer sensitive to emissions in the extreme-ultraviolet and far-ultraviolet wavelength range from 68 to 210 nm. The wavelength range covered

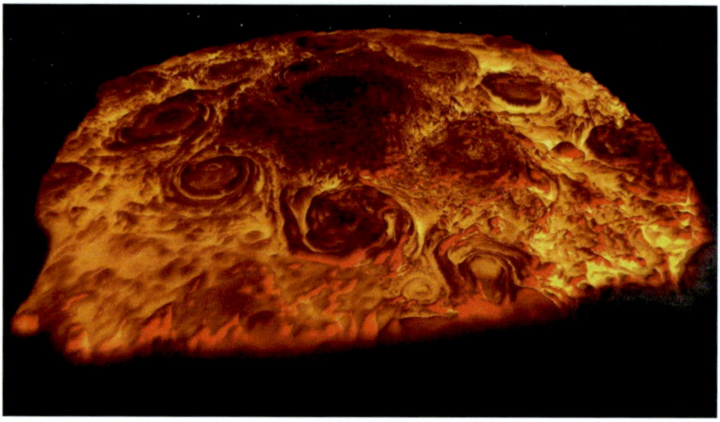

Fig. 9.21 Mosaic image of Jupiter's north pole from JIRAM experiment on Juno (Credit: NASA/JPL-Caltech/SwRI/ASI/INAF/JIRAM)

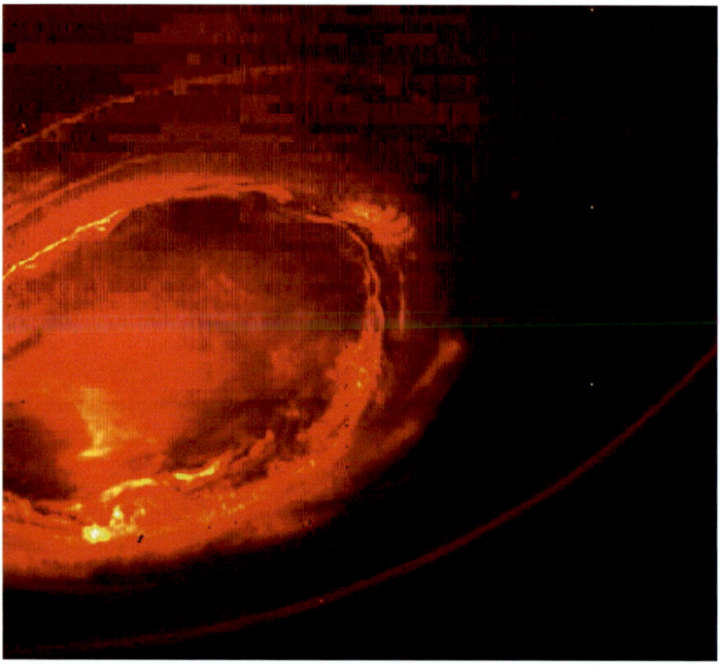

Fig. 9.22 Image of Jupiter's southern aurora from JIRAM experiment on Juno (Credit: NASA/JPL-Caltech/SwRI/ASI/INAF/JIRAM)

emissions from atomic hydrogen and molecular hydrogen in the auroras and absorption signatures of ammonia and other hydrocarbons in the atmosphere.

The ultraviolet imaging spectrograph was developed by the Southwest Research Institute (SwRI). The principal investigator was Randall Gladstone of the SwRI.

The instrument consisted of the Juno-UVS scanner and an electronic box. The UVS scanner was mounted on the bottom side of the upper shelf of the spacecraft with the field of view radial to the spinning spacecraft. The electronics box was mounted inside of the radiation-shielded vault.

Light entered the UVS scanner through an entrance baffle and struck a flat scan mirror. The scan mirror could be rotated ±30° in the plane of the spin axis of the spacecraft to examine specific targets. The scan mirror directed light to an off-axis parabolic mirror. A slit for the spectrometer was located at the focal point of the parabolic mirror. The long dimension of the slit was parallel to the spin axis. The width of the slit in the narrow direction had three contiguous widths. The fields of view projected onto the sky were 0.2 by 2.5°, 0.025 by 2.0°, and 0.2 by 2.5°. Light passing through the slit was directed at a toroidal grating.

The dispersed light from the grating was applied to an array of 2048 by 256 cesium iodide photocathodes.

A photocathode emits electrons when struck with a photon of sufficient energy. Electrons from the photocathode were amplified by a microchannel plate, and the resulting enhanced electron flux was attracted to a field of anodes. An electrical pulse was generated by the anode struck. The pulses were processed by local electronics before sending the pulse data to the electronic box for detailed processing.

Electronics Box

The electronics box included two redundant high-voltage power supplies for the microchannel plates and two low-voltage power supplies for the instrument. It also included a command and data handling (C&DH) function and electronics for the scan mirror and event processing. The C&DH used an 8081-equivalent microcontroller to control the operation of the instrument. Memory associated with the microcontroller included 32 Kbytes of PROM, 128 Kbytes of EEPROM, and 128 Kbytes of SRAM. The C&DH function received commands from the spacecraft and gathered, formatted, and sent housekeeping and engineering data to the spacecraft.

Results of Juno-UVS Experiment

Ultraviolet observation of Jupiter's auroras was conducted during approach to the planet by the UVS instrument. An image of the northern aurora taken by the Juno-UVS instrument was given previously in Fig. 9.17. The baseline total emitted power from the northern auroras over the time period of 3 June to 29 June 2016 was found to be about 3 terawatts (three trillion watts). The baseline total power emitted by the southern auroras during the same period was about 2 terawatts.

An important finding from the measurement of ultraviolet emissions from the auroras by Juno-UVS was the birth of early-morning brightening of the auroras at both the north and south poles. The brightening was intense but short-lived. It was found that the storms are born on the nightside of Jupiter. As the planet rotates, the storm moved into the dawn region and combined with the normal aurora. The short-lived dawn brightness increased energy to about ten times that of normal levels.

The Juno-UVS instrument detected a bright flash at ultraviolet wavelengths from a meteor entering the atmosphere of Jupiter. Entry was at an altitude of 225 km, and from the UVS data, the size of the meteor was estimated to be between 1 and 4 m.

Juno Waves Experiment (WAVES)

The WAVES experiment was designed to measure radio and plasma waves in Jupiter's magnetosphere. Of particular interest were radio waves associated with the auroras. Plasma waves accelerate charged particles in the magnetosphere that contribute to the auroras. The instrument was designed and built by the University of Iowa. The principal investigator for WAVES was William Kurth of the University of Iowa.

The WAVES instrument measured both electric fields and magnetic fields. The sensor for electric fields was a dipole antenna with the two elements in a "V" configuration. The sensor for magnetic fields was a search coil made up of 1000 turns of wire around a core 15 cm long. The electric and magnetic sensors fed four receivers having different frequency ranges. The receiver outputs were processed by a digital processing unit.

Electric Fields

The sensor for electric fields was a dipole antenna with two rodlike elements 2.73 m long. The elements were arranged in a V configuration with a 45-degree angle between elements. The antenna was mounted under the lower shelf of the spacecraft. The mounting aligned one of the elements parallel to the spacecraft +X-axis. A line between the tips of the elements was parallel to the spacecraft Y-axis.

The output of the antenna was applied to a wideband preamplifier. The amplified signal was sent to the electronics box where it was applied to four different receivers. One receiver, referred to as high-frequency receiver, covered the frequency range of 100 kHz to 41 MHz. A second receiver, referred to as high-frequency waveform receiver, covered the same frequency range of 100 kHz to 41 MHz. It was essentially the same as the high-frequency receiver and served as an active backup receiver. A third receiver, referred to as low-frequency receiver (high), covered the frequency range of 10–150 kHz. The fourth receiver, referred to as low-frequency receiver (low), covered the frequency range of 50 Hz to 20 kHz.

The high frequency receiver contained two paths. One path developed digital signals for spectrum analysis, and the other path detected the amplitude of the signal for waveform analysis. The frequency spectrum analysis path first passed the input signal through a bandpass filter and then to a quadrature mixer to convert a selected frequency band in the wideband input signal to baseband. The reference signal for the quadrature mixers was derived from a frequency synthesizer that had a step size of 250 kHz. The reference frequency was stepped through the desired frequency band, and the resulting quadrature baseband signals were processed in turn.

The two quadrature mixers were followed by low-pass filters with a passband of 500 kHz. The quadrature baseband signals were amplified and sampled at a 1.3 MHz rate for input to a 12-bit analog-to-digital converter. The most significant 8 bits of each quadrature channel were sent to the digital signal processor. The quadrature digital signals could be applied to fast Fourier transform (FFT) processing on board to develop frequency spectrums, or the quadrature baseband data could be downlinked for fast Fourier transform (FFT) spectral analysis on Earth.

The waveform analysis path in the high-frequency receiver passed the signal from the wideband preamplifier at the output of the antenna to a low-pass filter with a passband of 3 MHz. The bandlimited signal was applied to a logarithmic amplifier followed by a detector. The resulting voltage waveform was sampled and digitized by a 16-bit analog-to-digital converter for a digital representation of the waveform. Only the eight most significant bits are sent on to the signal processor.

The digital signal processor used a Y180 microprocessor for real-time processing of data and control of the instrument. It handled input commands, data output functions, and orchestrated data acquisition and analysis tasks. Another avenue of processing in the digital signal processor used a field programmable gate array (FPGA) to perform analysis tasks, including fast Fourier transform analysis of data from the receivers. Communications with the spacecraft for command and output data used RS-4221 data buses.

Magnetic Fields

The sensor for magnetic fields consisted of 10,000 turns of insulated copper wire wrapped around a 15-cm-long rod of high-permeability material. The sensor was mounted to the bottom of the lower deck with the axis of the coil parallel to the spin axis of the spacecraft. The sensor responded to magnetic fields in the frequency range of 50 Hz to 20 kHz. The output of the sensor was applied to a low-noise preamplifier.

The amplified signal from the preamplifier was applied to the low-frequency receiver in the electronics box. That receiver had a 50 Hz to 20 kHz channel for magnetic field measurements along with a 50 Hz to 20 kHz channel for electric field measurements. The magnetic waveform output of the receiver was applied to the digital signal processor.

Results of Waves Experiment

Examples of electric and magnetic field measurements are given in a study by J. E. P. Connerney et al. (2017b). Graphics from that study show strongest radio emissions near the poles of Jupiter with frequencies between 5 MHZ and 20 MHz and amplitudes up to 80 dB above background noise. Graphics of magnetic field measurements show magnetic fields strongest near the poles with frequencies from 300 Hz to 3 kHz and strength up to 10^{-4} nT^2/Hz.

Gravity Science

Variation in the gravity field of Jupiter affected the motion of the orbiting spacecraft. Variations in the gravitational field were detected by tracking the velocity of the orbiting Juno spacecraft very accurately. By using extremely accurate two-way Doppler processing of microwave signals, variations in gravity of Jupiter were sensed as Juno traveled over the planet.

Two-way Doppler measurements were made using the normal X-band communications equipment and by using dedicated Ka-band equipment. The high-gain antenna had feed provisions for both X-band and Ka-band. A special Ka-band translator on Juno received a Ka-band carrier at 34,367 MHz from the Goldstone Deep Space Network (DSN) station. The translator used a phase-locked loop to translate the frequency by a ratio of 3360/3599 to a carrier frequency of 32,085 MHz. The Ka-band frequency translator was furnished by the Italian Space Agency.

The translated carrier was amplified by a 2.5-W solid-state power amplifier and sent to the Ka port of the high-gain antenna for transmission to Earth. The downlinked carrier was received by the Goldstone DSN station on Earth, and the two-way Doppler shift was accurately measured. The two-way Doppler frequency at Ka-band was about 4.27 higher than at X-band. The higher Doppler frequency yielded higher precision of velocity measurements.

If the uplink Ka-band carrier signal was not available, a second means of generating the Ka-band downlink signal for the gravity science experiment used a reference signal from the telecommunications system transponder. The reference signal was frequency multiplied by a factor of four to 32,088 MHz. That Ka-band signal could be switched to the input of the 2.5-W Ka-band amplifier in the Ka-band translator instead of the translated input.

Results of Gravity Experiment

Results of the gravity experiment show that Jupiter's gravity field is nearly symmetric about its rotation axis but the field is very asymmetrical in the north-south direction.

References

Adriani, Alberto, et al., *JIRAM, the Jovian Infrared Auroral Mapper*, Space Science Reviews, Vol. 213, November 2017

Adriani, A., et al., *Two-Year Observations of the Jupiter Polar Regions by JIRAM on Board Juno*, Journal of Geophysical research: Planets, Vol 123, Issue 6, e2019JE006098, 2020

Bolton, S. J., and Connerney, J. E. P., *Editorial: Topical Collection of the Juno Mission Science Objectives, Instruments, and Implementation*, Space Science Reviews Vol. 213, Issue 1–4, 2017

Bolton, S. J., et al., *The Juno Mission*, Space Science Reviews Vol. 213, Issue 1–4, 2017

Connerney, J. E. P., et al., *Jupiter's magnetosphere and aurorae observed by the Juno spacecraft during its first polar orbits*, Science, Vol 356, Issue 6340, May 2017a

Connerney, J. E. P., et al., *The Juno Magnetic Field Investigation*, Space Science Reviews Vol. 213, Issue 1–4, 2017b

Evans, Ben, *How Juno Unmasked Jupiter*, Astronomy, Vol. 50, No. 8, August 2022

Gladstone, G. R., et al., *Juno-UVS approach observations of Jupiter's auroras*, Geophysical Research Letters Vol. 44, Issue 15, 2017a

Gladstone, G. R., et al., *The Ultraviolet Spectrograph on NASA's Juno Mission*, Space Science Reviews, Vol. 213, 2017b

Janssen, M. A., et al, *MWR: Microwave Radiometer for the Juno Mission to Jupiter*, Space Science Reviews Vol. 213, Issue 1–4, 2017

Mauk, B. H., et al., *Juno observations of energetic charged particles over Jupiter's polar regions: Analysis of monodirectional and bidirectional electron beams*, Geophysical Research Letters, 2017a, https://doi.org/10.1002/2016GL072286

Mauk, B. H., et al., *The Jupiter Energetic Particle Detector Instrument (JEDI) for the Juno Mission*, Space Science Reviews Vol. 213, Issue 1–4, 2017b

McComas, D. J., et al., *The Jovian Auroral Distributions Experiment (JADE) on the Juno Mission to Jupiter*, Space Science Reviews Vol. 213, Issue 1–4, 2017

NASA, NSSDCA/COSPAR ID: 2011-040A, *Juno*, updated 28 October 2022

NASA/JPL, *Juno – Jupiter Missions*, https://www.nasa.gov/juno, 2017

NASA/JPL, *Juno Launch Press Kit*, August 2011

NASA/JPL, *Juno Mission to Jupiter,* NASA facts, September 2015

NASA/JPL, *Jupiter Orbit Insertion Press Kit*, June 2016

Ryan, Mukai, et al., Descanso Design and Performance Summary Series, *Juno Telecommunications*, NASA, October 2012

Chapter 10
Future Missions

Four impressive new missions to the Jupiter system are planned with arrival times a few years from now. Two of the missions have already been launched. They are the European Space Agency (ESA) *Jupiter Icy Moons Explorer (JUICE)* and the United States *Europa Clipper*. JUICE has an ambitious mission of orbiting Jupiter, conducting flybys of the icy moons Europa, Ganymede, and Callisto, and finally orbiting Ganymede for in-depth study. JUICE was launched on 14 April 2023, and it is expected to arrive at Jupiter in July 2031.

The second mission already launched is Europa Clipper, developed by the United States. The mission includes an elliptical orbit around Jupiter that is planned to make 49 flybys of the moon, Europa. The orbit will take the spacecraft as close as 25 km to Europa during some flybys. Europa Clipper was launched on 14 October 2024, and it is expected to arrive at Jupiter in April 2030.

A third mission, still in development by the United States, is the Dragonfly rotorcraft which is planned to fly in the dense atmosphere of the Saturn moon, Titan. Dragonfly is planned to be launched in 2028 and arrive at Titan in 2034.

A fourth mission, under development by China, is the Tianwen-4 spacecraft. Tianwen-4 is planned to orbit the Jupiter moon, Callisto. It is possible that the mission will include a second spacecraft that would conduct flyby of the planet Uranus. Launch of Tianwen-4 is projected to be in September 2029 with arrival at Jupiter in 2035.

Jupiter Icy Moons Explorer

Jupiter Icy Moons Explorer (JUICE) is a large, capable spacecraft developed by the European Space Agency. It is presently traveling on an 8-year journey from Earth to the Jupiter system. An artist's rendition of JUICE in orbit around Jupiter, as given in NASA NSSDCA/COSPAR ID: 2023-053A, is shown in Fig. 10.1.

T. Lund, *Spacecraft Exploring Outer Planets*, Springer Praxis Books,
https://doi.org/10.1007/978-3-031-96544-9_10

Fig. 10.1 Artist's rendering of JUICE spacecraft (NASA file)

JUICE was launched by an Ariane 5 launch vehicle from Europe's launching site near Kourou, French Guiana, on 14 April 2023. It is expected to arrive at Jupiter in July 2031. The rather long transit time reflects having to use gravitational assists from Venus and Earth to boost the heavy spacecraft on a trajectory to Jupiter. Gravitational assists will be obtained from a lunar-Earth flyby in August 2024, flyby of Venus in August 2025, flyby of Earth on September 2026, and a final flyby of Earth in January 2029.

The objectives of the JUICE mission are to characterize the icy moons of Jupiter: Ganymede, Europa, and Callisto. These moons are suspected of harboring vast oceans beneath their surfaces. Detailed measurements, including by ice penetrating radar, will be made of each moon during a series of flybys. Ganymede will receive extra scrutiny by placing JUICE in orbit around the moon after several months of flybys of the three moons. A graphic showing the four larger moons of Jupiter, referred to as the Galilean moons, is shown in Fig. 10.2.

Background

The JUICE mission was selected by ESA in May 2012 as the first large-class mission within an ESA Cosmic Vision program. Study and planning led to the selection of ten scientific instruments for the spacecraft in February 2013. ESA selected Airbus Defence and Space SAS in France as the prime contractor for the JUICE spacecraft in July 2015.

The JUICE project manager at Airbus was Cyril Cavel. The design of the spacecraft progressed, and a critical design review (CDR) was held between December

Fig. 10.2 The Galilean moons, Io, Europa, Ganymede, and Callisto, shown left to right in order of distance from Jupiter (NASA graphic)

2018 and March 2019. The CDR was successfully concluded in March 2019, and full development and testing of JUICE began.

The spacecraft was assembled at Airbus Defence and Space GmbH in Friedrichshafen, Germany. The spacecraft was subjected to extensive environmental testing at the European Test Services test center in Noordwijk in the Netherlands. Following the successful completion of testing, it was readied for launch and transported to Europe's launching site near Kourou, French Guiana. Launch by an Ariane 5 launch vehicle occurred on 14 April 2023.

Spacecraft Details

JUICE is a large spacecraft. When all elements were stowed for launch, JUICE was 4.09 by 2.86 by 4.35 m in size. When all elements were deployed in orbit, the extent was 1.68 by 27.1 by 13.7 m. The mass without fuel, but including the adapter to mate to the launch vehicle, was 2420 kg. It carried 3650 kg of propellant. The total mass of science instruments was 280 kg. A high-gain antenna, 2.5 m in diameter, was mounted to one side of the spacecraft.

Electrical Power

The main electrical power source for JUICE is solar arrays. The solar arrays are made up of five panels, each 2.5 by 3.5 m in size, in a cross configuration on each side of the spacecraft as shown in Fig. 10.1. The total area of the array is 85 m^2, and the power developed at Jupiter is expected to be about 850 W. A bank of five lithium-ion batteries will supply power when the spacecraft is in the shadow of Jupiter or one of its moons.

Propulsion

The main propulsion engine, which produces 424 newtons (95 pounds) of thrust, is used for trajectory correction and orbit insertion. It uses monomethyl hydrazine (MMH) fuel and oxides of nitrogen (NTO) oxidizer. The fuel and oxidizer ignite upon contact. Two redundant sets of six thrusters with 10 newtons of thrust each are used for attitude control of the spacecraft, and two sets of four thrusters with 22 newtons of thrust each are used to maintain alignment of the spacecraft when the main engine is firing and to make small velocity corrections. The thrusters use the same fuel and oxidizer as the main engine.

Attitude control of the spacecraft uses four reaction wheels. Attitude reference is supplied by an inertial measurement unit, sun sensors, a star tracker, and a navigation camera for use around Jupiter.

Control and Data Management

Control of the spacecraft and data management are performed by the control and data management subsystem. This subsystem provides processing and extensive memory resources to enable the spacecraft to perform autonomously for much of its mission. It interfaces with the propulsion subsystem to control the flight of the spacecraft. It interfaces with the communications system to process and distribute commands uplinked from Earth, and it collects and formats science and engineering data to be downlinked.

Communications

JUICE communicates with the European Space Operations Centre (ESOC) via ESA's tracking station network, ESATRACK. ESOC is located in Darmstadt, Germany. The heart of the JUICE telecommunications system is two redundant deep space transponders. Each transponder has an X-band receiver section, an X-band transmitter section, and a Ka-band transmitter section. The X-band uplink frequency band is 7145–7190 MHz. The X-band downlink frequency is 8420.43 MHz. The Ka-band uplink band is 34,200–34,700 MHz, and the Ka-band downlink frequency is 31,997.64 MHz.

The transponders contain phase-locked loops that lock to the carrier of the uplink signal and provide a coherent reference carrier for the downlink signal so that two-way Doppler processing could be done on the ground. The uplink signal also contained a range signal that was retransmitted and processed on the ground to measure the range to the spacecraft.

The transmitter sections are modulated by telemetry data. The selected transmitter section can drive two redundant 52-W X-band traveling wave tube amplifiers (TWTAs) and two redundant 27-W Ka-band TWTAs. One or the other X-band TWTAs can be selected to drive one of the four selectable antennas: high-gain antenna, medium-gain antenna, and one of the two low-gain antennas. One or the other Ka-band TWTAs can be selected to drive either the high-gain antenna or the medium-gain antenna. The same antennas supply uplinked command signals to the selected receivers.

The high-gain antenna is mounted on the −X side of the spacecraft. It operates both X-band and Ka-band. It has a gain of 43.6 dB at the X-band uplink frequency and a gain of 45.0 dB at the X-band downlink frequency. It has a gain of 57.2 dB at the Ka-band uplink frequency and 56.6 dB at the Ka-band downlink frequency. The medium-gain antenna also operates at both X-band and Ka-band. It has a gain of 28.7 dB at the X-band uplink frequency and a gain of 30.1 dB at the X-band downlink frequency. The gain is 39.8 dB at the Ka-band uplink frequency and 39.2 dB at the downlink frequency.

Science Instruments on JUICE Spacecraft

JUICE carries ten scientific instruments to investigate Jupiter and its moons. A list of instruments, summary of purpose of each, and the principal investigator for each are given in Table 10.1.

Europa Clipper

Europa Clipper is a large sophisticated spacecraft developed by the United States to assess the habitability of Europa, an icy moon of Jupiter. The mission includes an elliptical orbit around Jupiter that is planned to make 49 flybys of the moon, Europa. The spacecraft was launched on 14 October 2024, and it is expected to arrive at Jupiter in April 2030. Europa Clipper is the largest and most capable planetary spacecraft ever launched by the United States. An artist's rendition of Europa Clipper over Europa is shown in Fig. 10.3.

Europa Clipper was launched by a Falcon Heavy launch vehicle from Launch Complex 39A at Kennedy Space Center in Florida on 14 October 2024. The trajectory was designed to take Europa Clipper to Mars for a gravitational assist flyby on 1 March 2025 and then return to Earth for a gravitational assist flyby in 2026. With velocity gained by the two gravitational assists, the spacecraft will enter a trajectory to intercept Jupiter. Arrival at Jupiter is expected in April 2030. Europa Clipper will be placed in an elliptical orbit around Jupiter with periapsis as close as 25 km from the surface of Europa during some orbits.

The objectives of the Europa Clipper mission are to use its nine science instruments, including an ice-penetrating radar, to characterize Europa in detail and determine if conditions on the moon would support life. Europa is suspected of harboring a sizable ocean beneath its surface. The mission will measure the thickness of the moon's icy shell and investigate interaction with its subterranean ocean.

Table 10.1 Scientific instruments carried by JUICE

Experiment	Purpose	Principal investigator
JANUS—Camera system	Image global, regional, and local morphology of Jupiter's moons and map clouds on Jupiter	P. Palumbo, Universita degli Studi di Napoli, Italy
MAGIS—Moons and Jupiter imaging spectrometer	Characterize ices and minerals on the surface of icy moons. Observe cloud features and minor species on Jupiter	F. Poulet, Institut d'Astrophysiqe Spatiale, France
UVS—UV imaging spectrograph	Characterize the composition of exosphere of the icy moons, study Jovian aurorae, and investigate composition and structure of upper Jovian atmosphere	R. Gladstone, Southwest Research Institute, USA
SWI—Submillimeter wave instrument	Investigate temperature, composition, and dynamics of Jupiter's stratosphere and troposphere, and the exospheres and surfaces of the icy moons	P. Hartogh, Max-Planck-Institut für Sonnensystemforschung, Germany
GALA—Ganymede laser altimeter	Use the laser altimeter to develop topography of the moons. Study the tidal deformation of Ganymede and the morphology and topography of the icy moons	H. Hussmann, Institut für Planetenforschung, Germany
RIME—Radar for icy moon exploration	Use an ice-penetrating radar to study the subsurface structure of the icy moons down to 9 km depth	L. Bruzzone, Università degli Studi di Trento, Italy
J-MAG—Magnetometer instrument for JUICE	Characterize the Jovian magnetic field and its interaction with the magnetic field of Ganymede to study subsurface oceans of the icy moons	M. Doughjerty, Imperial College London, UK
PEP—Particle environment package	Characterize the plasma environment of the Jovian system by measuring the density and fluxes of positive and negative ions, electrons, exospheric neutral gas, thermal plasma, and energetic neutral atoms	S. Barfabash, Swedish Institute of Space Physics, Kiruna, Sweden
RPWI—Radio and plasma wave investigation	Measure DC electric field vectors up to a frequency of 1.6 MHz by using Langmuir probes. Use high-frequency receivers and antennas to measure electric and magnetic fields at frequencies from 80 kHz to 45 MHz	J. E. Wahlund, Swedish Institute of Space Physics, Uppsala, Sweden

(continued)

Table 10.1 (continued)

Experiment	Purpose	Principal investigator
3GM—Gravity and geophysics of Jupiter and Galilean moons	Use Doppler shift of the Ka-band transponder signal to study the gravity field at Ganymede and determine the extent of internal oceans on the icy moons. Investigate the structure of neutral atmospheres and ionospheres of Jupiter and its moons	L. Iess, Università di Roma "La Sapienza," Italy
PRIDE—Planetary radio interferometer and Doppler experiment	Use telecommunication system of JUICE and very long baseline interferometry to make precise measurements of spacecraft position and velocity to investigate the gravity fields of Jupiter and the icy moons	L. Gurvits, Joint Institute for VLBI in Europe, the Netherlands

Fig. 10.3 Artist's rendering of Europa Clipper over the moon, Europa (NASA/JPL graphic)

Background

The Europa Clipper mission had a lengthy and sometimes contentious path to approval. Interest in Europa grew after images from the Jupiter orbiting spacecraft Cassini showed plumes of water erupting from the surface of the moon. Cassini orbited Jupiter from December 1995 to September 2003. Formal studies of a Europa mission began in 1996. A mission to investigate Europa was recommended by the National Research Council in December 2013. Several mission configurations were investigated before the present approach of Jupiter orbit with multiple flybys of Europa was selected. The Europa Clipper program was approved in May 2015. Nine science instruments were also selected at that time.

The Europa Clipper program was managed by the Jet Propulsion Laboratory (JPL) under the direction of NASA Science Mission Directorate. The program executive at NASA's Science Mission Directorate was Dave Lavery, and the program scientist was Curt Niebur. JPL formed an alliance with Johns Hopkins University Applied Physics Laboratory (APL), and APL developed and built a large portion of the spacecraft and some of the science instruments. The project manager for Europa Clipper was Jordan Evens at the Jet Propulsion Laboratory, and the assistant project manager was Thomas Manager at Johns Hopkins Applied Physics Laboratory. The project scientist was Robert Pappalardo at the Jet Propulsion Laboratory.

Preliminary design began in February 2017, and detailed design and construction began in August 2019. A critical design review was held in May 2020. Assembly and testing of the spacecraft began in March 2022. The large propulsion module and an RF module were assembled by APL, and the two modules were integrated and tested.

In March 2022, the propulsion module was shipped to JPL in California for continued integration and testing. A picture of the propulsion module after arriving at JPL is given in Fig. 10.4. Assembly and test of the complete spacecraft were completed in early 2024.

The spacecraft was shipped to the Kennedy Space Center in the spring of 2024 where the solar arrays were integrated and the entire spacecraft was tested. Launch from KSC occurred on 14 October 2024.

Fig. 10.4 Propulsion module, built by APL, at JPL (NASA/JPL image)

Spacecraft Details

A picture of Europa Clipper, complete except for the attachment of solar arrays, is shown at JPL in Fig. 10.5. The spacecraft is shown with deployable items in the stowed position as the spacecraft would have to fit in the nosecone of the launch vehicle. Europa Clipper is a large spacecraft. The body of the spacecraft is about 4.7 m high, 3.0 m wide, and 4.0 m deep. When the solar arrays are extended, they span 30.5 m tip to tip. The mass of the spacecraft at launch was 5892 kg. Of that, 2750 kg was for propellants for the propulsion system.

The coordinate system sited the X-axis along the long direction of the solar arrays, the −Y-axis along the boresight of the high-gain antenna. The +Y-axis is normally positioned along the nadir to the surface of Europa when gathering science data. The Z-axis is along the longitudinal axis of the spacecraft.

The spacecraft was made up of two modules: the propulsion module, developed and built by APL, and the avionics module, developed and built by JPL. The propulsion module was the main structure of the spacecraft. The avionics module and other spacecraft systems are attached to it. The avionics module has been mounted to the top of the propulsion module in Fig. 10.5.

The high-gain parabolic antenna, which was 3 m in diameter, is attached to one side of the propulsion module, and the communications subsystem is attached

Fig. 10.5 Europa Clipper in test at JPL (NASA/JPL image)

behind the antenna. Smaller medium-gain and low-gain antennas are attached at various points on the structure. Deployable rodlike antennas for the REASON radar, one of the science instruments, were mounted to the solar arrays.

A boom holding three fluxgate magnetometers was stowed in a canister mounted to the propulsion module. When extended, the boom is 8.5 m long. A large vault structure to shield electronics from the fierce radiation around Jupiter was also mounted to the propulsion module. The rectangular vault, built by JPL, had sides of aluminum-zinc alloy 9.2 mm thick.

Electrical Power

The Europa Clipper spacecraft requires about 620 W of electrical power. The main source of electrical power is solar arrays. The solar arrays are deployed on both sides of the spacecraft as shown previously in Fig. 10.3. When unfolded and deployed, each of the two solar arrays was 14.2 m long and 4.1 m high. The total area of the array is $116 m^2$, and the power developed at Jupiter is expected to be about 700 W. A photograph of the solar arrays being installed to the spacecraft is given in Fig. 10.6.

Three lithium-ion battery modules, with a total capacity of 540 Ah at 28 V, will supply power when the spacecraft is in the shadow of Jupiter or one of its moons.

Fig. 10.6 Installation of solar arrays on Europa Clipper (NASA image)

Propulsion

The propulsion module contained tanks for fuel and oxidizer and tanks for helium pressurant for the fuel system. The fuel was monomethyl hydrazine, and the oxidizer was dinitrogen tetroxide. The fuel and oxidizer ignited upon contact within the thrusters. The propulsion module contained 24 thrusters.

There were four assemblies of thrusters located on short booms located near the bottom of the spacecraft. Each assembly contained four main thrusters pointed aft (towards the bottom in Fig. 10.4) and two smaller thrusters pointed parallel to the +Y- or −Y-axes. The total of 16 main thrusters was divided into two redundant banks of eight thrusters each. Only one bank of eight main thrusters fired at a time. The main thrusters fired together for trajectory correction and orbit insertion. NASA estimates that a burn of 6–8 h will be required to slow the spacecraft for orbit insertion. The eight smaller thrusters are used as part of attitude control of the spacecraft.

The primary means of attitude control of the spacecraft is four reaction wheels. Attitude reference is supplied by an inertial measurement unit, sun sensors, a star tracker, and a navigation camera for use around Jupiter.

Computer and Control

Europa Clipper is controlled by two redundant Europa compute elements (ECEs). The ECE uses a PPC RAD750 radiation-hardened processor to run flight software, manage commands received from the ground, gather and format data from science instruments and engineering sensors, and manage propulsion and navigation modules. The ECEs communicate with the communications system to downlink large volumes of data to Earth.

Communications

Europa Clipper communicates with Earth through the very capable Deep Space Stations of NASA's Deep Space Network. Three main Deep Space Stations are used for the Juno mission. They are Goldstone located near Barstow California; Robledo near Madrid, Spain; and Tidbinbilla near Canberra, Australia.

The heart of the communications system for Europa Clipper is two redundant Frontier Radios designed by APL-JHU. Frontier is an advanced radio that uses software to accomplish many radio functions historically done in hardware. The radio contains an X-band receiver, X-band exciter, Ka-band exciter, and a digital signal processor. The X-band exciter output drives one of the two redundant traveling wave tube amplifiers (TWTAs) with an output power of 20 W. The Ka-band exciter drives two redundant Ka-band TWTAs with an output power of 35 W.

The output of the selected Ka-band TWTA is applied to the Ka-band feed horn of the parabolic high-gain antenna. The high-gain antenna has a gain of 57.5 dB and beamwidth of 0.12 degrees at Ka-band. The boresight of the high-gain antenna is along the −Y-axis of the spacecraft.

The output of the selected X-band TWTA can be applied through switches to one of eight different antennas. The antennas are: high gain with gain of 46 dB, beamwidth of one degree, and boresight along the −Y-axis; medium gain with gain of 17.9 dB, beamwidth of 32 degrees, and boresight parallel to the −Y-axis; three low gain with gain of six dB, beamwidth of 180 degrees, and boresights along the +Y-, −Y-, and −Z-axes; and three fan beam with gain of 12 dB and beamwidths of 30 degrees in the narrow direction and 180 degrees in the wide direction.

Science Instruments on Europa Clipper

Europa Clipper carried nine scientific instruments to investigate Europa. A list of instruments, summary of purpose of each, and the principal investigator for each are given in Table 10.2.

Dragonfly

Dragonfly is a sophisticated rotorcraft developed by the United States to fly in the dense atmosphere of Titan, the largest moon of Saturn. The goal of the Dragonfly mission is to assess the habitability of Titan while gathering information about the atmosphere and surface properties of the moon. Dragonfly will land on the surface, gather science data, take off and fly to different location, and repeat the operation several times during its mission. Launch of the spacecraft is planned in July 2028, and it is expected to arrive at Titan in 2034. An artist's rendering of Dragonfly in flight above Titan is shown in Fig. 10.7.

Background

Saturn's large moon, Titan, is of particular interest to scientists because it is the only moon in the solar system with a dense atmosphere, and its surface appears rich in prebiotic chemistry. Titan is a large moon, about 5150 km in diameter. It is slightly larger than the planet Mercury. It orbits Saturn at a distance of 1.2 million km with a period of 16 Earth days. Titan is tidally locked in synchronous rotation to Saturn, so the same face of Titan always faces Saturn, and a Titan day, T-sol, is also 16 Earth days. The lower atmosphere is composed of 95% nitrogen and 5% methane and trace amounts of hydrogen. The atmospheric pressure at the surface is 147 kPa or 1.45 times that of Earth. The density of the atmosphere is about 4.8 kg/m^3, about four times that of Earth. The gravity is 1.35 m/s^2, about 13.8% of Earth's gravity.

Table 10.2 Scientific instruments carried by Europa Clipper

Experiment	Purpose	Principal investigator
Europa imaging system (EIS)	Image Europa in detail by a narrow field-of-view and a wide field-of-view camera. Map about 90% of Europa with the wide-angle camera at a resolution of about 100 m/pixel	Elizabeth Turtle, Johns Hopkins University APL
Europa Clipper magnetometer (ECM)	Use fluxgate sensors to measure vector magnetic fields. Observe time varying inductive response at Europa to Jupiter's magnetic field	Margaret Kivelson, University of California, Los Angeles
Europa thermal emission imaging system (E-THEMIS)	Map daytime and nighttime temperatures of Europa to characterize the ice shell and identify heat anomalies	Philip Christensen, Arizona State University
Europa ultraviolet spectrograph (Europa-UVS)	Use an ultraviolet spectrometer to determine the composition of vapor plumes above Europa's ice shell	Kurt Retherford, Southwest Research Institute
Mass spectrometer for planetary exploration/ Europa (MASPEX)	Collect and ionize gases around Europa. Use transit time through instrument to determine the mass of various ions. The mass identifies identity of molecules	James Burch, Southwest Research Institute
Mapping imaging spectrometer for Europa (MISE)	Analyze infrared light reflected from Europa to develop a map of spectrums of amplitude as a function of wavelength. The spectrum will identify the composition of the surface	Diana Blaney, Jet Propulsion Laboratory
Plasma instrument for magnetic sounding (PIMS)	Use four Faraday cup sensors to study the density, temperature, and flow of plasma near Europa. The plasma originates at the volcanic moon Io, and it is carried around by the magnetic field of Jupiter. Knowledge of the plasma is important to compensate for its effect on other instruments	Adrienn Luspay-Kuti, Johns Hopkins University APL
Radar for Europa assessment and sounding: Ocean to near-surface (REASON)	Use radar operating in high-frequency (HF) and very-high-frequency (VHF) bands to penetrate the ice shell of Europa. Measure ice thickness and determine characteristics of an underlying ocean	Donald Blankenship, University of Texas, Austin
Surface dust analyzer (SUDA)	Gather dust grains during travel of the spacecraft and measure speed and trajectory of the grains. Measure charge of ions formed by impact of the particles to determine particle size	Sascha Kempf, University of Colorado Boulder

Titan is believed to have an icy crust over a layer of salty water. Much was learned about Titan from the Cassini mission and the Huygens probe that descended through the atmosphere of Titan and landed on its surface. The Cassini mission and the Huygens probe are described in Chap. 6 of this book. A picture of the surface of Titan taken by the Huygens probe was given in Chap. 6, and it is reproduced in Fig. 10.8.

Fig. 10.7 Artist's rendering of Dragonfly rotorcraft above Titan (NASA/Johns Hopkins APL/Steve Gribben image)

Concepts for exploration of Titan were published by Lorenz in 2000. He considered several vehicle concepts and concluded that a helicopter would be an ideal vehicle to achieve science objectives. Interest in Titan was heightened by the Cassini mission that orbited Saturn from 1995 to 2003 and dispatched the Huygens probe to land on Titan in January 2005. Funded studies on concepts to explore Titan continued and led to the Dragonfly concept.

In June 2019, NASA selected Dragonfly to become the fourth mission of the New Frontiers program. New Frontiers is a science program involving medium-sized spacecraft missions with high scientific priority and value. It would be the fourth mission of the New Frontiers program following New Horizons, Juno, and OSIRIS-Rex.

Johns Hopkins Applied Physics Laboratory (JH/APL) was awarded the contract to develop the Dragonfly spacecraft in April 2020. The principal investigator for the program is Elizabeth ("Zibi") Turtle of APL. The Dragonfly project manager is Peter Bedini of APL. Turtle has generated many technical papers on Dragonfly.

A preliminary design review of Dragonfly was conducted and passed in March 2023, and detailed design began. Final design is underway at the time of this writing in 2025. The critical design review was scheduled for April 2025. Launch of the spacecraft is planned for July 2028 with scheduled arrival at Titan in December 2034.

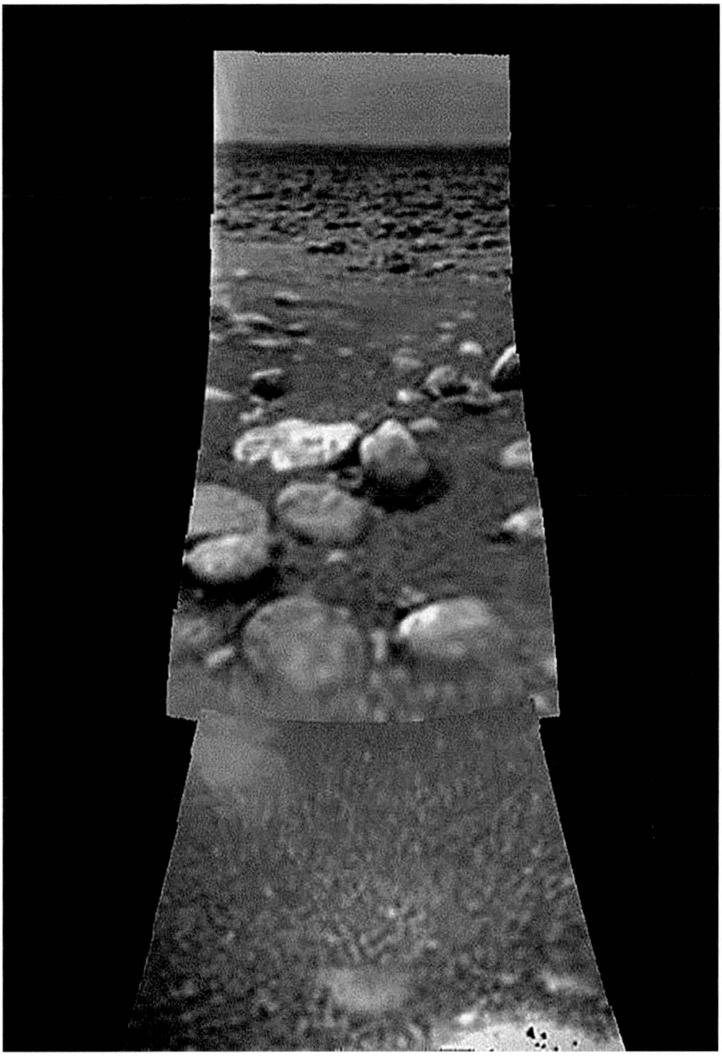

Fig. 10.8 View of the surface of Titan taken by Huygens after landing (NASA image cropped by author)

Rotorcraft Details

Dragonfly is a medium-size rotorcraft with four sets of counter-rotating blades for propulsion. The rotors and associated electric motors protrude from the sides on short booms as illustrated in Fig. 10.7. A JH/APL drawing of the rotorcraft that shows some details of the vehicle, including deployment of the high-gain antenna, is shown in Fig. 10.9.

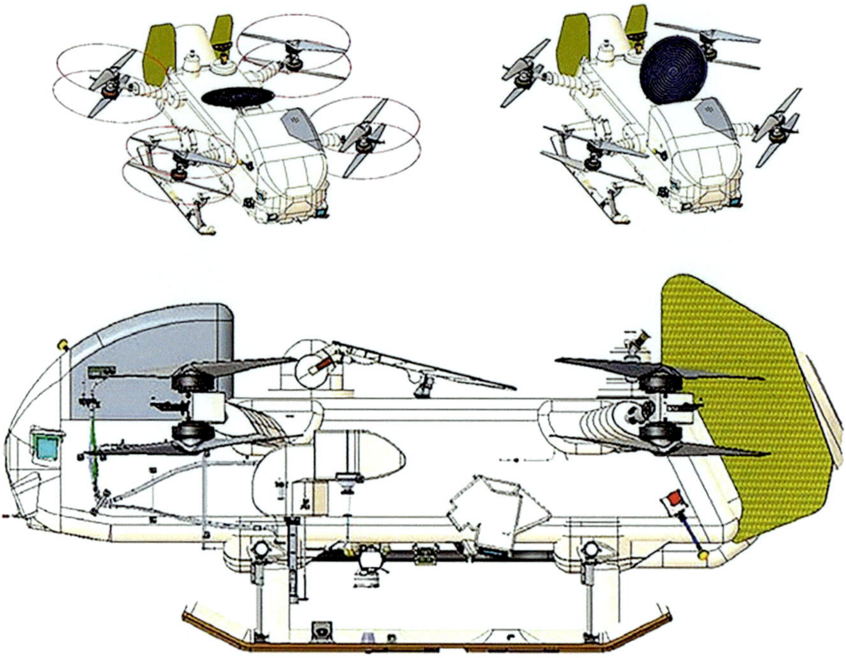

Fig. 10.9 Drawing of Dragonfly rotorcraft (NASA/JPL/APL graphic)

Dragonfly is about 3.9 m long, 3.9 m wide across the tips of the rotors, and 1.8 m high. It weighs about 420 kg. Each rotor has a length of 1.35 m. The rotorcraft has dual short rudders on the rear of the craft, and it lands on skids.

Electrical Power

Electric power is required for eight electric motors that drive the rotors, for systems in the rotorcraft, and for science instruments. The primary source of power is a multi-mission radioisotope thermoelectric generator (MMRTG). RTGs generate electrical power by applying heat from radioactive decay of a material such as plutonium-238 to a series of thermoelectric couples. Plutonium-238 decays into uranium-234 by emitting alpha particles. The half-life of the decay process is 87.7 years. The kinetic energy of the alpha particles is converted to heat when the particles strike the surrounding material.

The MMRTG contained a total of 4.8 kg of plutonium dioxide and a total of 768 thermocouples. The amount of power generated was 110 W at the beginning of life and will be about 72 W after 17 years of operation. The MMRTG is a cylindrical unit with heat-radiating fins aligned lengthwise around the cylinder. The diameter to the tips of the fins is 64 cm, and the length of the device is 66 cm. The width of the fins was shortened somewhat for the Dragonfly application. The unit weighs 45 kg. The MMRTG was mounted in the cylindrical enclosure at the rear of the rotorcraft as may be seen in Fig. 10.7.

The output power of 72 W from the MMRTG is much too low for full operation of Dragonfly, so an energy conversion system is used where the MMRTG is used to charge a bank of lithium-ion batteries during the long Europa night. The bank of batteries has a capacity of 11.5 kWh. A day on Europa, known as a T-sol, is 384 h or 16 days on Earth. The moon has daylight for about 192 h and darkness for 192 h.

The amount of power required to fly the rotorcraft above Europa is much less than on Earth because the force of gravity is only about 13.8% of that on Earth and the density is about four times that of Earth. Flight at optimum speed of 10 m/s (36 km/h) is estimated to require about 2 kW of power. The battery is required to supply power for a flight of perhaps an hour, operate the science instruments, and transmit data to Earth during one Europa daylight period.

Communications

Dragonfly will communicate with Earth through the very capable Deep Space Stations of NASA's Deep Space Network. The heart of the communications system for Dragonfly is a Frontier Radios designed by JH/APL. Frontier is an advanced radio that uses software to accomplish many radio functions historically done in hardware. The radio contains an X-band receiver, X-band exciter, and a digital signal processor. The X-band exciter output is used to drive a traveling wave tube amplifier (TWTA) with an output power of 100 W at X-band.

The output of the TWTA can be applied through switches to one of the three different antennas: high gain, medium gain, and low gain. The high-gain antenna is a deployable array antenna mounted on the flat region of the top of the vehicle. The medium-gain antenna and low-gain antenna are mounted on short booms on the flat area near the rear of the craft.

Drill for Acquisition of Complex Organics

The drill for acquisition of complex organics (DrACO) is a sample acquisition and delivery system that will provide samples of Titan's soil to the Dragonfly mass spectrometer instrument for analysis. DrACO includes two redundant rotary percussion drills, referred to as the port drill and starboard drill. The starboard drill is located on the right landing skid, and the port drill is located on the left landing skid.

The drills include a drill head that imparts both rotary and percussion motion to a hollow drill bit. The apparatus is capable of drilling up to 6 cm into the surface be it soil, ice, or rock. A pneumatic transfer system sucks up particles around and within the drill bit and transports the particles up to two sections of the mass spectrometer referred to as the laser desorption mass spectrometer and gas chromatograph mass spectrometer.

The particles are first delivered to a rotating carousel that contains two rows of sample cups. There are 40 sample cups for the gas chromatograph mass spectrometer and 18 sample cups for the laser desorption mass spectrometer. Valves allow selection of deposition of particles to cups of either spectrometer. The sample cups are moved around, and contents are distributed to a selected spectrometer for analysis.

Science Instruments on Dragonfly

Dragonfly carried four scientific instruments to investigate Titan. A list of instruments, summary of purpose of each, and the principal investigator for each are given in Table 10.3.

Entry and Descent to Landing on Titan

Planning for the descent and landing of Dragonfly onto Titan is being conducted by NASA Ames Research Center. The planning was led by Michael Wright, Dragonfly EDL Phase Lead.

Table 10.3 Scientific instruments carried by Dragonfly

Experiment	Purpose	Principal investigator
Mass spectrometer, DraMS	Measure mass of molecules in atmosphere and subsurface samples to identify organic compounds. DraMS functions in three modes: Laser desorption, gas chromatography, and atmospheric enrichment.	Melissa Trainer, Goddard Space Flight Center
Gamma-ray and neutron spectrometer, DraGNS	Detect gamma rays emitted from radioactive elements in the surface of titan. Measure intensity in the spectrum of radiation to determine elemental compositions in the surface and shallow subsurface of Titan	Patrick Peplowski, JHU APL
Geophysics and meteorology package, DraGMet	Measure atmospheric temperature, pressure, wind speed, and methane humidity. Measure properties of the surface including temperature, texture, and composition. Use seismometers to search for tectonic activity and probe a possible water layer beneath the surface	(TBD) JHU APL
Camera suite, DragonCam	A suite of eight cameras includes two cameras mounted on the steerable high-gain antenna for navigation and positioning, two forward-looking cameras for navigation, and two downward-looking, microscopic viewing cameras to view soil samples	Michael Malin, Malin Space Science Systems

A cruise vehicle, which may not yet have been selected, will transport Dragonfly to the vicinity of Titan. Dragonfly will be enclosed in an aeroshell with a protective cover. The aeroshell will be a 4.5-m-diameter, 60-degree cone. In preparation for separation of Dragonfly from the cruise vehicle, the cruise vehicle will be spun up to 2 rpm for stability and oriented for Dragonfly entry into the atmosphere of Titan. Dragonfly, in its protective aeroshell, will be separated from the cruise stage about 10 min before entry into the atmosphere.

Entry into the atmosphere will occur at an altitude of 1270 km. The velocity at entry will be about 7.3 km/s, and a peak deceleration of about 11 g is expected a few minutes after entry. The aeroshell will slow Dragonfly to allow a drogue parachute to be deployed about 6 min after entry at an altitude of 143 km. A main parachute will be deployed about 88 min after entry at an altitude of 4.8 km, and the aeroshell will be jettisoned. The lidar will be activated after separation of the aeroshell, and the landing skids of the rotorcraft will be deployed. Lidar lock on to reflected signals from the ground is expected at an altitude of 1.8 km, about 132 min after entry.

The Dragonfly rotorcraft will be separated from the structure holding the parachute 137 min after entry at an altitude of 1 km. The velocity will be about 2.9 m/s at the time. Dragonfly will transition to powered flight, fly over the surface, and descend to a landing. It will use its cameras to select an appropriate landing area.

The landing site is planned to be in the Shangri-La dune field region, east of the Huygens probe landing site. An artist's conception of Dragonfly on the surface of Titan is shown in Fig. 10.10.

Exploration of Titan

The baseline Dragonfly mission at Titan is planned to last for 61 Titan days (2.7 Earth years). Flights to another location will occur about every other Titan day. A day on Titan, known as a T-sol, is 384 h (16 days on Earth). Titan has daylight for about 192 h and darkness for 192 h. Dragonfly will use its science instrument suite to characterize the surface, subsurface, and atmosphere of Titan at several sites during its mission. It will repeatedly fly to a site, perform science investigations, and then fly to the next site.

The MMRTG electrical power source will be used to charge lithium-ion batteries that provide power for flight and to operate the science instruments while resting on the surface. The batteries will normally be charged during the long Titan night. Flight to a destination and investigation of Titan's surface and atmosphere will be conducted during daylight hours.

Typical exploration will involve a leapfrog flight operation where Dragonfly will fly over a site to verify it is safe to land before landing there. The process begins with a scouting flight with cameras operating and sending pictures to Earth. Dragonfly will then return and land at the original site. Photographs of the terrain will be analyzed on Earth, and a landing site is selected. On the next flight, Dragonfly

Fig. 10.10 Artist's rendering of Dragonfly on the surface of Titan (NASA image)

will fly past the selected landing site and scout ahead for the second landing site before returning and landing at the first selected site. This leapfrog operation is repeated on the traverse across Titan.

A typical flight will travel about 24 km, consisting of a 16 km forward leg followed by an 8 km return leg to the landing site. Typical flight time will be 30–40 min. After landing at each site, the science instruments are activated, and the skid-mounted rotary percussion drills will be used to sample material below the surface and bring it up for analysis. The mass spectrometer will be the primary instrument used to determine the composition of surface samples exposed by two redundant rotary percussion drills.

Special Testing

An operational half-scale model of the rotorcraft has been successfully flight tested over sand dunes near Yuma, Arizona. Test flights verified flight control algorithms, navigation hardware and software, and image processing. In addition, the rotors for the Dragonfly rotorcraft have been tested in the large wind tunnel at NASA Langley.

Launch of Dragonfly to Titan is planned for July 2028, and it is expected to arrive at Titan in 2034.

Tianwen-4

Tianwen-4 is a potential Chinese mission to investigate the Jupiter moon, Callisto, and perhaps include a separate small spacecraft to fly by the planet Uranus. Scientists from the China National Space Administration (CNSA) indicate a possible launch date of September 2029 with a trajectory obtaining gravitational assists from Venus and Earth and arriving at Callisto in December 2035. Once in the vicinity of Jupiter, a small spacecraft would be sent on to make flyby of Uranus, and the main spacecraft would orbit Callisto.

No details of Tianwen-4 or its mission have yet been found in the open literature. The spacecraft may be launched by a Long March 5 launch vehicle from the Wenchang Space Launch Site on Hainan Island.

References

Airbus, *JUICE Spacecraft Design Report*, JUI-ADST-SYS-DD-000122, Issue 4, 2022

Barnes, Jason W., et al. *Science Goals and Objectives for the Dragonfly Titan Rotorcraft Relocatable Lander*, The Planetary Science Journal, Vol 2, 2021

eoPortal, *Satellite Missions Catalogue, Other Space Missions*, JUICE (Jupiter Icy Moon Explorer), 2025

ESA, *ESA Science and Technology – JUICE*, https://sci.esa.int/s/w06IDBB, 2023

ESA/Science & Exploration, *Juice Launch Kit*, ESA/Science & Exploration/Space Science/Juice, 2023

Goddard Engineering and Technology Directorate, *What is NASA's Europa Clipper spacecraft?*, https://etd.gsfc.nasa.govour-work/eruropa-clipper/

Jet Propulsion Laboratory, *Europa Clipper Press Kit*, https://www.jpl.nasa.gov/press-kits/europa-clipper/, 2024

JHU/APL, *Dragonfly Web Site*, https://dragonfly.jhuapl.edu

Lorenz, R. D., *Post-Cassini Exploration of Titan: Science Goals, Instrumentation and Mission Concepts*, Advances in Space research, Vol. 36, Issue 2, 2005

Lorenz, Ralph D., *Dragonfly: A Rotorcraft Lander concept for Scientific Exploration at Titan*, Johns Hopkins APL Technical Digest Vol 34 No 3, 2018

NASA, NSSDCA ID: DRAGONFLY, 2025

NASA, NSSDCA/COSPAR ID: 2023-053A. *Jupiter Icy Moon Explorer*, 2023a

NASA, Science Editorial Team, *NASA's Dragonfly to Proceed with final Mission Design Work*, 2023b

NASA Science Editorial Team, *Why is NASA Sending Dragonfly to Titan? Here are Five Reasons*, 2020

NASA, *Dragonfly's Journey to Titan*, https://www.nasa.gov/emd/nepa/dragonflys-journy-to-titan

NASA, *NASA's Dragonfly Will Fly Around Titan Looking for Origins, Signs of Life*, 2019

NASA, *NASA'S Europa Clipper - Instruments*, https://europa.nasa.gov/spacecraft/instruments

Paczkowski, Brian, et al., *Europa Clipper Instrument Summaries*, Jet Propulsion Laboratory document, 2022

Pappalardo, Robert T., et al., *Science Overview of the Europa Clipper Mission*, Space Science Reviews, 220:40, 2024

Pensado, Alejandro R., et al., *Dragonfly Mission Entry and Descent Modeling and Simulation Overview*, AIAA 2025-2237, AIAA 2025-2237, 2025, https://doi.org/10.2514/6.2025-2237

Planetary Society, *Juice, exploring Jupiter's icy moons*, https://www.planetary.org/space-missions/juice, 2023
Srinivasan, Jeffrey M., et al., *Europa Clipper Flight System Overview*, Space Science Reviews 221:14, 2025
Turtle, E. P., et al., *Dragonfly: In Situ Exploration of Titan's Organic Chemistry and Habitability,* 50th Lunar and Planetary Science Conference, 2019
Turtle, E., et al, *Dragonfly Update to the NAS Committee on Astrobiology and Planetary Sciences*, 2024
Wright, Michael, *The Dragonfly Entry and Descent System*, UPPW 2019; Oxford, UK, 2019
Zacny, K., *Drill for Acquisition of Complex Organics (DrACO) for Dragonfly Mission*, 53rd Lunar and Planetary Science Conference, 2022

Index

T. Lund, *Spacecraft Exploring Outer Planets*, Springer Praxis Books,
https://doi.org/10.1007/978-3-031-96544-9

MIX
Papier aus verantwortungsvollen Quellen
Paper from responsible sources
FSC® C105338

If you have any concerns about our products,
you can contact us on
ProductSafety@springernature.com

In case Publisher is established outside the EU,
the EU authorized representative is:
Springer Nature Customer Service Center GmbH
Europaplatz 3, 69115 Heidelberg, Germany

Printed by Libri Plureos GmbH
in Hamburg, Germany